MW00563555

Praise for *The Power of Statistical Thinking: Improving Industrial Processes*

"Finally, a one-step reference which links statistical methodology with practical applications in the workplace. The authors have succinctly revealed that success will come through knowing your processes and using them to provide the value demanded by customers today. Statistical methods provide that knowledge."
> —Mary D. Dolan
> Director – Quality Improvements
> Campbell Soup Company

"The strength of the book is the use of case studies that illustrate applications of statistical thinking using tools the authors have introduced."
> —James L. Hess, Ph.D.
> Leader, Board for Quality & Process Control
> DuPont Engineering

"The authors distinguish themselves from others by writing from the perspective of those who have been there. After having taught and worked with companies in virtually every industry the authors are able to offer advice which is not only statistically correct, but also helpful!"
> —Jeff Peters
> Vice President and General Manager
> Harris Corporation

"The book's strength is in documenting the methods of application which could aid managers and technical people in making better and more appropriate application of control charts for process improvement. It talks directly to issues through examples."
> —William F. Fulkerson
> Staff Analyst
> John Deere & Company

"I am convinced the ability of an organization to survive and serve its customers with increasingly better products and services will depend on the 'power of statistical thinking.' This book effectively communicates statistical and process knowledge needed by managers and technicians."
> —David L. Beal
> Operations Manager
> Lake Superior Paper Industries
> A Company of Consolidated Papers, Inc.

"This book does an excellent job of teaching the fundamentals of continuous improvement tools. The recommended approach has been proven effective over and over again in various types of organizations. I particularly appreciated the extensive use of real case studies."

—Roger Hoerl
    Manager of Quality Methods and Information
    Scott Paper Company

The Power of
Statistical Thinking

Engineering Process Improvement Series

John W. Wesner, Ph.D., P.E., Consulting Editor

Lou Cohen, *Quality Function Deployment: How to Make QFD Work for You*

William Y. Fowlkes/Clyde M. Creveling, *Engineering Methods for Robust Product Design: Using Taguchi Methods® in Technology and Product Development*

Maureen S. Heaphy/Gregory F. Gruska, *The Malcolm Baldrige National Quality Award: A Yardstick for Quality Growth*

Jeanenne LaMarsh, *Changing the Way We Change: Gaining Control of Major Operational Change*

William J. Latzko/David M. Saunders, *Four Days with Dr. Deming: A Strategy for Modern Methods of Management*

Mary G. Leitnaker/Richard D. Sanders/Cheryl Hild, *The Power of Statistical Thinking: Improving Industrial Processes*

Rohit Ramaswamy, *Design and Management of Service Processes*

Richard C. Randall, *Randall's Practical Guide to ISO 9000: Implementation, Registration, and Beyond*

John W. Wesner/Jeffrey M. Hiatt/David C. Trimble, *Winning with Quality: Applying Quality Principles in Product Development*

The Power of Statistical Thinking

Improving Industrial Processes

Mary G. Leitnaker
Richard D. Sanders
Cheryl Hild

Addison-Wesley Publishing Company
Reading, Massachusetts   Menlo Park, California   New York
Don Mills, Ontario   Wokingham, England   Amsterdam   Bonn
Sydney   Singapore   Tokyo   Madrid   San Juan
Paris   Seoul   Milan   Mexico City   Taipei

Many of the designations used by manufacturers and sellers to distinguish their products are claimed as trademarks. Where those designations appear in this book and Addison-Wesley was aware of a trademark claim, the designations have been printed with initial capital letters.

The publisher offers discounts on this book when ordered in quantity for special sales.

For more information, please contact:

Corporate & Professional Publishing Group
Addison-Wesley Publishing Company
One Jacob Way
Reading, Massachusetts 01867

Library of Congress Cataloging-in-Publication Data
Leitnaker, Mary G., 1953-
     The power of statistical thinking: improving industrial processes / Mary G. Leitnaker, Richard D. Sanders, Cheryl Hild.
          p.   cm.
     Includes bibliographical references and index.
     ISBN 0-201-63390-6 (hardcover)
     1. Process control —Statistical methods. I. Sanders, Richard D., 1935-   . II. Hild, Cheryl, 1963-   . III. Title.
TS156.8.L44 1995
658.5'01'5195—dc20

95-39002
CIP

0-201-63390-6

1 2 3 4 5 6 7 8 9-MA-98979695

First printing, November 1995

Engineering Process Improvement Series

Consulting Editor, John W. Wesner, Ph.D., P.E.

Global competitiveness is of paramount concern to the engineering community worldwide. As customers demand ever-higher levels of quality in their products and services, engineers must keep pace by continually improving their processes. For decades, American business and industry have focused their quality efforts on their end products rather than on the processes used in the day-to-day operations that create these products and services. Experts across the country now agree that focusing on continuous improvements of the core business and engineering processes within an organization will lead to the most meaningful, long-term improvements and production of the highest-quality products.

Whether your title is researcher, designer, developer, manufacturer, quality or business manager, process engineer, student, or coach, you are responsible for finding innovative, practical ways to improve your processes and products in order to be successful and remain world-class competitive. The **Engineering Process Improvement Series** takes you beyond the ideas and theories, focusing in on the practical information you can apply to your job for both short-term and long-term results. These publications offer current tools and methods and useful how-to advice. This advice comes from the top names in the field; each book is both written and reviewed by the leaders themselves, and each book has earned the stamp of approval of the series consulting editor, John W. Wesner.

Key innovations by industry leaders in process improvement include work in benchmarking, concurrent engineering, robust design, customer-to-customer cycles, process management, and engineering design. Books in this series will discuss these vital issues in ways that help engineers of all levels of experience become more productive and increase quality significantly.

All of the books in the series share a unique graphic cover design. Viewing the graphic blocks descending, you see random pieces coming together to build a solid structure, signifying the ongoing effort to improve processes and produce quality products most satisfying to the customer. If you view the graphic blocks moving upward, you see them breaking through barriers—just as engineers and companies today must break through traditional, defining roles to operate more effectively with concurrent systems. Our mission for this series is to provide the tools, methods, and practical examples to help you hurdle the obstacles, so that you can perform simultaneous engineering and be successful at process and product improvement.

The series is divided into three categories:

Process Management and Improvement   This includes books that take larger views of the field, including major processes and the end-to-end process for new product development.

Improving Functional Processes   These are the specific functional processes that are combined to form the more inclusive processes covered in the first category.

Special Process Topics and Tools   These are methods and techniques that are used in support of improving the various processes covered in the first two categories.

Contents

Preface

The objective of this book is to provide present and future managers and engineers of industrial organizations with the knowledge and information needed for the statistical study of the processes of an organization. To remain viable in an increasingly competitive world marketplace, organizations must be continually engaged in the study and improvement of the effectiveness and efficiency of their operations. It is the premise of this book that knowledge of variation is and will continue to be critical to the effective execution of these responsibilities. Crucial competencies required of managers and engineers in the future will be the ability to characterize the organization's processes with performance indicators that serve as guidance for improvement and to direct the organization in improvement of these measures. A conscious recognition of variation, an informed interpretation of the messages it contains, and an understanding of the effects variation has on process performance will help define the different behaviors and decisions that will characterize the future role of the manager and engineer.

The topics in this book are intended to provide the statistical background necessary for the manager or engineer to assume these responsibilities. Presentation of the statistical methods and tools is supported by demonstration of correct use, practice, interpretation, and guidance for implementation. Clearly, a knowledge of statistics is insufficient by itself to provide the type of knowledge and guidance needed for process improvement. A critical component of improvement work is the management and technical knowledge brought to bear on determining where and how to employ the statistical techniques. The required knowledge of chemistry, physics, and of a host of applied engineering methods are not covered in this book. There is nothing directly on "people management skills." Instead, the book contains advice, based on observation and experience, as well as numerous case studies on the integration of statistical methods with management and engineering knowledge, which can lead to organizational improvement.

Just as the roles of managers and engineers are changing in organizations, so are the uses of statistical methods in organizations. This change in the manner in which statistical methods are being used is reflected in recent writings by W. E. Deming (1986) and Ronald D. Moen, Thomas W. Nolan, and Lloyd P. Provost (1991) on enumerative and analytical methods in statistics. Enumerative methods, although primarily intended for studying a fixed population of items, have formed the bulk of statistical techniques taught at the undergraduate and graduate levels in statistics courses for managers and engineers. A perusal of the table of contents of most statistical textbooks designed for these audiences would verify this statement. Analytical statistical methods, on the other hand, are intended to provide insights into the causal mechanisms affecting the outcomes of ongoing processes. Processes occurring over time are subject to changes in environment, materials, methods, equipment, and other circumstances. It is the study of process behavior and how it is affected by these changing circumstances that is addressed by analytical studies.

The demands for increased statistical skills for process investigation in the industrial world has had an effect on the content of statistical courses in universities and colleges and on professional education. There are a growing number of competent statistical books that present the basic formulas of statistical process control. In general, those books focus on the use of statistics as strict control or feedback mechanisms; they provide little demonstration of the use of these methods for the analytical studies required for identifying sources of variation. This book is intended not only to teach statistical techniques useful in such analytical studies, but also to illustrate how these techniques can be used in practice to understand the causal mechanisms that determine the output characteristics of processes and provide insights into the improvement of these processes. Some statistical theory is covered in an appendix. When used as a course text, the instructor can increase the value of the book by demonstrating the utility of theory in approaching a new situation, in searching for the correct technique, and by demonstrating and discussing the constraints often found in unspoken mathematical and process-based assumptions.

ACKNOWLEDGMENTS

The examples and ideas in this textbook are primarily the result of the authors' association with the University of Tennessee's Management Development Center. Although too numerous to name, managers and engineers from industry who have attended U.T.'s executive education courses have provided numerous ideas, insights, and critiques of the thoughts in this book. The authors would like to express their appreciation to John Riblett, the center's director, for his considerable efforts in supporting the growth of faculty and ideas through his management of the center. Also, without the advice, support, and insights from our faculty colleagues at the center this book would not have been possible.

1 Introduction to the Use of Statistical Methods in Strategic Organizational Improvement

During the past ten to fifteen years, significant changes have occurred in the way companies organize and perform work. The content of this change is not uniform across different industries nor even across diverse firms within an industry. Among the many changes that could be enumerated and described, there are three elements that are relevant to this book. These three elements have had and continue to have a persuasive influence across a broad range of businesses. They are

1. a renewed interest in the customer/consumer,
2. required changes in the behaviors of senior managers, and
3. at the operations level, continual evaluation of the process means by which products and services are provided.

These categories of change are not independent of each other. The characteristics favored by customers provide the metrics against which process performance is evaluated. This information is used to provide criteria by which to guide changes in processes. Senior managers decide what is to be provided to the customer, how it is to be provided, at what volume, and on what schedule. These decisions determine the internal measures against which process performance is evaluated. Process performance is continually improved in order to more efficiently provide products and services.

These changes have dictated that a primary responsibility of managers and engineers is process management. Process management in this book is defined as

- the alignment of process objectives with identified customer values,
- knowledge of the current process configuration and capability to attain process objectives, and
- the selection and direction of needed process improvements.

A past practice for carrying out responsibilities for obtaining process improvements has been to rely on new processing technology or innovations. This practice is insufficient; a dominating responsibility is to take the new technology or innovation and improve what it currently provides and how it is provided. A key competency in carrying out this responsibility is the use of statistical techniques to study and manage the variation in processes. The statistical techniques and supporting tools provided in this book have been proven to be useful in describing processes, assessing and understanding process performance, and making and evaluating process changes. This book is not intended to be a text on statistical methods. Instead, the overall objective is to describe how to use statistical methods for more effective and efficient management of processes. Its content is based on a context of continuously providing improved customer value, thus ensuring financial rewards and job security.

Three important concepts that have been introduced deserve further elaboration and justification:

1. process definition,
2. statistical techniques for process study, and
3. an organizational context for process management.

Together, these three ideas provide the framework within which the statistical methods discussed in this book are presented.

1.1   PROCESS DEFINITION

Several definitions of a process can be found in any moderately large dictionary. These definitions often include the following: A *process* is

1. a series of actions, operations, or functions leading to an end result or outcome;
2. a set of inputs, one or more transformations, and outputs;
3. the steps of a prescribed procedure.

These single definitions are obviously very general and can serve to define almost any activity in any circumstance as a "process." In industrial and business environments, an additional criterion is imposed that any process must have a purpose; process results and outputs, which are derived from purpose, must serve the organization's strategy. The complexity of strategy, purpose, results, and methods means that industrial processes are typically not linear, as the preceeding definitions might appear to imply.

Numerous processes of various levels of complexity are identifiable at any business or industrial site. These processes, as well as their elements, are typically interdependent so that the improvement of any one process or element of a process cannot be done without reference to how it improves the whole. Processes cannot exist in a vacuum or be treated as separate, individual entities because they cannot independently reach objectives or provide results. To capture the important aspects of processes, the methods and discussions provided in this book are based on a definition of "process" that integrates the previous single definitions:

> A **process** *is a structured set of interrelated operations, activities, and tasks that produces a set of specific outputs through the application of skills, methods, and practices specified by organizational protocol.*

Figure 1.1 provides an overview of an order/entry process. This graphic provides a visual illustration of the cross-functional nature as well as the non linear nature of an industrial process. Understanding the cross-functional nature of most industrial processes is critical to the application of the material in this book. The order/entry process does not reside within a specific, single function or department. An order/entry requires prior approval from Finance. The order cannot be scheduled without commitments from Manufacturing and Purchasing. Only the most perfunctory definition of order acceptance would not make reference to

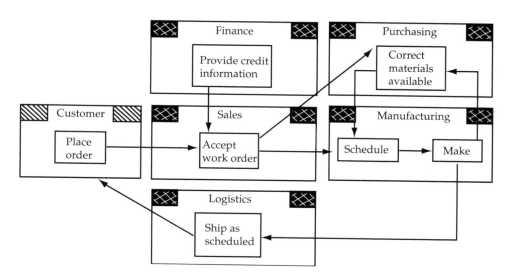

FIGURE 1.1 *Overview of an Order/Entry Process*

specific tasks within Warehousing and Logistics. As is generally true of most processes, no one manager has responsibility for the complete process.

As illustrated by the process in Figure 1.1, process results are created over a period of time by related operations and activities. The operations and activities within a process typically cross time, location, and functional area. The definition we gave earlier helps us to visualize work and how results are created. The definition appreciates order and relationships and recognizes that the creation of process outputs is not through haphazard, chaotic efforts. Process elements should exist because they are required for a process to achieve its objective. Their rationale for being is to contribute to the process outputs necessary to provide customer value.

The use of a process viewpoint and the realization of the relationships between operations and activities demand that individuals who work within a process clearly understand their contributions to process results. Process engineers must understand what relationships should be developed or exploited in order to create the required products or services. Managers must realize the importance of identifying those process elements that are critical in providing customer value and must prioritize work efforts and align tasks and responsibilities to achieve value. For a given list of products and services, the organization is made stronger and more competitive by improving processes in ways that better contribute to customer value.

A process orientation also provides the basis for determining measurements by which the performance of work efforts are judged. As previously stated, each process has an objective; it is required to provide specific outputs with certain characteristics. When these requirements are defined from the process customer's perspective, they become the criteria by which process effectiveness and efficiencies are measured. The process becomes characterized and evaluated based on crucial process parameters as opposed to inadequate and irrelevant task characteristics. For example, in a metal stamping operation, prior to viewing the operation as a set of interrelated tasks, the primary measure by which work was evaluated was the number of pieces processed and/or produced at each workstation. As expected, this measure resulted in high throughput figures for the individual stations as well as for the operation as a whole. However, when customer values were determined, when outputs and results were defined based on these values, and when the individual tasks and activities were viewed as components of a larger process, these measures were quickly identified to be insufficient. Two process measures were identified as the deviation between promised and actual ship dates and the proportion of defective parts prior to rework. These two measures revealed that optimizing individual throughputs without regard to the quality of the parts resulted in an overall process turn time that was too long to meet promised ship dates due to extensive rework.

As illustrated in the previous example, measurements on process outputs can be used to evaluate process performance against customer requirements. A process viewpoint also encourages the measurement of process inputs to provide information on ways to modify or optimize process operation. The process for designing a product is an input to a manufacturing process. The design of the product heavily influences manufacturing quality, cycle time, and manufacturing cost. An evaluation of the effects of the input provided by design gives direction for improving these inputs. Critical process characteristics of design and of manufacturing are identified, how these parameters are managed is known, and study or change is undertaken, as required.

In the course of presenting and explaining statistical methods and tools for process study, this book contains examples of specific processes from various types of organizations and companies. At times, the reference is to a high-level process that contains many smaller subprocesses, tasks, and activities. For example, a manufacturing process includes vendor-provided items and services; hence, the ordering activity and the checking of incoming materials are both contained within the larger manufacturing process. At other times, a lower level process, such as a machining or filtration process, is used for discussion and illustration of statistical procedures. It is important to remember that large-scale improvements are made by improving the critical processes of the organization and that tasks, activities, and operations only become the focus of improvement efforts after the critical processes are identified and studied.

1.2   STATISTICAL TECHNIQUES FOR PROCESS STUDY

Because the statistical techniques presented in this book are aimed at process study, they differ from those methods emphasized in most statistics books. Traditionally, statistical study has been based on the concept of taking samples from populations in order to make inferences concerning certain population characteristics. For example, a sample of incoming materials is taken in order to characterize the entire lot. In traditional statistics courses, hypothesis tests are taught as a means to make decisions about population characteristics based on sample results. Process study requires a different approach. The collection of data is necessary in order to understand the ongoing behavior of processes. Sampling is from an ongoing process; a finite population is not available to sample. The data are not used to characterize a population, but rather to help identify and understand the root causes of variation in a process. Measurements are focused on describing and understanding process performance over time as opposed to describing a set of data collected at a fixed point in time. The analysis of data for process improvement requires a different set of statistical techniques to aid managers and engineers in understanding the factors that impact the process and its outcomes over time.

The methods in this book rely heavily on the Shewhart control chart developed at the Bell Laboratories in the early 1930s. The use of these methods in the context of process management means that

- process measurements are consistently taken over time to evaluate a process,
- data plots over time aid in converting the numbers into information, and
- distinctions between ongoing variations and episodic variations support decisions about the correction or improvement of process operation.

It must be understood that process improvement does not simply refer to the stabilizing of process results and consistently providing the required results; process improvement, as previously described, is improving that which is currently provided and the means by which it is provided. Process results, summarized by control charts, provide a check on process management by focusing on the causes of variation and the reduction or elimination of the effect of those causes.

The statistical techniques and other tools included in this book have been selected because of their power and capability in supporting process managers, engineers, and operators in the performance of their respective duties. We emphasize that these are supporting techniques because the duties and responsibilities of these individuals go far beyond the use of tools and techniques. Statistical methods are not ends in themselves; they merely facilitate better and more effective fulfillment of process responsibilities. Throughout this book, a process manager is responsible for

1. understanding and disseminating information on what the customers of the organization value;
2. knowing what and how the process contributes value to its customers;
3. developing and deploying practices and methods that provide the intended results and confirming that these methods can be implemented;
4. knowing the past and current performance levels of the process in achieving its objectives;
5. assessing current process performance against requirements; and
6. making appropriate, constructive, and verifiable process changes that enable the process to improve significantly its contribution to enhancing customer value.

The unique advantage of the statistical techniques presented in this book over traditional statistical methods is their ability to aid the process manager in carrying out these responsibilities.

1.3   MANAGING PROCESSES WITHIN AN ORGANIZATIONAL CONTEXT

A process is a composite of functions, activities, and tasks performed by personnel using various skills and methods to provide required products and services with specific, required characteristics. These product characteristics (such as density, smoothness, thickness) and service characteristics (such as product availability, flexibility of product and process characteristics, reliability of supply) are determined by the strategy of the organization. The senior leadership of an organization must identify how its strategy is supported by process management. The importance of considering the role of strategy in describing process management objectives and hence in considering the context in which statistical methods in this book are used can be seen by reviewing the recent history of the use of these methods in organizations.

1.3.1   History of the Use of Statistical Methods in Industry

Since the early 1980s, the use of statistical methods in process work has been widely recommended. For many individuals, a powerful rationale for the use of statistical methods was found in an interpretation of Dr. W. Edwards Deming's contribution to the industrial success of several Japanese business firms. The message espoused by Deming had broader objectives, with statistical methods serving as a tool for attaining some of these changes. Deming was attempting to move managerial thought and practice from one of administrative oversight, managerial control, and obsession with managing financial outcomes to one that was more concentrated on *how* the organization provides products and services. The most powerful aspect of his message was that the senior managers of an organization must take responsibility for managing the means, not just the outcomes, of the systems of the organization. The work and teaching of Deming is responsible for the current focus on management and managers and their responsibility to the organization and customer. Statistical methods were the simplest aspect of what he taught; many in his audience chose to use the statistical methods to the exclusion of the first-order elements necessary for the successful practice of these methods. An overly simplistic interpretation of his emphasis on variation and its unsatisfactory effects on satisfying customer needs also served to push the interpretation of his message to the solely statistical.

Statistical methods have been proffered as a means for improving results and gaining and ensuring competitive advantage. Often, the advice was profuse and the promised returns lavish. Many organizations devoted large amounts of effort and resources to education and training in statistical methods, usually in the name of improving quality and productivity. Ten or so years after these initial efforts, it has become clear that the promised returns were often not realized. An understanding of the reasons for these often disappointing results is important for an understanding of the context in which statistical methods can be effective. Two primary reasons for the early ineffectiveness of the introduction of statistical methods follow:

1. *There was an erroneous belief that statistical methods were the means by which organizational improvement would occur.* This interpretation is misguided. Statistical methods are only a tool or technology. Although the use of these methods is a necessary skill, the more critical element of their successful use is the consideration of where they are applied. The presentation and explanation of their use must be placed within the context of the rationale and purpose for which they will be used in carrying out the responsibilities of process management.

2. *There was an overwhelming emphasis on training the manufacturing operator, supervisor, or technician.* The focus on the process technician or operator was often justified by "that is where the process knowledge is resident." Although the truth of that statement cannot be denied, it does not serve as a reason for focusing organizational efforts at this level of the organization. The deficiencies in that focus are many; four of the most critical ones are described here:

    a. Implementation of statistical methods uncovered problems and process deficiencies that could not be addressed by individuals at the bottom of the organizational chart; these individuals did not have access to the necessary resources nor control over accountable individuals. For example, they could not require obvious product or process design changes. The delegation of the expectation for "continuous improvement" to the factory floor imposed severe limitations on what issues could be addressed, the magnitude of improvement made, and the ability to maintain realized gains.

    b. If process improvement is to provide significant benefits, there must first exist knowledge of what market-valued improvement is. This knowledge is guided by organizational strategy, defined customer value, and information on market acceptance and customer evaluation and feedback. When this knowledge is not resident within the group charged with "process improvement," confusion and uncertainty as to what to work on and how to prioritize work efforts result.

c. The delegation of "process improvement" to supervisors and operators had no chance of overcoming years of practice that failed to
address customer value, that misunderstood quality and what it could
provide in the market, that valued schedule and volume over legitimate customer needs, and that used crude productivity measures as
ineffective proxies for efficiencies.

d. Improvement results were often overemphasized at the expense of
understanding and improving the process methods by which the
results were achieved. When there is little or no emphasis on building
process knowledge, the means by which improved results are
achieved often cannot be transferred to the next process change,
redesign, or innovation.

These comments on the realized benefits of using statistical methods for
process improvement are somewhat negative. However, learning from these
early results provides substantial recommendations about the organizational
context within which process management should occur and consequently
informs as to the circumstances under which the methods can be most
effectively used.

1.3.2   A Recommended Context for Process Management

Early experiences in the use of statistical methods for improving organizational
effectiveness were lacking in two major respects; the focus was on individuals
at the lower levels of an organization without a clear definition of the role of
senior and line management, and the intent of the work was expressed by the
vague, undefined goal of "improving quality and productivity." Ideas
concerning the management of industrial organizations over the past 10 or so
years have addressed these early deficiencies. It is not the intent of this book to
present a full treatise on this body of management thought. However, the
major foundations of these ideas are presented next in order to provide a
description of the context in which process management must fit. Three key
concepts capture recommended management practice.

> ***Concept One:*** *Senior management must understand and accept their responsi-
> bility for determining what is currently valued by customers and what will be of
> more value in the future.*

A perusal of current management literature would indicate that this is the
decade of "providing value." But, as with most phrases, the content of this
one can be cheapened or truncated. Customer value must have a much

broader interpretation than the price a customer is willing to pay for goods of a certain quality. Instead, value is determined by the customer as some complex evaluation of the realizations (that which is experienced or provided) and sacrifices (that which is given up in establishing a relationship that provides products and services). For example, realizations of value for commercial customers may be confidence that a vendor can provide a product or service that upgrades and sustains their production line efficiencies, that customer product and process changes are known and anticipated by the vendor, and that knowledge exists to provide for seamless changeovers required by the customer. As another example, customers may value machines that require less setup time with less complexity in what must be done and that provide information on internal reliability and performance. In both of these examples, the desired realizations go well beyond product and service features.

> **Concept Two:** *The cross-functional systems of the organization that provide customer value must be defined, understood, and specifically managed by designated senior managers.*

The cross-functional systems of an organization are often referred to as *suprasystems* since these systems transcend the functions and departments of an organization. These suprasystems are the means by which the organization provides that which is valued by the customer. For example, one of the desired customer values discussed previously was the ability of the organization to provide goods that sustain the customer's line efficiencies. Meeting this customer value would require a system that integrates the processes that design, equip, and manufacture the goods required.

> **Concept Three:** *Line managers' responsibilities for aligning processes with the strategic, organizational systems must be understood and executed.*

The alignment of process objectives with the objectives of the systems that provide customer value carries with it two responsibilities for process managers:

1. to provide suggestions for systemic change, and
2. to identify and implement appropriate improvements in process operation.

These later changes must be in agreement with process goals and objectives, support systemic management, and cumulatively build process capability and results. The alignment of process management initiatives with systemic requirements and strategic objectives is crucial for continuous improvement work to be successful. Any change in a process is not desirable; customer value and the business strategy are the foundations from which critical processes are identified and managerial attention and ownership are determined. The overall strategy provides direction for improvement efforts with regard to service characteristics, product features, and improved efficiencies and reduced costs.

It is obvious that process selection for study, control, and improvement has a direct impact on long-range success. The intent of process study is to understand the behavior of processes, to gain insights into the mechanisms that affect the outputs of the processes, and to guide the efforts toward process improvement or redesign. Obviously, process study requires prioritizing work efforts. Those processes identified as critical should be the first among those selected for improvement efforts. These processes are typically high-level processes that often provide outcomes to the external customer. Processes at this level have many elements that interact and interface with suppliers and customers in order to successfully attain the process purpose. Critical processes are typically cross-functional in that their major elements are located across several departments and functions. Therefore, no one manager usually has responsibility for the complete process. Processes identified as critical for fulfilling the plant's responsibilities should be assigned to individual mangers for ownership and accountability. Operations and tasks within these critical processes, as well as other supporting processes, are selected for study and improvement based on their criticality in the maintenance and improvement of the identified critical processes.

1.4   THE IMPACT OF VARIATION ON ACHIEVING PROCESS OBJECTIVES

Any manager or engineer who has been required to report such numbers as weekly line efficiency, overtime hours, shift downtime, machine breakdowns, line stoppages, or yield knows that there is variation exhibited by such numbers. Also understood by managers and engineers is that numerous strategies exist for affecting such numbers; in other words, some understanding exists about the sources affecting the variation in the numbers. This section illustrates how the *depth* of understanding of variability can be used to guide process management. Knowledge of the nature of the variation in such numbers, the causes and sources of variation in the numbers, and the effect that different engineering or management practices would have on the numbers will be useful in realizing how improvement in the numbers might be effected. As the depth of understanding about the variation in process outcomes increases, so does the range of viable and effective choices for improving process performance.

Variability results from the interaction of causes found in the procedures, people, material, and equipment used to create a product or service; variation of a given magnitude is a consequence of a way of doing business. In general, variation in results is not deliberately created by an organization; the sources of variation are not always known or appreciated. For example, variability in the cost or performance of a complex mechanical or electrical part may be partly due to manufacturing and assembly capability and largely due to materials selection and the actual design itself, the selection and implementation of the manufacturing process, or the specification and procurement of part and materials. Knowledge of the effects of variation, in terms of cost, capability, or performance, would be essential in deciding if changes were necessary or potentially beneficial. Perhaps because variability is always present, the effects of variation are not always well known or understood.

1.4.1   Fill Weights

Fill weight for a granular product shipped in containers with a given label weight provides an example for considering the impact of variation. There is a target value for container content. The target has been set based on the need to have content greater than label weight and by some understanding that process variability exists. If the filling process were managed such that it produced stable variation in fill weights, then managers could confidently determine the target, taking into account the existing, known magnitude of variation in fill weight. When this knowledge is lacking, target selection is influenced by the organization's lack of "confidence" in its ability to perform in a predictable fashion over time. Because a minimum value must be maintained, erratic, unpredictable, variability generally means that an even larger target value is set in order to ensure that the lower specification is met. Figure 1.2 provides a graphical illustration of the high target value that must be set when variation in fill weights is erratic.

Advantages of predictable and decreased variation in fill weight are easy to understand by considering graph B in Figure 1.2. Stable variation provides knowledge as to an appropriate target value. In addition, with knowledge of the amount of variation present in fill weights, estimates of the cost associated with a given level of variability can be calculated with confidence. Stable variation thus provides the opportunity to consider possible benefits from working to reduce the variability. Decreased variability in fill weight, as in graph C of Figure 1.2, provides management with the possibility of lowering the targeted fill weight, thus reducing costs and increasing efficiency and throughput time. This option, however, calls for additional knowledge as to what sources affect the average, by how much, and in what direction.

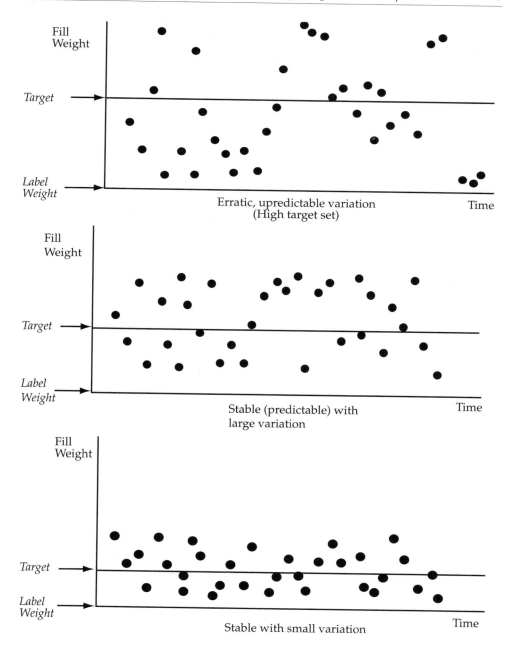

FIGURE 1.2   *Three Possible Descriptions of Fill Weight Measurements*

The advantages of predictable and decreased variation in fill weights are easy to recognize. The preceding discussion also implies certain types of actions and analysis by process managers. Process managers should be measuring and analyzing variability in actual fill content. To be able to effect sustained improvement, managers would have to know those operating practices, as well as those material, equipment, and environmental properties that affect variation and its stability over time, and would have to be willing to act on that knowledge. They would have to accept the charge that it is their job to do these things.

1.4.2   Capsule Wall Thickness

The importance of management's understanding of the implications of variation is discussed in this example.

For fill weight, large, though predictable, variation in net content meant a financial loss due to the practice of overfilling to maintain a lower specification, a realization that is immediately and easily understood and appreciated by the financial people. The relationship of variation to costs and benefits is not always as easily understood. Gelatin capsules, widely used in the pharmaceutical industry, have a product characteristic that offers a different perspective on what variability might mean to management. An important property for gelatin capsules is wall thickness. The current target value for wall thickness has been specified in order to provide the material in the capsule with the necessary protection from the environment and to ensure compatibility of the capsule with the customers' filling equipment. There is no need to change average wall thickness; indeed there are strong reasons for not changing the average. Because average wall thickness must remain unchanged, there is no opportunity to lower the average and save material after achieving decreased variation in wall thickness. However, suppose that current process variability consistently produced capsules having wall thicknesses outside of lower and upper specifications. This large variability has several consequences; the capsules must be sorted to remove those of an incorrect size, thus additional material is used to manufacture capsules that cannot be shipped, and machine and other resources are used to produce product that cannot be used. The costs associated with these losses are neither easily identified by current cost accounting philosophies nor searched for by current management practices. The process may, for example, be on budget because standards allow for a given amount of waste. If the costs are not seen to exist, then the benefits may also not be recognized.

Internal benefits can be traced to the decreased variability. After the improved variation has been documented and proven repeatable, there will be less inspection and cycle times will be increased. The financial gains may appear to be modest. However, other, and possibly more significant, benefits are to be found external to the business. With shipments of capsules having known, predictable values within a specified range, the customer can have greater assurance that a shipment of capsules will run on his or her equipment without machine stops, leading to higher equipment utilization and improved cycle times with attendant economies. This assurance in supporting high efficiencies by verified improvement in material properties can provide a competitive advantage. In addition, there may be a competitive advantage in having gained the process knowledge demonstrated by the ability to reduce variation. This improved knowledge allows the manufacturer of the capsules to be better able to respond to new information on customer needs in wall thickness, as well as to become a valued supplier of the customer.

It is significant that the benefits of decreased variation in wall thickness do not result in an immediate return to the manufacturer of capsules. Again, the manager must have understood the implications of variability in this process parameter; a narrow attitude will not uncover opportunities for gaining new customers or securing old ones. In this instance, the management group needs more than process knowledge. They must know what it is that their customers might value or where diminished sacrifice in use would offer potential gains. They must be able to correlate these attributes with internal systems and they must assume the responsibility for the stabilization and improvement of those systems. Experience indicates that financial audits are usually focused on the more narrow, internal evaluation of costs and benefits, thus directing attention away from the range of possibilities to be had from decreased variability. Reliance on existing financial models will not identify the financial gains that can accrue from managing the variation in processes that supply what the customer wants. Because of the complexity involved in thinking about and considering the effects of variation, it is imperative that a manager work through what excessive and erratic variation might mean in the fulfillment of system responsibilities. These system responsibilities include knowing and working to learn further about customer concerns. For a particular process, the determination of the costs or disadvantages associated with variation must be determined by the manager. By beginning to understand the effects of variation, the manager is in a position to make an objective analysis regarding the benefits of reducing variability.

1.4.3   Throughput Rates

An example is given here of the need to understand the sources of variation and their impact in order to choose improvement or redesign strategies.

Understanding the sources of variation in an output result is only part of the story. There must also be a motivation for first developing and then using that understanding. It is necessary that the manager understand clearly why it is that a particular system exists. Without this knowledge the manager is in no position to make informed judgments about variability and the benefits of its reduction, movement, or elimination. Understanding the sources of variation and the effect of variation in the context of system analysis will affect the way in which a manager approaches work.

Consider a situation in which an increased throughput rate is desired. The rationale for that objective is not questioned here; rather, the needed increase will serve as a starting point for a short discussion on tactics for obtaining the increase. Increased throughput could be attained in a variety of ways. It is the selection from among the alternatives and the rationale for choosing an alternative that forms the centerpiece of this simple example. Typically, the expectation that throughput be increased is expressed forcibly at the plant level. However, the management and staff may not have realistically considered how to achieve the requested gains. Some tactics for achieving the increase may be stated very specifically, such as working to eliminate a known bottleneck in one or more of the processes producing product or requiring that the throughput rate of each unit in the facility be increased. However, the expressed focus may not be consistent with the actual system sources of variation that deliver the current throughput levels. An overemphasis on the result, the throughput itself, may promote practices contrary to other expectations and needs required of the business. For example, under pressure to increase throughput, a department may release poor-quality material. By making this choice, the throughput rate may increase while the yield stays the same. As a further consequence, the opportunity to achieve improved throughput by increasing the ability to produce high-quality product consistently at each stage of operation is foregone because of the concentration on the schedule rather than on other system parameters. There are other losses. The capability of the people has not been increased because no process knowledge has been gained. The management group has not been strengthened by acquiring different behaviors by which to manage in the future. There may be no sustained experience in working to reduce defects; therefore, there is no assurance that the approach will yield appreciable benefits. There is no confidence that working systematically on the input side will result in improved throughput capability. Under pressure to attain or sur-pass schedule, equipment may be run without adequate maintenance, having an impact on quality and future plant abilities and well-being. Shifts, treated as if they are independent production units, are goaded toward quotas. The effect may be to delay appropriate maintenance or set aside the opportunity to better choose when to perform maintenance. If each shift is pressured to achieve a certain production quota, the effect is often to set one shift working against the others. Production knowledge is hoarded; appropriate work is shifted to other shifts or times by a variety of means. Each shift concentrates on getting its own quota, often emptying the line of all work in progress in order to achieve the objective. Thus a larger start-up job is left for the next shift. The idea of running the operation in a smooth, consistent fashion over the different shifts is not considered, nor are the benefits of improved throughput that might result.

A different practice for working to improve throughput levels would involve understanding the sources of variation contributing to the level and variability in throughput rates, evaluating those sources, and selecting where and when to make changes to improve the existing system. Implied in this statement is that there exists a multitude of sources that affect each other as well as throughput rate. The idea of searching for one, or the most prevalent, cause of deviations from a standard is to be replaced by the idea of understanding system behavior and the variations in components of the system that affect throughput levels. In examining the total system for making product, management might begin to look at specific activities in a different respect, finding opportunity in places pre-viously unexamined. Numerous setup changes and within-run modifications to accommodate raw material variation reveals that purchasing is not attached to production. At least two issues surface: the obvious impact on productivity and efficiency of frequent setups and modifications, and the evidence of system breakdown between purchasing and production. Examining the interface between purchasing and production offers large returns for management work.

1.5   THE SCIENTIFIC METHOD AND PROCESS MANAGEMENT

The different dimensions of process management are captured in Figure 1.3. Process managers need to translate what is valued by the customer into process requirements; process requirements need to be defined in terms of measurable properties of the process; and, measured properties of the process must be evaluated to assess capability to provide what is valued to the customer. As managers work to improve process operations, at the same time they work to identify solutions for the problems that plague the current operation. These two different efforts, process improvement and process problem solving, form two separate activities for successful process management that proceed concurrently with each other. The second activity, problem solving, should not be mistaken for process improvement. Although the continued operation of a process will, of course, require that problems be addressed, managers who rely only on apparent "problems" to identify work efforts will not be focusing on the dimensions reflected in Figure 1.3. The following section describes the limitations of relying only on problem solving to manage processes.

1.5.1   The Limitations of Problem Solving

A severe limitation of problem solving is that when work efforts are based solely on addressing apparent problems, the work is usually internally focused. Consequently, efforts are not spent on improving what the process provides to better meet customer values. An illustration of this phenomenon is provided by a firm that produces large medical devices for patient

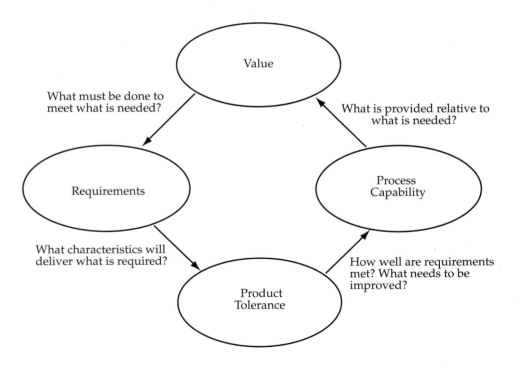

FIGURE 1.3 *Dimensions of Process Management*

Initial Run

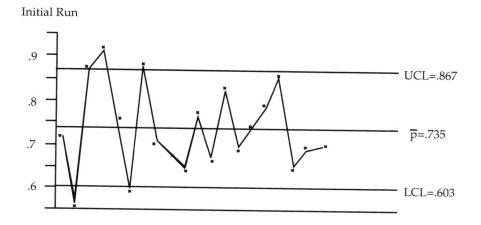

6 Months Later

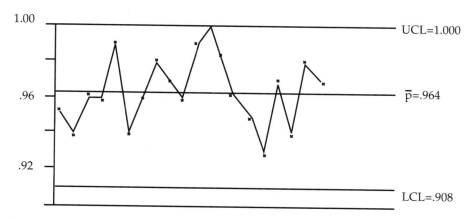

FIGURE 1.4 *Charts of Proportion of Assemblies That Meet Requirements on the First Pass*

diagnostics. Every one to two years an upgrade to the current model is made in response to improved technology. With each new upgrade, problems with assembly are experienced. Managers of the assembly process have consequently become quite expert at finding solutions to these problems. Typical results in assembly are captured in the graphic of Figure 1.4. The top chart in this figure shows that the proportion of assemblies that do not require rework behaves erratically (points fall outside of control limits) and the average proportion that did not require rework was 0.735. In other words, about 25% of the assemblies in the initial assembly run did not initially meet requirements. Six months later, information on the assembly process tells a different story. The proportion of assemblies that meet requirements without rework behaves more consistently, and the percentage that initially meet requirements is now very close to 100%.

Several additional points should be emphasized. First, the customers did not see the product before it was reworked. The organization verifies the quality of each device both before shipment and after installation. Second, the problem solving that was so effective with the device described in Figure 1.4 has been repeatedly used with the same success on each new device. Consequently, the limitations of problem solving in this manner may not be immediately apparent. (Certainly not to the managers of assembly who were being rewarded for their excellent problem-solving skills.)

However, a different perspective on the initial run control chart would be that this chart does *not* simply provide information on the assembly process; this chart also reports on the organization's ability to design and hand over to manufacturing and assembly the newest innovation. The fact that each new device displays similar results to the initial run chart of Figure 1.4 indicates that a critical element of process management is going unattended. Because the ability to initially produce devices is erratic, there will not be accurate knowledge of when devices will be available for the customer and promised delivery dates may not be met. The management group at this site would begin to work on quite different aspects of the assembly and design processes if the focus were on being able to provide customers with the new technology, without the added time and expense of reworking devices.

The preceding example provides an excellent contrast between problem solving and process improvement. Although there is no universally accepted definition of problem solving, here it is defined to be corrective activities that take place on recognition of a product, service, or process difficulty. Correcting these difficulties must be done, but that is not what is meant by process improvement. Process improvement takes place by changing process inputs or parameters to provide better results. Better results might be defined, as in our last example, as an improved ability to schedule to meet customer needs. At other times it might be reduced costs, improved resource use, improved product or service characteristics, etc. The distinction between problem solving and process improvement is an important one for process managers to understand. Management activity cannot remain focused on the reactive methods of problem solving. Time and energy must also be placed on improving what is provided to customers. Removing those things that create difficulty or dissatisfaction does not ensure that customers are being provided with the correct products and services.

1.5.2   Systematic Approaches to Process Improvement and Problem Solving

The scientific method is the foundation for most systematic process improvement and problem-solving strategies. The underlying principle of the scientific method is the collection of data (both facts and measurements) through observation or experimentation to verify (or not verify) conjectures or hypotheses about a given situation. A common representation of how this kind of systematic thinking can be applied to process study is the Plan-Do-Check-Act (PDCA) cycle, shown in Figure 1.5. The PDCA cycle is an expression of the discipline that should be used in process study. The planning phase of the cycle captures the need to have well-defined objectives prior to collecting data and certainly before implementing changes. Impacts of suggested changes are studied prior to permanently implementing those changes. Possibly the most powerful message conveyed by the PDCA cycle is the ongoing nature of the work to improve process operations.

Practicing the discipline of process study conveyed by the PDCA cycle is an

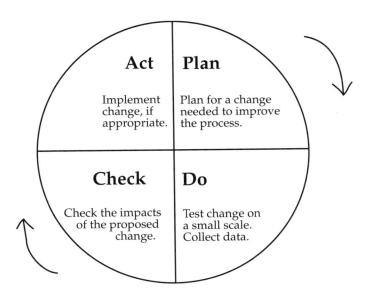

FIGURE 1.5 *The Plan-Do-Check-Act Cycle*

important aspect of process management. The use of the PDCA cycle implies the identification of critical process measurements, the collection of representative data, and the use of tools and methods to understand the process and analyze the impact of changes on those process measurements. The PDCA cycle is a helpful description of how to use sound reasoning to study a process, but it was never intended to serve as a road map of process management. It is woefully inadequate for this purpose. Missing from the study methods described by this cycle is the purpose for process study and how this purpose is aligned with identified process objectives.

Numerous organizations have found the development of a prescribed plan a valuable technique for promoting the correct practice of process management. A good example is provided by the AT&T methodology.* The seven steps of this methodology follow:

Step 1
    Establish process management responsibilities
Step 2
    Define process and identify customer requirements
Step 3
    Define and establish measures
Step 4
    Assess conformance to customer requirements
Step 5
    Investigate process to identify improvement opportunities
Step 6
    Rank improvement opportunities and set objectives
Step 7
    Improve process quality

There are two critical elements of this methodology. The first is the congruence of these steps with the objectives of process management. These objectives, as previously stated at the beginning of the chapter, are

- the alignment of process objectives with identified customer values,
- knowledge of the current process configuration and capability to attain process objectives, and
- the selection and direction of needed process improvements.

---

* Adapted with the permission of AT&T © 1987, 1988. All rights reserved. *Process Quality Management and Improvement Guidelines*, AT&T Bell Laboratories ©, 1987, 1988

The second critical element of this, or any organization's, methodology for process management is not that it is articulated, but that it be adopted and practiced.

Organizations and individuals may choose to develop their own prescribed plans for process management or adopt one like the one used by AT&T. In either case, the practice of process management articulated by these plans must reflect the responsibilities for process management. These responsibilities, though stated earlier, are summarized in the following list. The statistical methods discussed in the remainder of this book are tools that aid in the accomplishment of these responsibilities.

Responsibilities of Process Managers

Responsibility 1

Understand and disseminate information on what the customers of the organization value.

Responsibility 2

Know what and how the process contributes value to its customers.

Responsibility 3

Develop and deploy practices and methods that provide the intended results and confirm that these methods can be implemented.

Responsibility 4

Know the past and current performance levels of the process in achieving its objectives.

Responsibility 5

Assess current process performance against requirements.

Responsibility 6

Make appropriate, constructive, and verifiable process changes that enable the process to improve significantly its contribution to creating customer value.

2   Tools for Process Study

Chapter 1 discussed the value of managing process variation to improve the processes that provide customer value. This concept of managing variation was illustrated in Chapter 1 with the aid of several business examples, including

- the savings achieved from reduced variation and improved targeting of fill weights,
- the increased customer value provided by reduced variation in the wall thickness of gelatin capsules, and
- the importance of working to increase throughput rates.

In each of these examples, improvement in process operation was possible because the necessary process knowledge was in place prior to the analysis of process variability. Work to define and understand the process inputs, stages, outputs, and the interrelationships among these elements ensured that the prerequisites for process study were in place (see box).

Prerequisites for Effective Process Study

1. The critical characteristics of process operation that need to be studied are known.

2. The measurements made on the process provide correct information about the critical process characteristics.

3. The process parameters that affect the critical characteristics are identifiable.

These three pieces of process knowledge provide guidance and focus for the management of process variation. This chapter provides techniques that are useful to the manager and engineer in discovering and organizing these three types of process information. The construction of process flow diagrams is illustrated as a tool for identifying the critical characteristics of process operation. The importance of operational definitions, which specifically describe how measurements are to be made, and the usefulness of check sheets for collecting the data are described. Finally, the construction of Pareto diagrams for summarizing process data and the development of cause and effect diagrams for understanding causal factors are described. Cause and effect diagrams occupy an important role in process study, because they capture information about factors, or causes, that are known or thought to affect process parameters and outputs. Information on possible factors that affect process variation provides critical insights into possible means of managing process variation.

2.1   USE OF FLOWCHARTS FOR DESCRIBING PROCESS OPERATIONS

A process flowchart is a schematic picture of the sequence or relationships of steps or activities in the current operation of a process. Prior to considering the methods for constructing a flowchart, it is beneficial to consider some of the benefits that are gained from the construction of flowcharts (see box).

Information Provided from a Process Flowchart

1. Collection of knowledge on current process operation and an identification of areas where process knowledge is lacking, misunderstood, or not uniformly applied.

2. The identification of critical steps in the process, where the outcomes from these steps are critical to better process operation.

3. The identification of redundant or unnecessary process steps.

4. The development of an understanding of the relationships between inputs into the process, individual jobs and responsibilities, and the outcomes of the process.

5. The provision of a common reference point for team discussions.

6. A determination of key points in the process for study, including points for data collection.

To accomplish all of the purposes listed on the box, no one flowchart of a process is adequate. Instead, a series or collection of charts is often useful to capture the different kinds of process information needed. The following example on the construction of hornpad assemblies illustrates the use of more than one flowchart to describe a process while also describing the basic elements of flowchart construction.

2.1.1 Constructing Flowcharts: The Hornpad Assembly Process

A manufacturing plant produces hornpad assemblies, which are purchased by different automobile manufacturers for use in the steering wheels of automobiles. To remain a valued supplier of these assemblies, the plant has begun an initiative to continually improve its ability to supply high-quality assemblies on schedule within a reasonable cost structure. This focus led the plant to begin developing a process flowchart for the hornpad assembly process. A flowchart is developed to provide a reference point for individuals engaged in the process and to use for discussions of improvement opportunities. Figure 2.1 contains the initial flowchart of the hornpad assembly process. This initial flow is used as a baseline against which other views of the process flow can be evaluated and deviations from the designed flow can be understood. Each of the boxes in the flowchart of Figure 2.1 represents a process step. The arrows describe the direction in which the process "flows." In the hornpad process, these flows represent the movement of materials through the process. The ovals represent steps outside of the boundaries of the process. Even though the stages represented by ovals are outside of the actual manufacturing process, it is sometimes necessary to address problems with these stages in order to achieve the desired process improvements.

The information contained in a flowchart is useful only if the flowchart adequately represents process operation. The following guidelines for constructing a flowchart are useful to ensure the adequacy of this representation:

Guidelines for Flowchart Construction
Guideline 1
   Determine and define the purpose for the flowchart prior to constructing the chart. For the flowchart of Figure 2.1, the purpose is to provide a common reference point for understanding process operation. Thus, a general description is adequate. A different purpose may require more detail or a different type of flowchart.

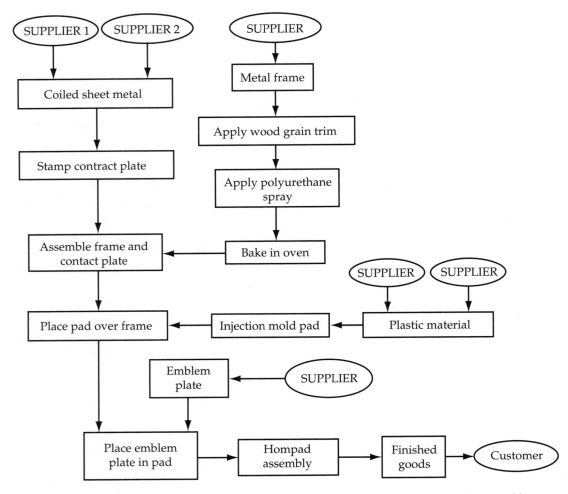

FIGURE 2.1 *Flow Diagram of Manufacturing Process for Hornpad Assemblies*

Guideline 2

Determine appropriate process boundaries. The following questions are helpful in determining the process boundaries:

- Where does the process start?
- What are the inputs into the process?
- Where does the process end?
- What are the outputs of the process?
- What departments or functions are involved in the process?

Guideline 3

> *Determine the appropriate persons to be involved in the flowcharting work.* It is important to have individuals from various shifts and different departments, since initially there are often different understandings of how the process operates.

Guideline 4

> *Include only as much detail as is needed to achieve the stated purpose for the flowchart.* The level of detail on the flowchart will vary according to the needs of the individuals using the chart, the purpose of the chart, and the maturity of the process study. For the hornpad assembly flowchart, the intent is to provide a common reference point for understanding process operation. Thus, a fairly general process description is provided. As different elements of the process are investigated, more detail will undoubtedly be required.

Guideline 5

> *Include all steps that are a part of the process. Chart the actual sequence of steps, not the sequence by design or procedure.* The earlier identification of the process boundaries provides a framework within which this question can be addressed. While constructing the flowchart, differences in the sequencing of the steps are often uncovered. For example, if different conditions exist on different shifts, different flowcharts are necessary for the different shifts. In other cases, one particular sequence of steps may give superior results. Then, further information about the effects of sequencing of steps on process outcomes is needed in order to decide which sequence should become the standard captured on the flowchart.

Guideline 6

> *Validate the flowchart.* To ensure that the process flow is adequately represented by the chart, those involved in the process should be consulted about any changes that are needed to improve its accuracy.

2.1.2   Adding Additional Detail to a Flowchart

The flowchart of Figure 2.1 provides an overview of the steps involved in the production of hornpad assemblies. However, this basic flowchart does not necessarily capture adequate detail about actual process operation. For example, the manufacturing plant desires information about how material flows through the process steps. However, the flowchart in Figure 2.1 only captures the sequence of operations performed on the hornpads. It does not capture the way in which material flows through

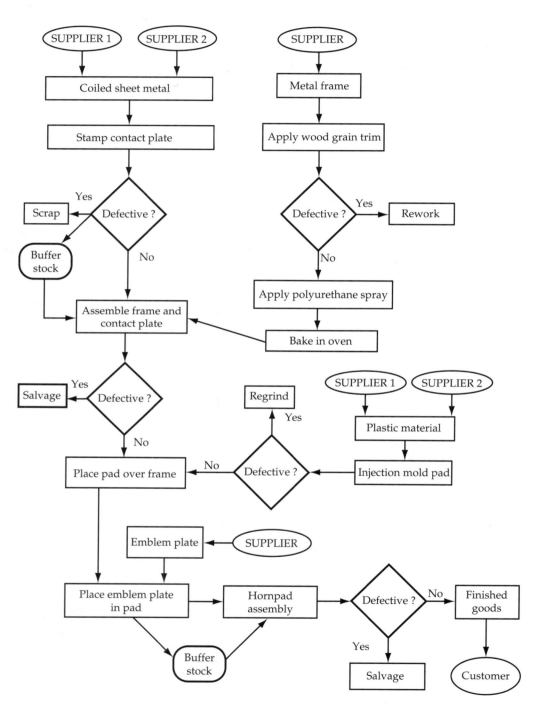

FIGURE 2.2 *Revised Flowchart of Hornpad Assembly Process*

the process. To better capture actual material flows, we can modify the flowchart of Figure 2.1 by providing more detail on how material actually moves through the various steps of production. The resulting flowchart is shown in Figure 2.2. Note that another symbol, the diamond, is used in Figure 2.2 to describe a decision point in the process.

The flowchart in Figure 2.2 provides further information on the material flow in the hornpad process as well as insights into a direction for process work. The steps outlined in bold capture points in the process that are primarily inspection points or storage of buffer stock. Of course, both inspection and buffer stock are currently necessary because of the inability to carry out each stage correctly in the correct amount of time. Identification and removal of the needs for the inspection and buffer stock points will allow the process to run in a more efficient fashion. Thus, future work will include developing the capability at each process stage to run the operation correctly in order to eliminate any scrap or rework and the need for costly inspection. Once the output from each stage is consistently reliable, it is possible to reduce, or possibly eliminate, the need for buffer stock. The decision to work on the capability of the process to deliver correct product requires that the characteristics defining correctness, or "quality," be determined. The next two sections on data collection and data summaries address this issue.

The flowchart in Figure 2.2 captures additional information about how materials flow through the process. Clearly, other additional types of information can be added to the original flowchart of Figure 2.1. Further valuable detail can be incorporated by including information on where samples are taken and measurements are made, on the number and type of equipment used at each step, on the times spent at each stage of production, and on other types of information useful to those working on or within the process. The flowchart is most useful as a working document, which is used to capture pertinent information about the various aspects of process operation.

2.1.3   Restructuring Process Flow

The construction of process flow diagrams often reveals redundant and non-value-added steps in a process. Redundancy occurs in a process when a function, task, or activity is being done more than one time; that is, when functions are duplicated in the process. For example, in the assembly process of manufactured buildings, the materials and parts are inspected as they are received into stock. Due to a past occurrence of defective materials getting past this initial inspection, some of the materials are tested again prior to final assembly. Thus, a redundant inspection activity is built into this process. Also, redundancies commonly exist because the numerous personnel involved with a process unknowingly repeat a procedure that is already being performed elsewhere. Non-value-added steps are activities or functions that could be removed from the process without taking away from the critical characteristics of the outputs if each step of the process operated correctly. Redundant and/or non-value-added steps are often deliberately built into a process in order to handle poor results at one or more stages of a process. Obviously, redundant steps do not add value because they would not be needed if the process operated correctly. In either case, exploration into the causes of redundancies and non-value-added activities provides useful direction for process study.

The flow diagram in Figure 2.3 describes the steps in the order process within the parts department of a heavy equipment manufacturer's service center. This example demonstrates the flexibility of process flow diagrams because the diagram in Figure 2.3 actually describes two process flows in the same diagram. The steps outlined in bold and the bold arrows show the flow for items that are designated as priority items, while the stages with non-bold outlines capture the order process when an item is not designated as a priority item. Those steps shared by both priority items and nonpriority items are outlined twice, both in bold and non-bold lines.

Process flowcharts, supported by information flows, decision points, and indicators of time and resource utilization, provide data for making changes to achieve specific process attributes. An examination of the process in Figure 2.3 led the manager of the parts department to conclude that the current order process contains redundant and non-value-added steps that cause significant delays and unnecessary costs in the ordering process. It is likely that, under previous departmental changes and business growth, steps have been added and changes have been made over a period of time without determining the long-term effects of the changes on process complexity or performance. To visualize a more efficient process and to provide a reference for explaining future order process operation to the purchasing and accounting groups, the manager worked with others involved in the order process to develop the flow diagram shown in Figure 2.4. However, before making any changes to the process, the manager realized the importance of establishing metrics and evaluating the current level of process performance. With this type of baseline data, it is easier to determine and report any improvements in effectiveness and efficiency obtained by the process changes.

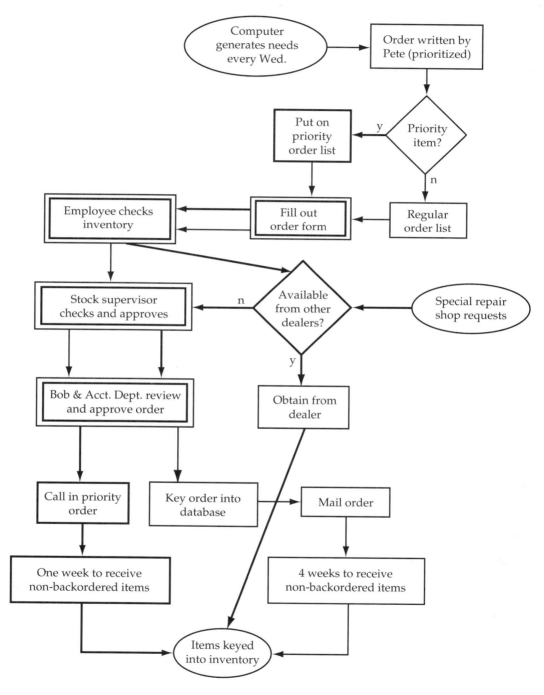

FIGURE 2.3 *Current Parts Order Process*

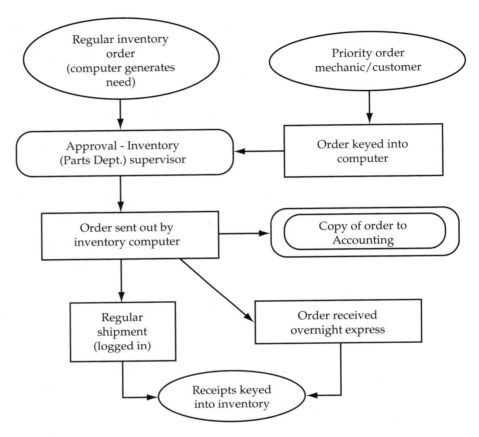

FIGURE 2.4 *Ideal Order Process*

2.1.4   Top-Down Flowcharts

The flowchart of Figure 2.1 captures the major steps in the hornpad assembly process. When trying to track the actual flow of material in the process, we saw that a more detailed chart was required. Other process issues in the hornpad assembly encourage the addition of different types of information in the chart of Figure 2.1. Although adding some detail to flowcharts is often necessary for gaining specific insights into process operation, too much detail (without reason) typically causes difficulties in gaining an overview of process operation. Top-down flowcharts are useful in organizing more detailed information in a clear manner. In a top-down flowchart, the major process steps are drawn in flowchart form across the top of the page. Then, underneath each of the major steps, more detailed information is captured. Figure 2.5 illustrates the construction of a top-down flowchart that describes the production, bottling, and packaging of chocolate syrup. This chart was

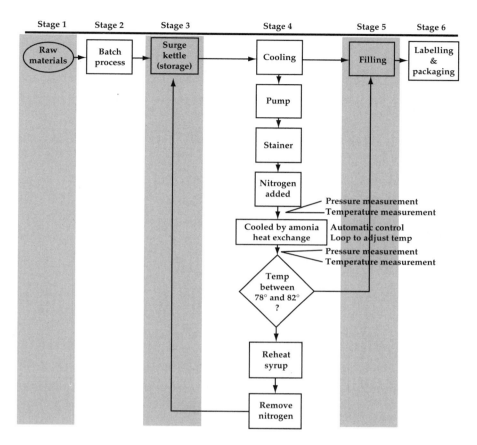

FIGURE 2.5 *Top-Down Flowchart of a Syrup Process*

developed to help a group focus on the bottleneck in the process being caused by a difficulty in cooling the syrup to the correct temperature before filling plastic bottles with it.

2.1.5   Summary

The construction of process flow diagrams is an essential early step in process study. As work on a process proceeds, the updating and detailing of flow-charts are performed. Used effectively, process flow diagrams provide valuable insights into critical process parameters to be measured and tracked, opportunities for reducing process complexity or processing time, and key processing relationships. Figure 2.6 summarizes the basic flowchart symbols— the circle, rectangle, and diamond—and their use. Appendix C provides a description of additional flowchart symbols that can be used when more

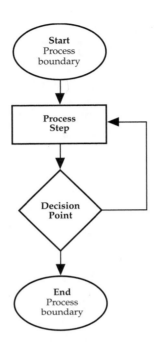

FIGURE 2.6   *The Basic Flowchart Symbols*

detailed information is needed in the process flow diagram. The lines and arrows indicate the sequence of process steps and/or the direction of materials flow.

2.2   PLANNING FOR DATA COLLECTION

A viable process description provides a common understanding of how a process actually works. It identifies key process characteristics and critical leverage points in the process, which are potential points for measurement. However, prior to initiating the actual measurement process, further investigation and planning are required to determine the specific data to be collected. Measurements are taken for a variety of reasons. Often, data are obtained on final product items in order to determine whether or not the product meets customer requirements. Specific characteristics are measured to compare the actual characteristic measurement with the specification limits. These types of measurements are used simply to describe process outputs numerically. Process improvement requires information on the relationships between causal factors and process outcomes and on the effects of critical variables. Data do not automatically provide information. Planning a data collection strategy involves determining the information desired from the data, identifying the appropriate point of measurement, and deciding on the type of data that will provide the desired information.

2.2.1   Types of Data

Even though data are commonly collected for documentation and reporting, the primary purpose of data collection should be to obtain information to support process changes that improve the future outcomes of a process. Given this purpose, the type of data examined dictates proper summary and analysis of the observations contained in the data set. The two major categories of data that are typically analyzed in process study are variables and attributes data. Variables data are commonly referred to as metrical or "measurement" data and attributes data are frequently called "count" data. These two types of data differ in their statistical natures because they provide different types of information and require different methods and statistical tools for analysis.

Variables data are obtained from a measurement process that results in continuous values. When the characteristic being measured can realistically obtain an infinite number of values, the actual measurements are referred to as "variables" data. For example, a measurement of the outer diameter of a rubber seal may result in the value of 5.44 mm. However, the outcome of this measurement is dependent on the precision of the measurement gauge. The actual diameter can possibly take on an infinite number of values within the range of 5.4351 and 5.4449. Other common examples of variables data are weights, times, dimensions, flow rates, temperatures, and other characteristics that can take on any numerical value on a continuous scale.

Attributes data usually result from a counting process. For example, a count of the number of blemishes on a bolt of fabric results in a discrete number (i.e., an integer). In general, a count of the number of defects results in a measurement that is referred to as an "attributes" data point. Attributes data also result from a count of the number of defective items that exist in a larger lot or group of items. Consider the final inspection of tubing for air conditioning units. Stainless steel tubes with burrs, scratches, or discolorations are considered to be defective. The inspection results are reported in terms of the proportion or percentage of tubes that are defective. Count data that is converted into a percent or a proportion is also attributes data, since the reported proportion cannot take on an infinite number of values. If 500 stainless steel tubes are inspected and the number of types with unacceptable burrs, scratches, or discolorations are counted, then only integer values will result for the counts. A count of 35 nonconforming tubes out of a lot of 500 will yield a proportion defective of .070, while a count of 36 tubes will result in a proportion defective of .072. It is impossible for the proportion of defective tubes in a lot of 500 tubes to take on the intermediate value of .071.

2.2.2   Operational Definitions

Attributes data are obtained to provide information on product characteristics that cannot be or are not chosen to be measured on a continuous scale. Anytime data serve as a basis for action on a product or process, consistent communication on product or process requirements and on decision criteria is necessary for effective process management. The information provided from the data usually refers to characteristics of the product or process that must meet certain requirements in order to obtain outcomes that are usable by subsequent process steps and by the final customer. Without the existence and implementation of an operational definition regarding a specified quality characteristic, the determination of whether or not a characteristic meets the necessary requirements is based on individual, subjective judgments. What results is inconsistency in meeting customer requirements. Operational definitions ensure consistency by providing clear and precise

---

Characteristics of an Effective Operational Definition

1. The levels of the quality characteristic that are considered to be acceptable are communicable. Hence, all persons have the same understanding of the criteria on which decisions are to be made.

2. There exists a consistent, agreed-on method for the evaluation or measurement of the quality characteristic. The equipment to be used to measure the characteristic, the method for sampling the product items, and the way in which the measurement is to be obtained are clearly defined.

3. The decision made from the evaluation of the quality characteristic is the same, irrespective of the person making the decision. The criteria yield a clear yes or no decision on whether or not the specifications are met.

Source: These characteristics of an effective operational definition are paraphrased from *Out of the Crisis,* by W. Edwards Deming, by permission of MIT and the W. Edwards Deming Institute, published by MIT, Cambridge, MA, 1986.

criteria concerning the process or product characteristic of concern and the measurement of that characteristic. Thus, a clear understanding of the requirements is a prerequisite to the development of an effective operational definition. The characteristics of an effective operational definition are listed in the box.

Management is responsible for making sure that practicable operational definitions are in place, that they correctly describe the requirements, and that they are practiced by the engineering, production, inspection, and user groups. The correctness and adequacy of an operational definition is determined through an agreement between the producer and the end user. Consider the situation in which bolts of fabric are inspected for dye imperfections. Without an operational definition to clearly specify what constitutes a discoloration, it is realistic to assume that different inspectors might make different judgments regarding the acceptance of the bolts of fabric. Figure 2.7 shows a possible operational definition for classifying dye imperfections as either acceptable or unacceptable. Such an operational definition provides for consistency in inspection results, in the data collected, and, hence, in the decisions based on the collected data. An operational definition that correctly describes the required product or service characteristics is essential for reaching agreements between the customer and supplier on the requirements and for ensuring that the product or service delivered to the customer consistently meets these requirements.

2.2.3   Developing Check Sheets

Information obtained from collected data and the effectiveness of decisions made from the resulting measurements depend heavily on the planning of the data collection strategy. To determine causes of variation in process outcomes and to make improvements to the process, the collected data should correspond to process parameters that impact process outcomes. In most organizations, a large amount of historical data is collected and stored. Often, historical data do not contain information that is meaningful or useful as a basis for decision making. To gain pertinent information from data, a data collection strategy should be developed that is based on an accurate description of the process and on potential causal factors that affect process outcomes. The data collection strategy dictates the quality characteristics to be measured, the point in the process where the measurement is to take place, and the type of data to be collected.

Once the desired information is defined and the data collection strategy is developed, implementation requires that the collected data be carefully and accurately recorded. Check sheets (i.e., data collection sheets) are a means

**Criteria for determining whether a 500 yard
bolt of fabric meets specificactions
regarding dye imperfections**

Measure a pice of fabric from the bolt that is 100 linear yards and
48 in. wide.

In this area, locate all possible changes in coloration.
Compare each possible change in coloration to Prototypes I and II.
If the discoloration is darker than the upper limit shown on
prototype I, the discoloration is marked as a potential defect.
If the discoloration is lighter than the lower limit shown on
prototype II, the discoloration is marked as a potential defect.
Otherwise, the discoloration is not marked as a defect.

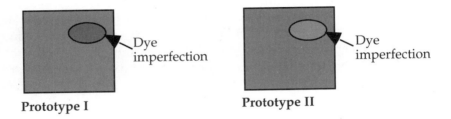

Prototype I   **Prototype II**

For each potential defect, using a ruler that measures in
millimeters, measure the range of discoloration at the two
most distant points. See figure below.

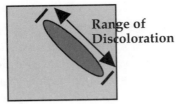

If the range of discoloration is greater than 0.12 mm, the
discoloration is to be classified as a defect.
If the range of discoloration is less than or equal to 0.12
mm, the discoloration is not to be classified as a defect.

FIGURE 2.7 *Operational Definition for Dye Imperfections on Bolts of Fabric*

Information to Include on a Check Sheet

1. The times of data collection and identification of the person(s) recording the data.

2. Any conditions or changes that may impact the results revealed by the measurements, such as changes in temperature, maintenance procedures, adjustments to machines, and line stops.

3. Auxiliary process information, such as suppliers of items being measured, lot numbers on incoming raw materials, machine numbers, and part numbers.

---

of organizing and simplifying the data collection process. A carefully designed check sheet not only makes the recording of the data easy and less time consuming, it also improves the accuracy of the information recorded and facilitates data interpretation. A check sheet should provide space for recording all information that provides insights into the ability of the process to consistently provide outputs with the required characteristics.

A check sheet that aids in obtaining information useful for process change is designed only after a clear understanding of the process and the possible causal factors are developed. It is likely that "items" will be deleted and added to the check sheet as process knowledge is gained, as certain issues are resolved, and as new customer issues are discovered and defined. Successful process work means that the type, nature, and frequency of defects and issues change over time. Thus, no single check sheet is relevant and adequate forever.

The desired information and the type of data to be collected dictate the structure of the data collection sheet. Therefore, it is not possible to provide a prototype of a check sheet that will be useful in all situations. However, the development of a check sheet for evaluating the quality of wooden plaques provides an illustration of the construction of a check sheet. To gain insight into the causes of different types of defects on the plaques, it is necessary to know not only the types of defects and their frequency of occurrence, but also the locations of the various types of defects. To obtain this information, a check sheet is designed so that the inspectors and workers can easily mark both the type and the location of the defect. The check sheet shown in Figure 2.8 is easily understood by management, engineering, and operators. It clearly conveys information on the types and locations of defects, and it points to possible process steps where defects

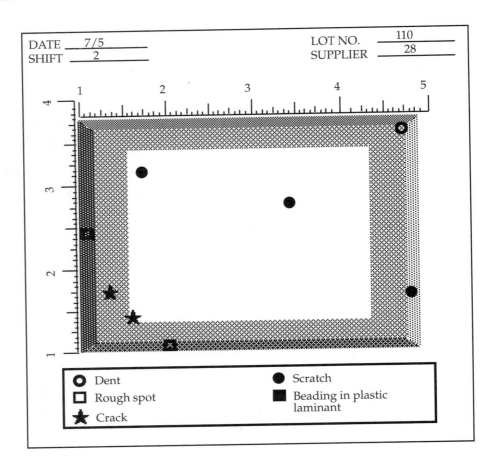

FIGURE 2.8 *Check Sheet for Location and Types of Defects on Wooden Plaques*

commonly occur. Of course, the judgment about where defects occur needs to be supported by a study of the process over time.

The recording of information on a check sheet enhances the ability to connect the times at which large numbers of defects occur with the causes of these defects. When data are to be collected and analyzed for statistical control, the check sheet should allow for the preservation of the time order of the data. Figure 2.9 shows a check sheet for maintaining inspection records of counts and types of the hornpad defects as the defects occur over time. This simple yet effective check sheet provides information on the machine, shift, and the types of defects. The dates of data collection are provided and space is given for additional comments. Additionally, space for totals and subtotals could easily be included on this data collection sheet.

Machine	Shift	6/3	6/4	Date 6/5	6/6	6/7	Comments
1	1	ADI	EA	AADF			Line stop
	2	FB	ADC	DE			
2	1	AEG	AI	EB			
	2	AEG	AED	EAAD			Change in spray container
3	1	DDI	EAB	D			
	2	AB	EF	DDB			

Machine	Shift	6/8	6/9	Date 6/10	6/11	6/12	Comments
1	1						
	2						
2	1						
	2						
3	1						
	2						

Types of Defects
A Defective urethane spray   **D** Surface defects   **G** Blow pressure
B Improper assembly   **E** Excessive flash   **H** Loose nuts
C Missing parts   **F** Bad trim plate   **I** Other

FIGURE 2.9 *Check Sheet for Hornpad Defects*

The design of this check sheet for defects on hornpad assemblies provides for the study of the stability of the number of all defects over time as well as for analyzing each type of defect individually. Associations can be made as to the source, type, and frequency of the defects in order to better understand potential sources of variation. Also, the data can be analyzed to determine if variation is caused by the use of different machines, by shift-to-shift differences, by different lots of materials, or by day-to-day changes and occurrences.

2.3   PARETO DIAGRAMS

From the check sheet in Figure 2.9, the data are easily consolidated to expose the relative magnitudes of the defect types or the potential causal factors. In the hornpad assembly example, the defects in nine different categories of defect types are counted. Counts of events in three or more categories are often summarized by the use of a Pareto diagram. The Pareto diagram is simply a vertical bar graph of the frequencies of events in each of several specified categories. This diagram provides a comparison of the relative magnitudes of the areas, problems, or causes of problems in terms of frequency of occurrence or the cost of each category. The major benefit of the Pareto diagram is its ability to display data in such a way that areas that need focused work are easily identified and the appropriate order of improvement efforts is portrayed.

The first step in the construction of a Pareto diagram is the determination of the categories to be used to summarize the counts of occurrences. Inspection of results against in-house specifications is a commonly used method for creating the categories. When this method for selecting categories is used, priority is directed toward the most frequently occurring of these violations. Prior to using such in-house specifications, it is necessary to consider the source of the specifications, their validity and relative importance, and their relationship with verified quality, performance, or reliability characteristics. When these considerations are not a part of the selection of categories for the Pareto chart, then the priorities identified for work may have no real or lasting value to the organization. Another possible means for identifying the categories for summarizing events is to select them based on a customer perspective. The categories are then determined to be events that impact the customer, e.g., line stoppages, increased setup times, use deviations, and other inconveniences for the customer.

Once the categories are selected, the counts are summarized according to the chosen categories and the categories are ordered, largest to smallest, based on the magnitude of the resulting counts. Infrequently occurring, unrelated categories are often combined into a category labeled "Other." The counts of

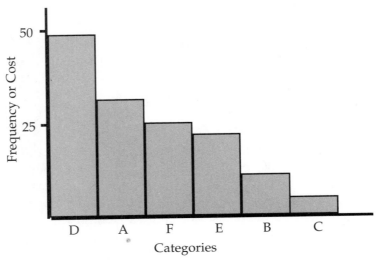

FIGURE 2.10 *Pareto Diagram*

events in each category are represented by the heights of the bars on a vertical bar graph. The vertical scale on the graph typically corresponds to frequency. However, a vertical scale based on dollars or time lost can often be useful in deciding the prioritization of work efforts. The bars are drawn above labels on the horizontal scale which identify the selected categories. Figure 2.10 shows the general form of a Pareto diagram.

The data recorded on check sheets (like the one in Figure 2.9) allows for the number of defective items found at final inspection to be categorized according to machine, shift, or type of defect. Table 2.1 shows the information from a set of check sheets summarized by the type of defect found at final inspection. The summary counts of each type of defect found during inspection of 1000 hornpad assemblies are shown in the Pareto diagram of Figure 2.11. These counts have been arranged in descending order of magnitude.

TABLE 2.1 *Number of Defects by Type Found During Final Inspection of Hornpad Assemblies*

Type of Defect	Number	Percent of Total
Defective urethane spray	65	25.0
Surface defects	64	24.6
Excessive flash	63	24.2
Improper assembly	24	9.2
Bad trim plate	12	4.6
Other	11	4.2
Blow pressure wrong	9	3.5
Missing parts	8	3.1
Loose nuts	4	1.5
Total	260	~100.0

The top three types of defects in the hornpad assemblies are urethane spray defects, surface defects, and excessive flash. Because the purpose for obtaining the data and constructing the Pareto diagram is to gain insight into how to focus work efforts in order to eliminate the causes of the defects, the results of the Pareto analysis lead to focusing on the stages of the production process at which these major defect types occur. Urethane spray defects almost always occur in the process step where the spray is applied to the hornpads. Both surface defects and excessive flash are defects that occur in the pad molding stage. To determine where in the production process work should begin, the data from the check sheets are reorganized according to point of origin of the defects in the process. A Pareto analysis showing the results from this reorganization of the data according to production stage is shown

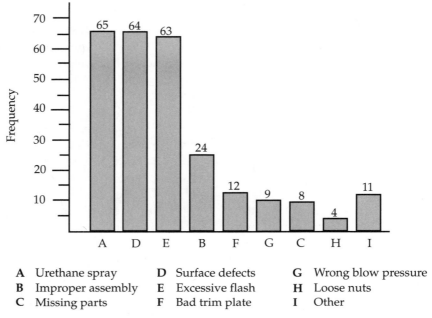

A   Urethane spray   D   Surface defects   G   Wrong blow pressure
B   Improper assembly   E   Excessive flash   H   Loose nuts
C   Missing parts   F   Bad trim plate   I   Other

FIGURE 2.11   *Pareto Diagram for Hornpad Assembly Defects by Type*

in Figure 2.12. This Pareto diagram effectively displays the evidence for focusing the work on the pad molding stage of the process in order to gain knowledge on root causes of the defects which occur in this process step and to make planned changes in order to reduce the frequency of the surface and excessive flash defects.

The prioritization of problems, their causes, and the areas in which they occur is useful as a guide for improvement efforts. However, several issues must be considered to prevent the expenditure of time and resources on efforts that do not provide beneficial results. One issue is that a stable process generates varying numbers of different types of defects or problems over time. Data collected over time allow for distinguishing between those types of problems that occur sporadically and those that occur consistently over time. This kind of information is not provided by the Pareto diagram. Control charts, as introduced and discussed in Section 2.5, provide a mechanism for identifying and eliminating erratic sources of variation acting on a process. Identification and removal of the problems that occur sporadically should be the first focus so that knowledge can be gained of the average rate of occurrence over time of the chronic, consistent problems. Once an area for focused work is chosen, the manner in which work proceeds within this area must be

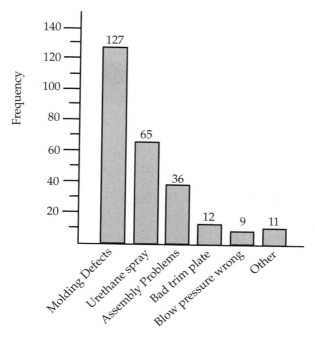

FIGURE 2.12 *Pareto Diagram of Hornpad Defects Categorized by Process Step*

determined based on a clear understanding of what the process is to provide in terms of customer value. These considerations require that managers provide direction and resources for the work to proceed effectively.

2.4   CAUSE AND EFFECT DIAGRAMS

Cause and effect diagrams are used to display possible causal factors affecting process outcomes or process characteristics (see box). Capturing this type of process information is a vital, central element in directing process study and improvement. The cause and effect diagram should capture as many potential sources of variation as possible, since managing process variation requires that the effects of causal sources be well understood. Cause and effect diagrams constructed by only one or two individuals typically do not contain complete, extensive information concerning the causes of variation, since individuals often have functional or personal biases. All persons who work within a process area, who are affected by the results of the process, and who impact the inputs into the process should participate in the construction of a cause and effect diagram. Additionally, people with different types of technical knowledge and experiential backgrounds are beneficial in obtaining thorough information on causal factors. Department managers, operators, process engineers, and design engineers are among those whose inputs are useful in the construction of a cause and effect diagram.

Uses of Cause and Effect Diagrams

1. To organize current process knowledge about potential sources of variation.

2. To highlight causal factors where effects are not well understood.

3. To consolidate the ideas of many employees into an organized description of causal factors.

4. To facilitate discussions about known causal factors and those that require further study.

Typically, a group begins construction of a cause and effect diagram with a particular characteristic (i.e., the "effect") that needs further study. Some examples of characteristics that might be studied are a type of defect in a manufactured part, the amount of impurities in a chemical product, or the times required to deliver a product. The construction of the diagram begins with a horizontal line in the middle of a page connected to a box at the right end of the line with the characteristic or "effect" written in the box. The group generates a list of the many factors that could possibly affect the characteristic being considered. These factors are organized under broad headings, usually called "Main Factors." Figure 2.13 illustrates the main features of a cause and effect diagram.

Although no one list of broad headings (i.e., Main Factors) is useful for every cause and effect diagram, the causal factors are commonly organized under six headings:

1. personnel
2. materials
3. equipment
4. methods
5. measurement
6. environment.

Another way to organize the causal factors is according to the major stages in the process flow that are responsible for creating the causal factors. Once

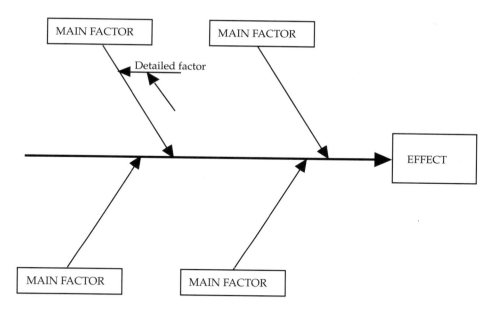

FIGURE 2.13 *General Form of a Cause and Effect Diagram (Fishbone Diagram)*

the headings are determined, the possible causes are organized as branches and sub-branches under each of the main headings. Although no specific rule exists for determining this organization, the intent is to capture relationships between the different factors.

A completed cause and effect diagram, constructed in a plant that manufactures deodorant, is provided in Figure 2.14. This diagram illustrates the organization of causal factors under the main factor headings. The deodorant bottles have polypropylene balls forced into the top openings. Deodorant is dispensed by inverting the bottle and rolling the ball across the skin. A common problem associated with this type of packaging is that the ball in the top of the bottle does not easily roll, making it difficult to dispense the deodorant. To understand the causes contributing to this effect, a group at the plant constructed the cause and effect diagram shown in Figure 2.14.

The cause and effect diagram is a useful method for capturing current knowledge and for entertaining informal speculation about the sources of variation affecting the identified quality characteristic. Cause and effect diagrams do not necessarily provide answers to questions, but the construction of these diagrams raises numerous questions:

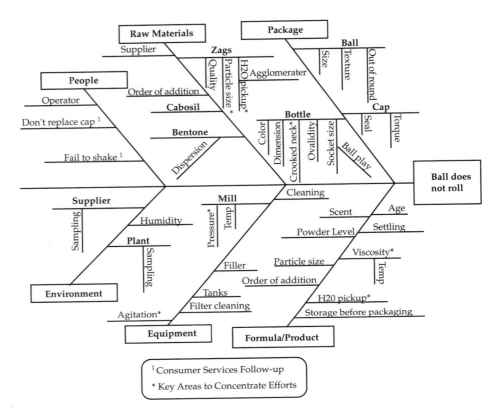

FIGURE 2.14 *Cause and Effect Diagram for Factors Affecting the Ball Not Rolling in Assembled Deodorant Bottles*

1. What is currently known about the causes listed? Which causes need more study? Which causes are the major contributors to the effect under consideration?
2. Which of the factors listed are currently controlled? For those factors being controlled, what has been the experience in controlling these factors? How can factors that are not currently controlled be identified? Which factors are technically or economically not feasible to control? On what basis has this decision been made?
3. What are the current operating levels of the factors listed? What are their averages and variation? Are their averages and variations stable over time? What limitations in the levels of these factors must be adhered to?
4. What is known about the relationships between causal factors? What is known about the effects of interactions between these causes on the quality characteristic?
5. How is the "effect" measured? Which of the causal factors should be measured? How should the causes be measured?

Cause and effect diagrams are effective tools for the improvement of process operation and performance when they are developed with the intention of providing answers to these questions. Otherwise, they contribute little to process improvement efforts.

2.5   INTRODUCTION TO THE CONCEPT OF STATISTICAL CONTROL CHARTS

A model that provides a powerful perspective on the variability in a series of measurements taken over time describes both common and special causes of variation. Common causes of variation are those system sources that affect each and every outcome and are exercised or experienced on an ongoing, consistent basis. Figure 2.15 is a plot that typically illustrates the behavior of measurements subject only to common cause sources of variation.

Special causes of variation are those that occur sporadically and affect only some of the results. Measurements subject to special cause sources of variation in addition to common cause sources might behave like those plotted in Figure 2.16. Given the level and the degree of variation in the series of values, the most recent value plotted in Figure 2.16 appears to be exceptional. It appears that something in addition to the common, consistent sources of variation is occurring. Establishing control limits for these data will confirm that a special cause for the sudden deterioration in these values must exist. Subsequent chapters discuss the statistical analysis required to determine when process measures are affected by special causes of variation. Special causes occur intermittently, arising from behaviors, methods, and equipment variations that are beyond usual practices and experiences.

Time

FIGURE 2.15   *Plot of Measurements Subject Only to Common Causes of Variation*

The concept and interpretation of common and special causes are guidelines to help identify and eliminate sources of erratic variability as well as to provide support for focusing on work to improve the underlying system. The common cause/special cause model does not suggest that a system subject only to common causes of variation is the ideal or perfect system. In fact, once the special causes have been identified and eliminated, the manager must evaluate the process relative to what is required and make judgments according to defined criteria. In definition, the concepts in the special/common cause model appear simple and straightforward. In practice, they require study, elaboration, and insight in order to understand the possible sources of variation and to guide improvement efforts.

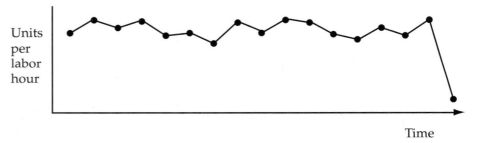

FIGURE 2.16   *Plot of Measurements Subject to Both Common and Special Causes of Variation*

It is common for organizations to use standards to indicate what is expected of a process and its outputs and to report periodic numerical indicators of performance against those standards. In Figures 2.17 and 2.18, the same time series depicted in Figure 2.15 is plotted along with a line indicating the standard of what is acceptable performance for the plant. Without the awareness of process history and the concept of stable variation about the average, each deviation from the standard is typically used as a separate measure of the performance of the process.

In Figure 2.17, the process is performing consistently and predictably at a level below the standard value. Because there are no special causes of variation acting on the process, the process is not capable of obtaining the standard level unless management makes design changes to the process. In Figure 2.18, the process has a stable average over time at the standard value. Due to the common causes of variation acting on the process, it is unreasonable to expect the units per labor hour to be exactly at standard value in each

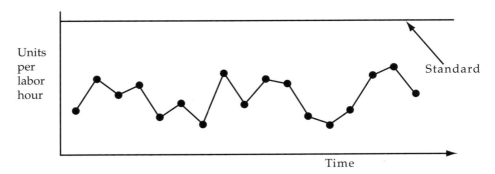

FIGURE 2.17   *Stable and Predictable Process Performance Below Standard Value*

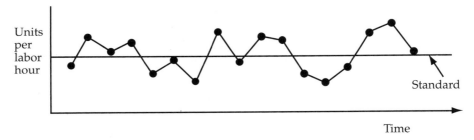

FIGURE 2.18   *Stable Process Average at the Standard Value*

time period. Expecting outputs over time to be exactly equal to a standard value assumes that absolutely no variation exists in the process.

The idea of using a standard to judge current performance is prevalent in many industries. Typically, a manager required to report such performance measures will also be asked to explain deviations from the standard. Of course, explanations will be sought and found, but the usefulness of this activity is questionable. Standards, of course, are set for a number of different reasons. In some instances, they are meant to describe the results of best practices. In other instances, they serve as a goal or as an attempt to change behaviors. In evaluating current results against standards, the purpose of the standard and its definition must be operative and be used in the evaluation. Further, the system's capability must be considered. If the establishment of the standard is not based on the known capabilities of the system, but on the hope that a high standard will promote greater productivity, then explaining each deviation from standard misdirects attention from studying the behavior of the process and determining causes of variation impacting that behavior.

The use of deviations from standard as a management signal has numerous limitations. Focusing on deviations from standards often causes an improper treatment of variation. If each deviation from a standard is treated as a one-time occurrence, then the common causes are addressed and treated as special causes when the process is subject only to common causes of variation. Hence, individuals and process areas may be held responsible for deviations from standard when the causal factors may reside in policies and practices at higher organizational levels or may result from decisions made in other divisions or places. In a process subject only to common causes, a measurement from that process may yield any one of several values for the deviation in a particular time period. This view offers the advantage of allowing history to help judge the effect of past decisions and practices on the process outputs and to serve as a guide for process modifications or changes.

When each result or outcome is considered a one-time event, it is difficult to appreciate the concept that work is a process. Yet, work is a process and a part of a system in which elements and behaviors interrelate and interact to provide products and services to the customers of the organization. Without a process point of view, deviations remain one-time events, carrying little insight into the behavior of the underlying processes and limiting potential work to improve the processes' ability to provide outputs of value to the customers. The ability to measure variation and understand its sources is necessary for the manager with process responsibility. Tracking process variation over time allows a process manager to know the size of the variation, to understand the impacts of changing conditions on the variability in the process outputs, and to learn about the causes and effects of critical process activities.

2.6   SUMMARY

This chapter provides an introduction to techniques useful for attaining current process knowledge that is necessary to improve process operations and outputs. Correct practice in organizing and working on process management involves an ongoing study of the process over time. Managers engaged in this practice with engineers and operators find it useful to construct verified *process flow diagrams* that indicate how, when, and where the process work is performed. Flowcharts ensure that all persons involved in process improvement efforts have a common understanding of the steps of the process and their interrelationships. Different types of flowcharts provide for the updating of the charts as work proceeds and for portraying additional, key information about process operation.

Once a representative process description is in place, key process characteristics and leverage points in the process are more easily identified. To obtain the desired information concerning these characteristics and leverage points, a plan should be in place for data collection. Planning a data collection strategy involves determining the desired information, identifying where measurements should be taken, understanding the appropriate type of data to collect, and establishing a consistent measurement process. The development of *operational definitions* ensures consistency of data collection, and it ensures that the process outputs consistently meet customer requirements. Because collected data should correspond to process parameters that impact process outcomes, the data collection strategy should ensure that the data obtained from the process are accurately recorded and easily interpreted. *Check sheets* help in obtaining information from data that can be quickly portrayed and studied using other tools such as *Pareto diagrams*. Pareto diagrams are useful for identifying areas on which to focus work and for determining the appropriate sequence of work efforts.

Critical process and product characteristics are measured with the intent of understanding how these characteristics are currently behaving and how they can and should be managed. *Cause and effect diagrams* portray and organize suspected sources of variation to be studied, and as work on the process proceeds, the cause and effect diagrams contain verified information about relationships between critical input variables and the outcomes of interest. The techniques in this chapter are designed to help managers and engineers gain process knowledge that serves as a foundation for the determination of critical measures of process performance and the initiation of work to eliminate the sources of variation impacting process operation. The three pieces of process knowledge discussed in the introduction to this chapter serve as prerequisites to the effective use of the techniques and methodologies discussed in the remaining chapters. These prerequisites and the methods that support the development of this knowledge are summarized:

- *Process flowcharts* provide information on the critical characteristics of process operation that need to be studied.

- *Operational definitions* and *check sheets* ensure that the measurements made on the process provide correct information about the critical process characteristics.

- *Cause and effect diagrams* and *Pareto diagrams* facilitate the identification of the process parameters that affect the critical characteristics.

3

Control Charts for Attributes Data: *p* and *np* Charts

As with all control charting techniques, the effective use of *p* or *np* charts for understanding and improving the processes of an organization requires a competent application of the statistical methods used to analyze the collected data. Mere knowledge of these methods is not sufficient for realizing the type of improvements discussed in Chapter 1. For control charting techniques to be effective, a number of issues must be addressed both before and during the use of these charts. These issues include

- the reason for developing the control charts,
- what process knowledge is recorded and known by the personnel working on the process prior to the collection and analysis of data,
- the use of this knowledge in developing sampling and subgrouping plans for data collection, and
- the management practices that must exist in order to realize benefits from the information provided by the statistical study of processes.

These issues are discussed in more detail after aspects of constructing and interpreting *p* and *np* charts are covered. Additionally, several case studies are presented to illustrate appropriate use of these statistical techniques in the continual improvement of processes.

3.1   DESCRIPTION OF ATTRIBUTES DATA REQUIRING THE USE OF *p* OR *np* CHARTS

The information gathered to study the behavior of a process is often obtained by counting the number of items that possess a certain characteristic or attribute. This type of attribute data, as discussed in Chapter 2, is called categorical data.

Each item examined is judged to fall in one of two categories—those that possess the specified characteristic and those that do not. In quality improvement work, these data generally arise by counting the number of items in a collection or subgroup of items which are judged to be nonconforming according to a set of criteria describing the requirements for use. Some examples follow of situations in which this type of attributes data is generated:

1. Counts of the number of damaged cans in lots of 1,000 cans. Each can is judged as being damaged or not damaged.
2. Counts of the number of shafts in samples of 200 whose diameters fail to meet stated specifications. Each shaft either meets or fails to meet the specifications.
3. Daily counts from each day's production on the number of assembled motors that fail a voltage test. Each motor is judged as either passing or failing the voltage test.
4. Weekly counts of the number of invoices that contain at least one error. Each invoice either contains one or more errors or it is correct.

Several evaluation objectives may exist that lead to the collection of a set of attributes data.

> **Objective One:** *One objective might be to collect such data for informational purposes.*

For example, a merchandiser records the number of dented cans in a lot of canned goods purchased from a food producer. This information may be needed to arrive at the number of dented cans for which the producer will reimburse the merchandiser. The interest in obtaining this information is only to describe the quality of the lot of cans inspected.

> **Objective Two:** *A second reason for collecting data on nonconforming items is to build a product or service profile that can be used to describe the product or service characteristics over time.*

At an engine plant, 10 engines are selected from each shift's production and subjected to extensive testing. The reason for this testing is to track the performance of the plant over time. In the test, 47 different characteristics are checked on each engine. If an engine fails on any 1 of the 47 characteristics, it is recorded as a reject. The numbers in Table 3.1 are the counts of the number of engines rejected by both shifts in a week (five days) of production for 21 successive weeks. When attributes data are collected over time, as in the present example, the consistency of process results can be evaluated. If these results appear to be consistent or stable, then a basis exists for predicting the future performance of the process with regard to the characteristic being evaluated.

> *Objective Three: Finally, data on nonconforming items may be generated as an aid to ongoing improvement work.*

Attributes data are then used to gain insight into the mechanisms of the process and the dynamics of the system that generated the product under study. In this instance, current knowledge of the process is used to guide why and how data will be collected. Analysis of the attributes data is then used to update the current process knowledge as well as to indicate process and system improvements.

TABLE 3.1 *Number of Rejected Engines by Week*

Week	Number Inspected	Number Rejected	Fraction Nonconforming
1	100	6	.06
2	100	8	.08
3	100	2	.02
4	100	3	.03
5	100	5	.05
6	100	10	.10
7	100	4	.04
8	100	7	.07
9	100	2	.02
10	100	12	.12
11	100	10	.10
12	100	5	.05
13	100	7	.07
14	100	8	.08
15	100	11	.11
16	100	3	.03
17	100	5	.05
18	100	9	.09
19	100	5	.05
20	100	7	.07
21	100	2	.02
	2100	131	

3.2   CONSTRUCTION OF *p* CHARTS

p and *np* charts are used for evaluating the level and the behavior of the variation in categorical data over time. The data in Table 3.1 are counts of the number of engines rejected in each week (five days) of production for 21 successive weeks. For this particular set of data, there are 21 subgroups which represent the 21 weeks over which the data were collected. Each subgroup consists of counts of the number of nonconforming engines in a sample containing 100 engines from one week's production.

Notation

---

k denotes the number of subgroups.

n denotes the number of items in a subgroup.

3.2.1   Plotting the Points on a *p* Chart

The construction of a *p* chart begins with a plot of the fraction nonconforming, *p*, for each subgroup. The fraction nonconforming is calculated for each subgroup by dividing the number of nonconforming items in the subgroup by the number of items contained in the subgroup. An examination of Table 3.1 shows that for the 100 engines tested during the first week, 6 were found to be nonconforming. So, for the first subgroup, $p = 6/100 = .06$. The *p* values for all 21 weeks are plotted on the *p* chart shown in Figure 3.1, which has the subgroup number along the horizontal axis and the fraction nonconforming along the vertical axis. The vertical axis is divided into equal increments from 0 to the largest values for fraction rejected thought to be likely. The data are recorded in the time order in which they were collected and this time ordering is maintained when the data are plotted. Therefore, the horizontal axis represents time; in this instance the time from the first week to the twenty-first.

3.2.2   Calculating the Centerline for a *p* Chart

The centerline is the average proportion nonconforming. To determine the centerline for a *p* chart, denoted by $\bar{p}$, the total number of nonconforming items in all subgroups is divided by the total number of items examined in all of the subgroups.

$$\bar{p} = \frac{\text{total number of nonconforming items}}{\text{total number of items inspected}}$$

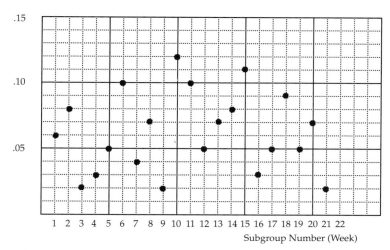

FIGURE 3.1 *Plot of the Proportion of Rejected Engines*

When the number, n, of items inspected in each of k subgroups is the same:

total number of items inspected $= k \times n$

For the data in Table 3.1, there are 131 rejected engines among the 2100 engines inspected. Therefore, the average proportion of rejected engines in the data collected is:

$$\bar{p} = \frac{131}{2100} = .0624$$

This value of $\bar{p}$ is used to position the centerline on the p chart in Figure 3.2.

3.2.3 Calculating the Control Limits for a *p* Chart

The upper and lower control limits define the amount of variation expected to occur in p values collected from a stable process when n items are inspected for each subgroup. (The insights provided by this definition are discussed after the calculation of these limits is illustrated.) The formulas used to calculate the upper control unit (UCL) and lower control limit (LCL) are, respectively:

$$\text{UCL}_p = \bar{p} + 3\sqrt{\frac{\bar{p}(1 - \bar{p})}{n}}$$

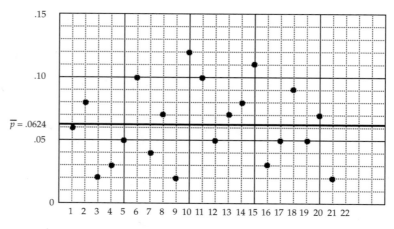

FIGURE 3.2 *Centerline on* p *Chart for Rejected Engines*

$$\text{LCL}_p = \bar{p} - 3\sqrt{\frac{\bar{p}(1 - \bar{p})}{n}}$$

where $\bar{p}$ and n are the average proportion and the subgroup size, respectively. For the data on the engine tests, the upper control limit is calculated as:

$$\text{UCL}_p = .0624 + 3\sqrt{\frac{.0624(1 - .0624)}{100}} = .1350$$

The calculations for the lower control limit, shown below, result in a negative number, −.0102:

$$\text{LCL}_p = .0624 - 3\sqrt{\frac{.0624(1 - .0624)}{100}} = -.0102$$

Since a negative fraction of nonconforming items is impossible, the convention adopted in this book is to say that there is no lower control limit. Thus, for the engine testing data:

$$\text{LCL}_p - \text{none}$$

Figure 3.3 displays the completed control chart with the upper control limit drawn on the chart.

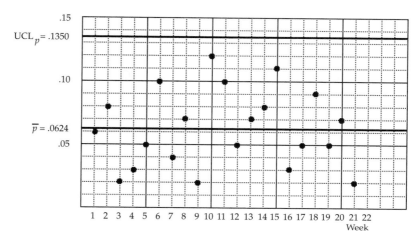

FIGURE 3.3   p *Chart for Nonconforming Engines*

3.2.4   Summary

Construction of *p* Charts

---

k denotes the number of subgroups and *n* refers to the number of items in a subgroup.

1. Classify each item in a subgroup of *n* items as either conforming or nonconforming. Record the number of nonconforming items for each subgroup.

2. Calculate *p* (fraction nonconforming) for each of the *k* subgroups.

$$p = \frac{\text{number of nonconforming items in the subgroup}}{n}$$

3. Calculate the centerline for the chart.

$$\bar{p} = \frac{\text{total number of nonconforming items}}{\text{total number of items inspected}}$$

Construction of p Charts (continued)

---

4. Calculate the upper and lower control limits.

$$\text{UCL}_p = \bar{p} + 3\sqrt{\frac{\bar{p}(1 - \bar{p})}{n}}$$

$$\text{LCL}_p = \bar{p} - 3\sqrt{\frac{\bar{p}(1 - \bar{p})}{n}}$$

If the resulting calculation for the lower control limit is below zero, then it is concluded that there is no lower control limit.

3.3   THE CONCEPT OF STATISTICAL CONTROL

The p chart constructed from the data on engine tests provides information about the process producing these engines. The values of p vary from the smallest value of .03 to the largest of .12. What can be learned about the process from this observed variation? The largest value of p, .12, occurs for engines made during the tenth week. Is it useful to try to identify what was different about the tenth week of operation? The lowest value of p, .03, occurs during both the fourth and sixteenth weeks. Does this value indicate that the work performed in these 2 weeks was better than the work performed in the other 19 weeks? The answers to these questions require consideration of what is meant by a stable process (i.e., one that is in a state of statistical control), how much variation is expected in the outcomes from a stable process, and how one interprets data collected from a process.

3.3.1   Types of Variation Acting on a Process

A common/special cause model of process behavior provides a valuable way of understanding process behavior. This model ties the variation produced in outputs of a process to the way in which this variation occurs in the process. An inescapable fact is that the number of nonconforming items produced hour to hour, day to day, and week to week varies. The common/special cause view of process behavior is that, for a given data collection strategy, the amount of variation observed in the measurements produced by a stable process is of a predictable size, whereas that from an unstable process is unpredictable. The causes of variation in a stable process, referred to as "common causes," are those causes acting on all outcomes. An unstable process is not only affected by common cause sources of variation, but also by "special causes," which affect the process at some times but not at others. A process subject to only common cause sources of variation is said to be "in statistical control." Special causes act on some outcomes to produce additional variation in the outcomes. This additional variation may be sporadic or episodic; referring to this additional variation as "special cause variation" means that it occurs at some times and not at others. For example, a line stoppage may be the special cause of a drastic shift in the number of nonconforming items produced by a process. Such a line stoppage might occur sporadically or on a regular basis. A common cause of variation in the number of nonconforming items might be temperature changes that occur consistently over time.

The separation of variation into these two types, common and special cause, provides a basis for understanding the root sources of variation affecting a process. This understanding will require a consideration of how process factors are likely to create variation in results. In the engine example, common causes of variation in the fraction nonconforming would be those causes that were present throughout the time in which the data were gathered. These causes may stem from the capability of the equipment used for engine manufacture, the maintenance practices for this equipment, the original design of the engine and alterations to this design, and methods used throughout the production of the engine. For example, the hardness of the metal used to cast the engine block is of varying degrees from block to block. This variation in the hardness means that machining operations require varying amounts of time. Sometimes in the production process, adequate machining is not done because of the difficulty in accommodating the machining equipment to the varying amounts of hardness. Consequently, some engines are improperly machined, possibly resulting in oil leakage or inadequate combustion characteristics. This type of problem occurs at times throughout the manufacture of the engine and is, therefore, common to some degree in all outcomes.

As stated previosly, special causes affect only some of the process results. For example, another cause of variation in the fraction of nonconforming engines produced might be the type of cutting tools or grinding wheels used in the machining operation. These tools are purchased from several suppliers. Thus, the type used one week may not be the same as the type used in a previous week. Inadequate tools purchased from one supplier do not affect all outcomes of the process. At the point in time when an inadequate tool is used, a marked increase in nonconforming engines occurs. This additional variation in the level of nonconforming engines is identified as occurring because of a special cause. In the previous section, the question was asked about whether the *p* value of .12 in week 10 was large enough to indicate a difference between the way engines were produced in week 10 and the way they were produced in the other weeks for which data were collected. This question can be rephrased as: "Does the value of $p = .12$ differ so much from the other p values recorded that there is an indication of a special cause occurring in week 10?"

3.3.2   The Control Chart as a Tool for Distinguishing Between Types of Variation

Control charts aid a manager or process owner in classifying the variation exhibited in process measurements as due to special causes or due to common causes. The control limits on a control chart indicate the amount of variation expected from a stable process, one that is subject to only common cause sources of variation. An *in-control process* is stable and predictable over time. For an in-control process, all plotted points fall within the upper and lower control limits. An *out-of-control process* has an unpredictable amount of variation. Points that fall outside of the control limits indicate special causes of variation acting on the process. The use of control limits to aid in determining the state of control of a process is illustrated in Figure 3.4. Points falling on or outside the control limits are signals of the presence of assignable causes. Points on or above the upper control limit for p charts indicate the occurrence of an abnormality creating a larger than usual fraction of nonconforming items. Points on or below the lower control limit indicate lower values than "normal" for fraction nonconforming, perhaps signaling a sudden lowering of inspection standards or revealing unusually good results for reasons that deserve investigation and identification.

In addition to points falling outside of the control limits, distinct patterns in the data points are also taken to indicate the presence of special causes. These patterns in the plotted points might be an increasing or decreasing trend or a cyclical pattern. One approach for deciding whether or not there is a systematic pattern in the data are *runs tests*. A *run* is defined to be a sequence of one or more consecutive points either above or below the centerline. Figure 3.5 is an example in which five runs exist in a sequence of plotted points. The first run is of length 11 since there are 11 consecutive points above the centerline before the twelfth point, which falls below the centerline.

A large number of runs tests are available for testing data patterns. Appendix B provides a detailed description of many of these tests. However, the simultaneous use of many runs tests on a set of data is not recommended. Any time a run test is applied to a set of data, there is a small chance that the test will indicate the presence of a pattern in the data when in fact there exists no

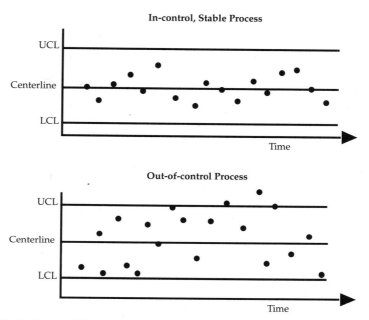

FIGURE 3.4 *Control Limits as Indicators of the State of Control of a Process*

special cause creating the observed pattern. As more and more tests are applied to a given data set, it becomes more and more likely that at least one of these tests will incorrectly indicate the presence of one or more special causes. As a general guideline, it is recommended that the analyst adopt one or two run rules and consistently apply these rules to baseline data sets and to ongoing process data.

A simple, effective run test, referred to as the "Rule of Seven," is used in this book. This test states that if seven or more consecutive points occur either

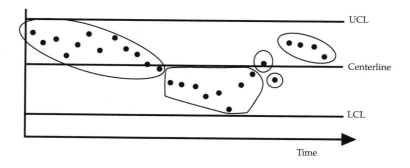

FIGURE 3.5 *Example of Runs in a Sequence of Data Points*

above or below the centerline, then there is evidence that a special cause is affecting the process. In Figure 3.5, there are two runs each containing more than seven data points that fall above or below the centerline. Hence, based on the Rule of Seven, we conclude that there exist special causes of variation acting on the process from which the data points were obtained. If an application of the Rule of Seven indicates the presence of an assignable cause, the process is concluded to be unstable, out of statistical control. In summary, a process is determined to be out of control if a systematic pattern in the data points exists or if there are one or more plotted points existing outside of the control limits.

The purpose for evaluating the state of control of a process is to better understand the sources of variation affecting a process. This increased understanding is useful for

1. determining whether a process is stable and therefore predictable,
2. monitoring a process for maintenance and stability,
3. determining the magnitude of the effects of changes made on a process, and
4. developing engineering or operating insights into sources of variation captured in the collected data.

Additional signals on control charts that can provide information on the sources of variation acting on a process are shown in Figure 3.6. To be useful, these signals require proactive study of the current process, the data collection strategy, and the possible causes of variation impacting the chosen

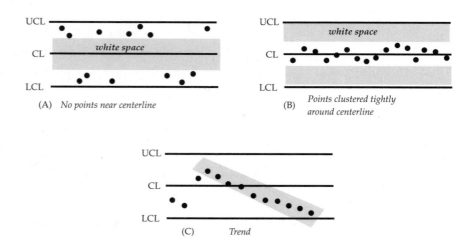

(A) *No points near centerline*

(B) *Points clustered tightly around centerline*

(C) *Trend*

FIGURE 3.6 *Additional Signals on Control Charts*

process measurement. For instance, a large amount of "white space" formed by the absence of points at or near the centerline is shown in plot A of Figure 3.6. White space of this type is typically created when entire subgroups are selected from one cause system and other, complete subgroups are taken from a different cause system. For example, two different machines, each subject to different causes of variation, may be used in a machining operation. If all items in some subgroups come from one machine and all items in others from a different machine, the pattern of points on plot A might occur. Different cause systems could also be the result of two different inspection systems being used to evaluate different subgroups.

Another type of white space, shown in plot B, has data points tightly clustered about the centerline. Patterns of this type are typically created by subgrouping plans that consistently mix in each subgroup output from two or more processes that operate at significantly different levels. For example, some proportion of sampled units in a subgroup may be produced by a machine operating at one level of rejected product, the remaining units in a subgroup come from another machine, operating at a different performance level. Aggregation of this type is often rationalized as giving a "fair" representation to all sources producing similar product. This is not a correct idea; subgrouping plans should separate the sources of the product when these sources are known or suspected to provide different results. Further discussion of the formation of subgroups in the presence of multiple cause systems is contained in Section 3.7.

3.3.3   Statistical Control as a Tool for Planning

When the outcomes from a process behave in a stable fashion over time, the future behavior of the process is considered to be predictable; the behavior will tend to replicate itself unless specific changes are made to the process. Conversely, the behavior of a process that is subject to both special and common causes of variation is unpredictable unless the special causes can be identified and their effects eliminated. Understanding the predictability of process outcomes is critical to effective business planning. Recall the p chart for nonconforming engines in Figure 3.3. From this chart, it is concluded that the proportion of nonconforming engines produced in a week is stable over time. Thus, it can be predicted that the proportion of nonconforming engines produced in a week will vary between 0 and .1350 and that, on average, 6.24% of engines produced will be nonconforming. This information is obviously useful in estimating the cost of production, in managing the schedule for engine production, and in forecasting the number of engines that can be supplied to the automotive assembly division.

The study of the state of statistical control of processes is an important part of business planning and quality improvement. Unfortunately, cost estimates and production schedules are often built without knowing whether or not process outcomes are predictable over time. When schedules are built without the ability to run the process in a stable manner, large discrepancies often occur between scheduled production and actual production. Plant personnel sometimes compensate for these discrepancies by costly means such as buffer stocks, overtime production, and holdups at the next stage of production. These costly ways of handling poor scheduling ability do not improve the process's ability to meet production schedules; instead, they add complexity and difficulty to the production process.

To improve process performance, knowledge concerning the stability of the process outcomes must be supported by information on the process mechanisms that affect these results. When the information collected from a process indicates that the process is unstable, the first step in working to gain process knowledge is to identify the special cause(s) creating this instability. If a special cause results in increased levels of nonconforming items, what can be done to remove this cause? If a special cause is acting to reduce the level of nonconforming items produced, what can be done to ensure that this special cause is made part of the ongoing operation of the process? Once all special causes are identified and either removed or made a part of the standard operations, then the process is subject only to common causes of variation. When a process is stable and subject only to common causes of variation, only process changes that impact one or more common causes will result in improvement. To improve a stable process, it is necessary to collect information that allows for the effects of common causes on the process outcomes to be studied. In the engine example, the data in Table 3.1 were simply acquired at the end of the production line producing the engines. At this level of aggregation, data are useful for high-level review; trends are revealed, consistency in operations can be seen, and the effects of specific changes can be evaluated. But these data provide little insight into the underlying causes of variation in the production process. To study and improve that process, additional data are needed that facilitate an understanding of the impact of the common causes of variation acting on the process.

In addition to providing guidance on actions that are appropriate to improve a process, distinguishing between common and special causes is also important in order to avoid actions that are inappropriate. Without knowing whether or not special causes are operating on a process, one might overreact to a particular result and proceed as if there is an indication of a special cause when, in fact, there is not. The completed *p* chart in Figure 3.3 shows that none of the points plotted falls outside of the control limits. Since the control limits on the *p* chart indicate the amount of expected variation in the plotted points if the process is stable, examination of this chart leads to the conclusion that there is no evidence of instability in the process. Given that the process is in control, reacting to the *p* value of .12 in week 10 as if it were a "special" cause will not likely result in an action or decision that is effective in improving the process. At worst, treating common causes of variation as special causes can, by overadjustment, produce more variation in the outcomes of the process.

As mentioned earlier, the data collected from the engine production process provide little direction for process improvement. The sampling strategy was not designed to provide information on possible process factors that contribute to engine defects. The data, as collected, can be used only to describe current levels of defects. A strategy for beginning work to reduce the level of defects is to begin identifying the potential factors contributing to defects and to use control charts as a means of studying these factors. This powerful use of control charts to support process improvement is illustrated by example in Section 3.6.

3.4   CONSTRUCTION OF *np* CHARTS

Both *p* and *np* charts can be used to evaluate the variation observed in data on nonconforming items collected from a process over time. When the fraction nonconforming, *p*, for each subgroup is plotted on a chart, then a *p* chart is used to analyze the data. When the subgroup size, *n*, is the same for all subgroups, the analyst may prefer to plot the *number* of nonconforming items, rather than the fraction nonconforming. An *np* chart is the appropriate technique for evaluating the stability of the number of nonconforming items. When each subgroup is based on the same number of items, the choice of analyzing the data by a *p* chart or by an *np* chart is merely a matter of convenience or preference on the user's part. The data on the number of rejected engines by week in Table 3.1 is used to illustrate the construction of an *np* chart. Instead of the fraction of nonconforming engines, the number of nonconforming engines is plotted in Figure 3.7.

3.4.1   Plotting the Points on an *np* Chart

Table 3.1 contains columns for the week of inspection, the number in a subgroup, and the number of nonconforming engines in each subgroup. On an *np* chart, the *number* of nonconforming items are plotted. These values, plotted in time order, are shown on the chart in Figure 3.7. On an *np* chart, the subgroup numbers again appear along the horizontal axis and the range of the number of nonconforming items appears along the vertical axis. The analysis of the information contained in this plot proceeds in the same fashion as for a *p* chart. A centerline is calculated; then upper and lower control limits are computed. The control limits, as with the *p* chart, describe the amount of variation that should occur in the plotted points if the data are obtained from a stable process.

3.4.2   Calculating the Centerline and Control Limits for an *np* Chart

On an *np* chart, the value used to draw the centerline is the average number of nonconforming items in a subgroup. This value, denoted by $n\bar{p}$, is the total number of nonconforming items in all subgroups divided by the number of subgroups.

$$n\bar{p} = \frac{\text{total number of nonconforming items}}{\text{number of subgroups}}$$

For the data in Table 3.1, there are 131 nonconforming engines in the 21 subgroups, so the average number of nonconforming engines is:

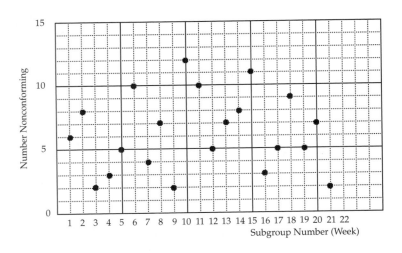

FIGURE 3.7   *Plot of the Number of Nonconforming Engines*

$$n\bar{p} = \frac{131}{21} = 6.2381$$

This value of $n\bar{p}$ is used to establish the centerline in Figure 3.8.

The calculation of the upper and lower control limits for an np chart requires that the value of $\bar{p}$ be calculated. The calculation of this value is described in Section 3.2.3. Alternatively, $\bar{p}$ can be calculated by dividing $n\bar{p}$ by n, the subgroup size.

$$\bar{p} = \frac{6.2381}{100} = .0624$$

The upper (UCL) and lower (LCL) control limit are then calculated as follows:

$$UCL_{np} = n\bar{p} + 3\sqrt{n\bar{p}(1 - \bar{p})}$$

$$LCL_{np} = n\bar{p} - 3\sqrt{n\bar{p}(1 - \bar{p})}$$

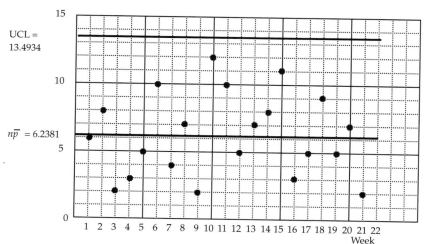

FIGURE 3.8 np *Chart of Number of Nonconforming Engines*

For the data in Table 3.1, the upper and lower control limits for the np chart are:

$$\text{UCL}_{np} = 6.2381 + 3\sqrt{6.2381(1 - .0624)} = 13.4934$$

$$\text{LCL}_{np} = 6.2381 - 3\sqrt{6.2381(1 - .0624)} = \text{none}$$

These values are used to draw the upper and lower control limits as shown in Figure 3.8. The completed chart shows no indication of a special cause operating. From the available data, it appears that week to week, throughout the time period studied, the proportion of nonconforming engines is stable with about 6.24% nonconforming engines being produced. An important observation is that the np chart on the number of nonconforming engines provides the same information as the p chart constructed on the same data in Section 3.2. Not only are all points within the control limits on both charts, but the pattern in the plotted points is exactly the same for the two charts. As a result, the same decisions about the behavior of the process producing engines will be made regardless of which chart type is used.

3.4.3   Summary

Construction of np Charts

---

k denotes the number of subgroups and n refers to the number of items in a subgroup

Note: np charts should not be used when the subgroup size varies.

1. Classify each item in a subgroup of n items as either conforming or nonconforming. Record the number of nonconforming items for each subgroup.

2. Calculate the centerline on the chart.

$$n\bar{p} = \frac{\text{total number of nonconforming items}}{\text{number of subgroups}}$$

3. Calculate the value of $\bar{p}$

4. Calculate the upper and lower control limits.

$$\text{UCL}_{np} = n\bar{p} + 3\sqrt{n\bar{p}(1 - \bar{p})}$$

$$\text{LCL}_{np} = n\bar{p} - 3\sqrt{n\bar{p}(1 - \bar{p})}$$

If the resulting calculation for the lower control limit is below zero, then it is concluded that there is no lower control limit.

3.5   CONSTRUCTION OF *p* CHARTS WHEN *n* VARIES

If a set of data on nonconforming items has varying subgroup sizes (i.e., *n* varies), then the appropriate chart for analyzing these data is a *p* chart. An *np* chart is not recommended when *n* varies. The data in Table 3.2 provide an illustration of the use of *p* charts for varying subgroup sizes. These data are from a label printing process in a custom packaging company. To determine a baseline of current process behavior with respect to number of defects, the manager of the process decided to collect data on the number of misprinted labels in each batch of printed packages. Because customers prefer to order customized packages in batch sizes of 250, 500, 750, or 1000, the subgroup sizes vary in the collected data.

In Table 3.2, the fraction nonconforming, *p*, has been calculated by dividing the number of misprinted labels in a batch by the number of packages contained in that batch. Although the definition of *p*, the fraction nonconforming in a subgroup, remains the same as defined in Section 3.2.1, remember that these fractions are based on varying amounts of inspected items. The values of *p* from Table 3.2 are plotted on the chart in Figure 3.9. Just as described in Section 3.2.2, the centerline for the *p* chart, $\bar{p}$, is calculated by dividing the total number of nonconforming items in all subgroups by the total number of items examined. In the 19 subgroups, there are a total of 950 misprinted labels. Thus:

$$\bar{p} = \frac{950}{12,500} = .076$$

The formulas for upper and lower control limits for *p* charts, as given in Section 3.2.3, are reproduced below. These limits are a function of *n*, the subgroup size. Another interpretation of this statement is that the amount of expected variation in the fraction of nonconforming items selected from a stable process depends on the number of items in a subgroup. Thus, when the subgroup sizes in a collection of data differ, it is not correct to use the same control limits to assess the variation in the fraction nonconforming found in each subgroup. Instead, the control limits used for each plotted point will depend on the number of items in that specific subgroup. Thus, for those *p* values that are based on a subgroup size of 250, the upper and lower control limits are:

$$UCL_p = \bar{p} + 3\sqrt{\frac{\bar{p}(1-\bar{p})}{n}}$$

$$UCL_p = .076 + 3\sqrt{\frac{.076(1-.076)}{250}}$$

$$UCL_p = .1263$$

$$LCL_p = .076 - 3\sqrt{\frac{.076(1-.076)}{250}}$$

$$LCL_p = .0257$$

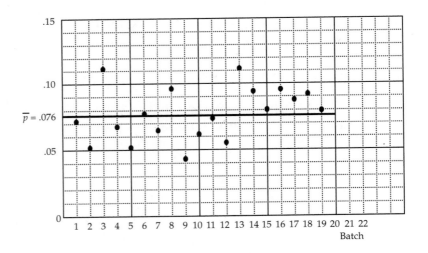

FIGURE 3.9 *Plot of Nonconforming Labels Printed on Packages*

TABLE 3.2 *Number of Misprinted Labels on Customized Packages*

Batch	Number of Packages	Number of Misprinted Labels	Fraction Misprinted
1	750	54	.072
2	750	39	.052
3	250	28	.112
4	250	17	.068
5	250	13	.052
6	1000	78	.078
7	1000	65	.065
8	750	72	.096
9	500	22	.044
10	500	31	.062
11	1000	74	.074
12	1000	56	.056
13	250	28	.112
14	1000	94	.094
15	1000	80	.080
16	500	48	.096
17	250	22	.088
18	750	69	.092
19	750	60	.080
	12500	950	

The control limits for those p values with subgroup sizes of 500, 750, and 1000 are calculated in the same manner using the differing values of n. These control limits are calculated to be:

n	UCL_p	LCL_p
250	.1263	.0257
500	.1116	.0404
750	.1050	.0470
1000	.1011	.0509

These limits have been used to construct the control chart in Figure 3.10. This chart shows no points outside of a control limit. However, there is a run of seven points above the centerline from batch 13 to batch 19. Hence, it is concluded that the process is out of control and that there is evidence of special causes affecting the proportion of labels misprinted in the printing process.

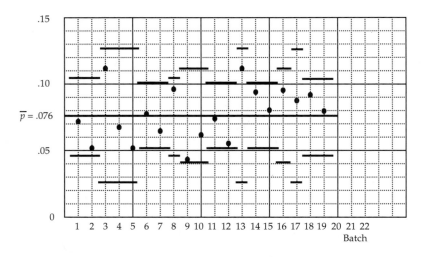

FIGURE 3.10 *p Chart for Fraction of Misprinted Labels*

The next phase of process investigation must focus on the identification of the special cause or causes impacting the level of the misprinted labels. Unfortunately, the data as collected provide little information to aid in the identification of possible special causes acting on the process. Because non-conforming labels due to many types of defects have been counted, many causes need to be investigated in order to determine which of them is creating the unstable condition noted on the chart in Figure 3.10. For instance, is there a particular type of misprint, such as crooked labels, that has increased in the most recent time periods? Was machine maintenance performed or new equipment added to the line? Without additional information on the types of misprints and information on the possible causes creating such misprints, improvement efforts are limited. The data collected for this study of a label application process provides, as did the data in the engine example, information to baseline current process behavior. Other information will be required. In particular, types and reasons for misprints could have been recorded in the present example. Large occurrences of certain types of misprints could then be used to prioritize further process work. Data collection could also be planned so that information on sporadic events occurring in the printing process are captured. Correlation between the occurrence of these events and changes in process outcomes could provide clues as to reasons for process abnormalities. The following case study illustrates the development of such a data collection strategy to support process improvement.

3.6   COLLECTION OF ATTRIBUTES DATA TO SUPPORT PROCESS STUDY: A CASE STUDY

A plant produces single-serving frozen meat and vegetable pies, called pot pies. Completed pot pies have a bottom and top crust enclosing a meat and vegetable filling. Figure 3.11 is a flowchart of the process that produces these pies. The stage of the process where the bottom crusts are formed (labeled A in the flowchart in Figure 3.11) has been a trouble spot. Although the equipment was designed to produce bottom crusts at a rate that would allow the filling operation to proceed in a consistent fashion, this has not been the actual experience in the operation of this line. The filling operation frequently has to slow or stop because there are not enough bottom crusts ready to be filled. Operators of the equipment that produces the bottom crusts report that a major reason for the inadequate production is that a large number of the crusts are incorrectly formed. Since these crusts are not suitable for filling, they are removed from the line. The dough that forms the crust is removed from the metal tin and added back to the dough mixture, which feeds the machine that forms the bottom crusts. A more detailed description of the formation of bottom crusts is shown in Figure 3.12.

Operators currently perform 100% inspection of the bottom crusts to remove any crusts with obvious breaks in the dough or that do not completely cover the metal tray. The line producing the bottom crusts ran two shifts each day. Initial data gathered at inspection from these two shifts indicated that the process was currently producing about 18% nonconforming crusts. The management group began evaluating the bottleneck caused by bottom crust formation. This group realized that the reported 18% nonconforming crusts was the result of an inadequate process being installed and that it was their responsibility to provide the means and operating procedures to improve the outcome. It was also understood that the plant's inability to produce good crusts without rework resulted in additional expense, since machine and operator time were used to produce unusable product. Further, other characteristics, such as the taste and texture of the crust, deteriorated during the process of reworking the dough. Therefore, rework was not an appropriate solution.

To support the work on the pie crust line, management began to gather current process knowledge. Two members of the management group and several operators who worked on the pot pie line met and brainstormed possible causes that were active in producing nonconforming pie crusts. Lengthy discussions about the possible causes resulted in the cause and effect diagram shown in Figure 3.13. After discussing the many possible causes for nonconforming crusts, it was decided that the 18% figure on nonconforming crusts

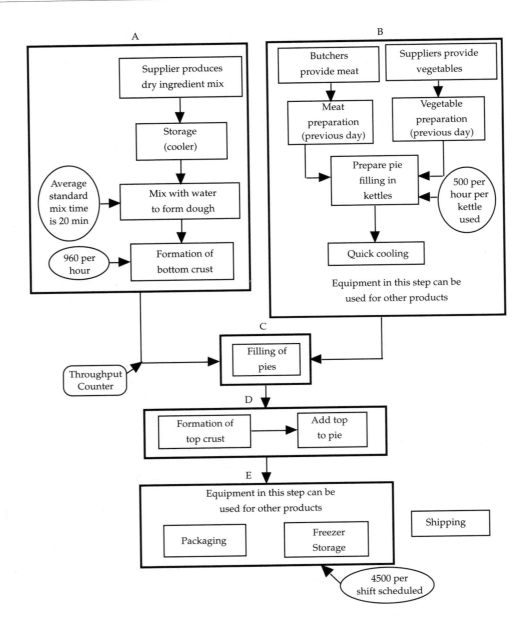

FIGURE 3.11 *Pot Pie Production Flowchart*

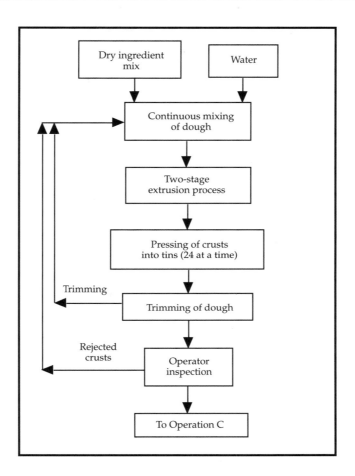

FIGURE 3.12 *Flowchart of Formation of Bottom Crusts*

provided incomplete information about the current capability of the pot pie line. In particular, there was little knowledge about whether more noncon-forming crusts occurred on some days and less on others. If this were true, the group should focus on studying those causes that were present on some days but not others. Also, the group had no knowledge about whether the number of nonconforming crusts was larger at some times of the day rather than at other times. Again, knowledge of the behavior of the process during a day would provide direction as to what might and might not be a factor in causing nonconforming crusts.

During these meetings, the point was made that a more thorough knowledge of how the process was currently operating would provide both information as to needed corrections as well as a baseline by which future changes to the process could be judged. Therefore, the management group decided to collect information on the number of crusts discarded throughout a day's operation. The flowchart in Figure 3.12 describes the current operation of the process where the bottom crusts are formed and shows the point in the process where data were collected. The production of pie crusts begins with mixing the dough in a vat. Next, the dough is extruded from the vat and pulled into a forming machine, where it is rolled into a thin sheet. The dough is then cut into wide strips by the forming machine. One strip covers 24 metal trays. The dough is pressed into the pie trays and any excess dough is trimmed. The forming of a set of 24 crusts is referred to as a cycle. The operator running the forming machine controls the times at which cycles occur. This same operator inspects the pie crusts after the forming operation for visible breaks in the dough and for metal trays that are not completely covered by dough. At this point in the operation, a large number of crusts are discarded. Therefore, the decision was made to collect data on nonconforming crusts at this process stage.

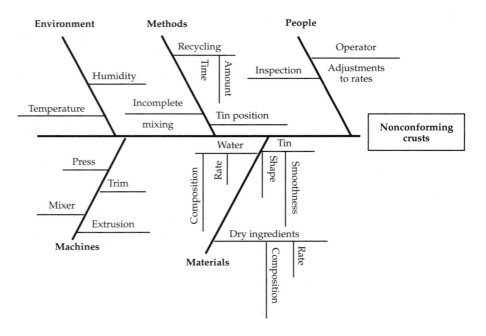

FIGURE 3.13 *Cause and Effect Diagram for Nonconforming Pie Crusts*

3.6.1   ## Analysis of Baseline Data on Nonconforming Crusts Using an *np* Chart

To study the current behavior of the pot pie line, data were collected over a period of several days on the number of nonconforming crusts being produced throughout each day. The operator inspecting the crusts was instructed to record the number of nonconforming crusts produced in four consecutive cycles once each hour. The operation of this line occurs during only one shift (the day shift) each day. The number of nonconforming crusts produced in these four cycles was counted for one week (five working days). These data are recorded in Table 3.3.

As explained in Section 3.2, a *p* chart can be used to evaluate the information contained in these data. However, the management group decided it would be simpler to plot the *number* of nonconforming crusts using an *np* chart rather than plotting the fraction nonconforming. In this case, the subgroup size is 96. (Recall that each subgroup was formed by counting the number of nonconforming crusts in four cycles of 24 crusts.) Because each subgroup contains the same number of inspected crusts, it does not matter whether a *p* or *np* chart is used to analyze the data. Both types of charts will reveal the same information about the data in Table 3.3.

Table 3.3 contains information on the time of inspection, the number of subgroups, and the number of nonconforming crusts in each subgroup. The management group plotted, in time order, the number of nonconforming crusts in each subgroup. These plotted values are shown on the chart in Figure 3.14. After plotting the values, the centerline and control limits were computed in order to describe the amount of variation that should occur in the plotted points if the crust formation process was stable.

To obtain the centerline for the *np* chart, the average number of nonconforming crusts in a subgroup was determined. For the data in Table 3.3, there were 579 nonconforming crusts in the 35 subgroups, so the average number of nonconforming crusts was:

$$n\bar{p} = \frac{579}{35} = 16.5429$$

This value of *p* was used to establish the centerline as shown in Figure 3.14.

TABLE 3.3 *Number of Nonconforming Crusts in Four Consecutive Cycles of 24 Crusts*

Date	Time	Number in Subgroup	Number Nonconforming
11–08	8:00 a	96	15
	9:00 a	96	16
	10:00 a	96	17
	11:00 a	96	20
	12:00 a	96	12
	1:00 p	96	19
	2:00 p	96	13
11–09	8:00 a	96	18
	9:00 a	96	27
	10:00 a	96	16
	11:00 a	96	21
	12:00 a	96	22
	1:00 p	96	20
	2:00 p	96	19
11–10	8:00 a	96	9
	9:00 a	96	14
	10:00 a	96	18
	11:00 a	96	23
	12:00 a	96	19
	1:00 p	96	12
	2:00 p	96	13
11–11	8:00 a	96	15
	9:00 a	96	16
	10:00 a	96	18
	11:00 a	96	18
	12:00 a	96	16
	1:00 p	96	12
	2:00 p	96	16
11–12	8:00 a	96	11
	9:00 a	96	17
	10:00 a	96	11
	11:00 a	96	18
	12:00 a	96	15
	1:00 p	96	14
	2:00 p	96	19
			579

To calculate the upper and lower control limits, the value of $\bar{p}$ was calculated by dividing $n\bar{p}$ by n, the subgroup size.

$$\bar{p} = \frac{16.5429}{96} = .1723$$

The upper (UCL) and lower (LCL) control limit were then calculated as follows:

$$UCL_{np} = 16.5429 + 3\sqrt{16.5429(1 - .1723)} = 27.6439$$

$$LCL_{np} = 16.5429 - 3\sqrt{16.5429(1 - .1723)} = 5.4419$$

These values were used to draw the upper and lower control limits as shown in Figure 3.14.

The completed chart showed no indication of a special cause operating. From the available data, it appeared that hour to hour, throughout the week studied, the proportion of nonconforming pie crusts was stable with about 17% nonconforming crusts being produced. At first, the management group felt like the *np* chart provided the same information as that contained in their initial statement about the pot pie line—that it was producing about 18% nonconforming. They still needed a direction for study of the pot pie line in

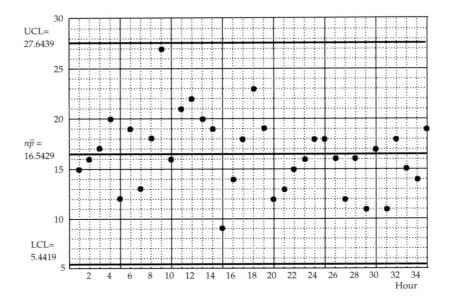

FIGURE 3.14 *Control Chart for the Number of Nonconforming Pie Crusts*

order to begin improving the current capabilities of the line. They began to look more carefully at the *np* chart to see if it could be used to indicate which direction might be most fruitful. They recalled that the points on the chart were collected hour to hour over a week's period; thus, the chart indicates that the process is stable when judged in this fashion. Consequently, they realized that those sources on the cause and effect diagram that act to create differences between days or between hours might not be the most useful areas for study at this point. Because the chart reported that hour to hour and day to day the process was behaving consistently, there was no evidence that a special cause was acting at some hours and not at others. Therefore, they decided that working to detect why some hours or days differ from others would not be useful. Instead, they believed that they needed to further explore those causes on the diagram that might affect nonconforming crusts throughout an hour. The cause and effect diagram was used to help identify which causes might be acting consistently within an hour to produce nonconforming crusts.

3.6.2 Selecting the Subgrouping Strategy to Gain Information on Causal Factors

In the study of the pie crusts, each subgroup on the *np* chart consisted of four groups of 24 crusts. These four groups came from four consecutive cycles of the machine, which formed 24 pie crusts at each operation. The 35 subgroups on the chart in Figure 3.14 were collected at various points in time throughout one day. By selecting the subgroups in this fashion, any changes in the process that were detected on the control chart as a special cause would be changes that occurred between subgroups; in other words, differences that occurred from hour to hour or day to day. On the other hand, causes active in each of the hours would not be indicated as special causes on the control chart. Thus, the way in which the subgroups are selected determines those causes of nonconforming crusts which can be detected as special causes on a control chart. Consequently, what is learned about the operation of the process from a set of data depends on how the data were gathered. The next stage in the study of the pie crust process illustrates this idea.

One of the possible causes identified on the cause and effect diagram for nonconforming pie crusts was tin location. The group studying the process believed that tins occupying an end position when the dough is placed over the tins are more likely to be defective. Several reasons were stated for this phenomenon. One reason is that the dough is thinner at the end positions than in the center. Secondly, if the dough is not properly rolled out it will not completely cover the pie tins. The final reason stated was that the dough dries more quickly at the ends than it does at the middle positions. If rejected pie crusts are truly associated with tin position, then they expected that different proportions of crusts would be rejected at the two positions. From this discussion, two questions arose:

1. Are the number of nonconforming crusts formed at the two locations consistent (or stable) over time?
2. How does the number of rejected crusts at end tins compare with the number nonconforming at the middle positions?

The previous subgrouping strategy was not designed to provide answers to these questions because each subgroup contained both middle and end crusts. To determine if the end positions do, indeed, produce a higher proportion of defective crusts, the management group realized that a different subgrouping strategy was necessary. Subgroups were selected so that some subgroups contained only "end" crusts and others contained only "middle" crusts.

Collecting data using this method required a much larger investment of time, since each crust had to be identified by its location in addition to recording whether or not it was nonconforming. Since the line was not to be stopped for this data collection, a check sheet was designed to simplify the recording of the data. The type of form used is shown in Figure 3.15. The circles on the form correspond to the 24 locations of the pie tins as they are covered with dough. The operators collecting the data were instructed to make an "X" on

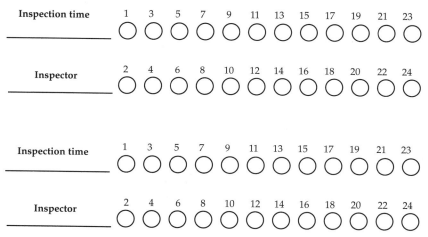

FIGURE 3.15 *Data Collection Format for Number and Location of Nonconforming Crusts*

the form to indicate the locations of nonconforming pie crusts. As before, this information was recorded on four consecutive runs every half hour.

The first set of completed forms is shown in Figure 3.16. Unlike the first set of data collected on the line, these data are summarized by position. The positions labeled 1 through 4 and 21 through 24 on the form are the end positions. The crusts in these eight positions from the first four consecutive runs form the first subgroup. The number of nonconforming crusts in this first subgroup is given in Table 3.4. The sample size for this subgroup is $n = 4 \times 8 = 32$. The second subgroup consists of the remaining $4 \times 16 = 64$ "middle" crusts from the same four consecutive operations from which the first subgroup was collected. The remaining data in Table 3.4 are recorded in a similar fashion.

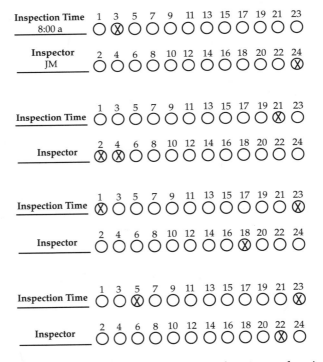

FIGURE 3.16   *Completed Data Collection Form for Nonconforming Crusts in Subgroups One and Two*

TABLE 3.4 *Number of Nonconforming Crusts by Location*

Subgroup	n	np	p	Subgroup	n	np	p
1	32	9	.281	2	64	2	.031
3	32	10	.313	4	64	9	.141
5	32	8	.250	6	64	5	.078
7	32	9	.281	8	64	4	.063
9	32	8	.250	10	64	9	.141
11	32	6	.188	12	64	9	.141
13	32	7	.219	14	64	6	.094
15	32	9	.281	16	64	6	.094
17	32	10	.313	18	64	6	.094
19	32	12	.375	20	64	1	.016
21	32	11	.344	22	64	7	.109
23	32	12	.375	24	64	5	.078
25	32	10	.313	26	64	7	.109
27	32	14	.438	28	64	3	.047
29	32	10	.313	30	64	7	.109
31	32	13	.406	32	64	2	.031
33	32	10	.313	34	64	8	.125
35	32	13	.406	36	64	6	.094
37	32	14	.438	38	64	5	.078
39	32	13	.406	40	64	7	.109
		208				114	

3.6.3 Construction of a *p* Chart to Analyze Data Under Current Subgrouping Strategy

The 40 subgroups were collected over one shift of operation. The odd-numbered subgroups were formed from end crusts and had subgroup sizes of 32. The even-numbered subgroups were formed from the "middle" crusts and had subgroup sizes of 64. Because this set of data has varying subgroup sizes (i.e., n is not constant), a p chart was used to analyze this data. The values for fraction nonconforming, p, were plotted by subgroup and the centerline was drawn on the chart as shown in Figure 3.17.

As described in Section 3.2.2, the centerline, $\bar{p}$, was calculated by dividing the total number of nonconforming crusts in all subgroups by the total number of crusts examined. In the 20 subgroups with 32 crusts there were 208 non-conforming crusts, and in the 20 subgroups with 64 crusts there were 114 nonconforming crusts. Thus:

$$\bar{p} = \frac{208 + 114}{(32 \times 20 + 64 \times 20)} = .1677$$

The rationale for computing one average for results from both end and middle positions is to first determine whether end and middle crusts behave as if they were affected by the same sources of variation or whether one is subject to additional, "special" causes of variation. If, in fact, there is no difference between positions, then there is only one average process value. However, if the ends differ from the middle, then the computed average, corresponding to both locations, describes the results from neither location and a control chart should indicate the inconsistency of crusts from end positions with those from the middle positions.

Because the amount of expected variation in the fraction nonconforming crusts measured from a stable process depends on the number of items in a subgroup, the same control limits could not be used to assess the variation in

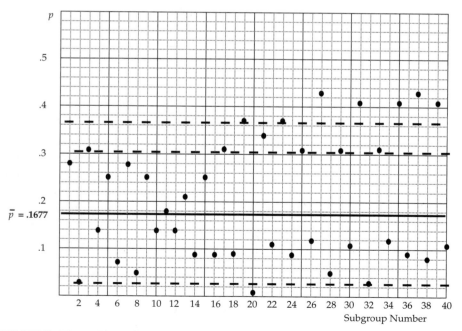

FIGURE 3.17 p *Chart for End and Middle Nonconforming Crusts*

the fraction nonconforming across all of the subgroups. Therefore, two different values of n, 32 and 64, were used to compute the control limits for the p chart. For those points based on 32 end crusts, the upper and lower control limits were computed to be:

$$\text{UCL} = .1677 + 3\sqrt{\frac{.1677(1 - .1677)}{32}}$$

$$= .3081$$

$$\text{LCL} = .1677 - 3\sqrt{\frac{.1677(1 - .1677)}{32}}$$

$$\text{LCL} - \text{none}$$

For those points which are the fraction nonconforming in 64 middle crusts, the upper and lower control limits were calculated as:

$$\text{UCL} = .1677 + 3\sqrt{\frac{.1677(1 - .1677)}{64}}$$

$$= .3081$$

$$\text{LCL} = .1677 - 3\sqrt{\frac{.1677(1 - .1677)}{64}}$$

$$= .0276$$

These limits were used to construct a p chart like the one in Figure 3.17. The odd-numbered subgroups have subgroup sizes of 32, so the upper control limit for these points is .3658 with no lower control limit. The remaining points have a subgroup size of 64. The upper and lower control limits for these points are .3081 and .0276, respectively. The constructed chart showed several points above the upper control limit. As seen in Figure 3.17, these points correspond to subgroups 19, 23, 27, 31, 35, 37, and 39. Also, the p value for subgroup 20 was below the lower control limit. Since all of the points above the upper control limit are for subgroups of "end" crusts, it appeared that "end" crusts were inconsistent when evaluated against "middle" crusts. In other words, the "end" crusts tended to have larger nonconforming rates than did the "middle" crusts.

The strategy used for detecting this difference in the nonconforming crusts by location is worth reviewing, because it provides a good introduction to using control charts for process study. A possible source of variation in nonconforming crusts was identified as tin position. To investigate this suspicion, a subgrouping strategy was used that consisted of "end" crusts in some subgroups and "middle" crusts in others. The points associated with end positions appeared to be inconsistent with those from the middle positions, as indicated by points outside of both control limits when the overall average was used as a centerline; this inconsistency was taken as evidence of a "special cause" acting by location. Because the chart indicated a higher nonconforming rate among "end" crusts, the decision was made to construct separate control charts for the two locations.

These findings led the management group into a discussion about how to understand the actual reasons for the higher proportion of nonconforming crusts produced on the ends rather than in the middle tins. They knew that the way in which this work would proceed was dependent on whether the number of nonconforming crusts when considered separately by type was stable over time or not. If the number of nonconforming "end" crusts was out of control, then they would attempt to identify those sources present at some times and not others that resulted in inconsistent levels of nonconforming crusts. On the other hand, if the process producing "end" crusts was stable, their work would include discovering and studying the effects of the common cause sources present throughout the production of "end" crusts.

Figures 3.18 and 3.19 contain the individual p charts constructed on the fraction nonconforming "middle" crusts and "end" crusts. It is left to the reader to verify that the centerline and control limits on these charts are correct. The behavior of the data points on the p chart for "end" crusts, shown in Figure 3.19, was particularly noticeable. It appeared as though the number of nonconforming "end" crusts was increasing over time since there were many points below the centerline on the left of the chart and more above the centerline in the later time periods. An application of the Rule of Seven indicated a run of nine points (points 1 through 9) below the centerline. Thus, they concluded that a special cause was present. Efforts should now be made to identify the cause or causes that act to produce the increasing trend in the number of nonconforming pie crusts. Of course, nothing on the chart indicates why this systematic pattern occurred. The ability, or "luck," in identifying such causes depends on the observation powers of the managers, technical people, and operators and on the ability to correlate these observations with the way process work is performed.

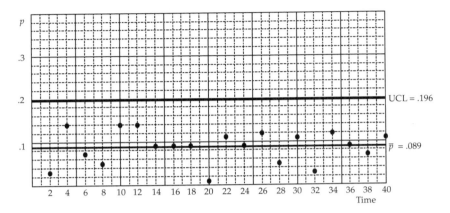

FIGURE 3.18 p *Chart for Nonconforming Middle Crusts*

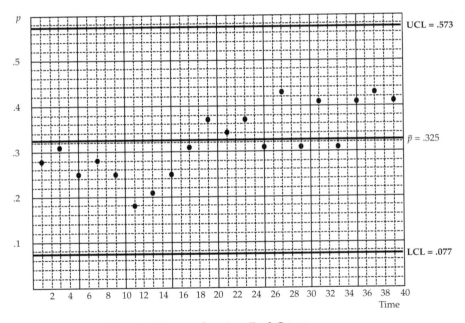

FIGURE 3.19 p *Chart for Nonconforming End Crusts*

3.6.4 Studying Cause and Effect Relationships

The ability to detect special causes is enhanced by having multiple persons from different levels of responsibility involved in data collection so that many types of information are collected from the process. When events such as changes in material, changes in operators, and line shutdowns are noted on a chart, guidance in the search for special causes is provided. Such possible causes are more easily noted when the individuals involved in the process identify, prior to data collection, process factors that potentially affect process outputs. Although development of a complete list of possible causes of unacceptable results prior to data collection is unlikely, a preliminary cause and effect analysis increases the chance of successfully identifying the actual causes. The collective knowledge of all persons who work with the process is invaluable in attempting to identify causes of variation acting on a process.

The management group responsible for the pot pie line began to focus attention on possible causes for a gradual increase in the number of nonconforming "end" crusts throughout a day. The following two causes were suggested:

1. *Changing environmental conditions*: At the start of each day, the moisture level in the dough is set by adjusting the rates at which dry and liquid ingredients flow into the mixing vat. The moisture level is critical, because dough that is too dry tears easily. As the temperature and humidity in the plant change throughout the day, the setup rates may not maintain the appropriate moisture level.
2. *Recycling of excess and scrap dough:* The operation of cutting the dough to fit the tins generates a certain amount of scrap dough. Current practice returns this dough to the mixture from which other pie crusts are rolled. Additionally, dough from nonconforming pie crusts is returned to the mixture. There is some concern that this practice causes deterioration of the dough throughout the day, resulting in increased numbers of nonconforming "end" crusts.

Although these two causes were identified as possible reasons for the trend in nonconforming pie crusts, it was not clear that either one of them was the true cause of the noted trend. Therefore, the next step in the process work was to study the actual effect of these two causes and to identify other possible reasons for the observed increase in nonconforming crusts. The effect of recycling of dough was to be studied by discontinuing the recycling practice, then observing whether or not the trend persisted after recycling ceased. If the trend did not persist, this process change may be attributed to the discontinuance of the recycling. However, questions would remain about the effect of the environmental conditions on the increasing trend in nonconforming end crusts. Studying the cessation of recycling under a variety of temperature and humidity conditions is now suggested. In addition, identifying other possible causes for the trend, in addition to the recycling of dough, would be useful for identifying those sets of conditions over which the process should be observed in order to better understand the process.

To begin developing the required knowledge of the effects of recycling on nonconforming end crusts, recycling was discontinued for a day. The number of nonconforming end crusts were counted in the same fashion as before. A summary of these data is contained in Table 3.5. Figure 3.20 is a p chart constructed from these data. The fact that no trend is noted on the chart in Figure 3.20 supports the assumption that the recycling of the dough causes an increase in the number of nonconforming "end" crusts. This result determined two important issues that should be addressed at this point:

Issue 1: *What kind of confirmation work should be done on this recycling study?*

TABLE 3.5 *Number of Nonconforming End Crusts Without Recycling*

Subgroup	n	Number Nonconforming	p
1	32	7	.2188
2	32	10	.3125
3	32	8	.2500
4	32	4	.1250
5	32	11	.3438
6	32	6	.1875
7	32	7	.2188
8	32	5	.1562
9	32	10	.3125
10	32	9	.2813
11	32	9	.2813
12	32	4	.1250
13	32	8	.2500
14	32	9	.2813
15	32	7	.2188
16	32	5	.1562
17	32	11	.3438
18	32	8	.2500
19	32	9	.2813
20	32	7	.2188
		154	

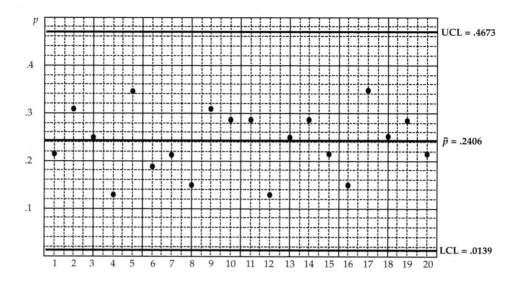

FIGURE 3.20   p *Chart for Nonconforming End Crusts Without Recycling*

Performing confirmation studies is important not only in the present context, but is a recurring issue in work directed at the continual improvement of a process. Confirmation studies are performed repeatedly under a variety of conditions to determine if the improvements noted in earlier studies are maintained under different conditions. Too often an effect noted at one point in time, such as the reduction in nonconforming crusts when recycling is halted, does not reoccur in later studies. The increased focus on a particular area may result in improvements in that area that are not sustained after attention is directed elsewhere. These improvements may be incorrectly attributed to a change made to the process rather than to the increased attention. Confirmation studies help to prevent this type of incorrect conclusion. Confirmation studies are required for other reasons. As previously discussed, other causes may be acting on the pie crust line to cause a trend in the number of nonconforming "end" crusts. The improvement noted may be due to one of these less understood causes. Just as confirmation studies help detect improvements that come simply from increased attention to a process, they are also useful in determining if an improvement occurs for some reason other than the one originally proposed. If confirmation studies do not show that the improvement can be maintained by ceasing recycling, then additional work to identify other causes that could be affecting the process is indicated. In the pie crust example, confirmation studies might consist of continuing to collect data over a larger span of time, which includes a greater variety of operating conditions. Performing similar studies on other shifts and other lines can help the process manager determine if the recycling is actually responsible for the increasing level of nonconforming pie crusts.

> ***Issue 2***: *Given that stopping the recycling showed that the number of non-conforming crusts did not increase throughout the observed shift, what kind of process improvements should follow?*

The answer to this question depends on further study of the process. These are some of the points that need to be considered:

What characteristics of the returned dough contribute to the problem?

How much dough is currently being recycled? At what rate is it recycled?

Why is recycling performed? Is it a cost-saving measure or a way of improving the scheduling of the operation?

If the dough is recycled, can it be reconstituted so that it does not have a detrimental effect on end crusts?

What are the effects of recycling on dough characteristics, such as taste and texture?

Would the same type of improvement occur if the amount of recycling was reduced but not stopped?

Answers to these questions require technical process knowledge, information on the current operating procedures of the process, and an understanding of the basis for the managerial decisions made concerning the process. This need for a wide range of information inherently dictates that management lead the improvement efforts.

3.7   SUPPORTING IDEAS FOR THE EFFECTIVE USE OF *p* AND *np* CHARTS

The ideas discussed in the previous case study provide the reader with some ideas on how control charts (in particular, *p* and *np* charts) are effectively integrated into a program for continually improving business processes. These ideas are important to ensure sustained process improvements; these suggestions, of course, may not be applicable in all situations. However, the critical idea that the practitioner must understand as much as possible about a process in order to decide how information should be collected is a necessity common to all process investigations. Such process knowledge provides a clear understanding of the values and limitations of the information gained from the analysis of data using control charts. Additional ideas and recommendations for the effective use of process data are captured in the following text paragraphs.

> ***Idea 1***: *A process flow diagram should be in place prior to data collection.*

The benefits are as follows:

1. The process is understood in terms of how it affects and is affected by other processes and systems.
2. Alternatives for process improvement become apparent.
3. The consistency of proposed improvements with other priorities is encouraged.
4. Potential process improvements can be more easily evaluated in terms of their impact on other aspects of the process. It is important not to induce larger problems in other areas of this process as a result of a proposed improvement.

Given such an important role in process improvement, a flowchart that simply lists the order in which operations are intended to occur is usually insufficient. The flowchart should describe how the process currently operates. As focus is directed toward a process and work proceeds to improve that process, the chart should be revised and updated to incorporate new process knowledge and process changes.

> ***Idea 2***: *Operational definitions of the specifications, characteristics, and other attributes under study should be in place.*

The benefits are as follows:

1. Management has set the priority for studying certain process characteristics based on known information about attributes necessary to deliver customer value.
2. The team working on the process is strengthened when the members understand the importance of the consistent application of critical operational definitions.
3. The validity of the measurements is strengthened when training in the operational definitions has been performed appropriately.
4. The value of process descriptions, conclusions, and recommendations resulting from data collection is increased.

Procedures should be in place to update or revise the definitions and specifications as the product is changed based on careful evaluation of customer response (in terms of changing expectations or better information). Operational definitions need to be checked and tested on a regular basis.

Also, inspection practices should be checked and verification provided that inspection is being properly and consistently performed by the different shifts and departments. Proper attention to operational definitions is vital. If inspection procedures are inconsistently applied or if records are inexact or incomplete, then nothing useful can come from the application of a p chart.

> **Idea 3**: *Plans for using the various types of information gained from a* p *chart must be in place. Work on identifying, understanding, and preventing special causes requires management and supervisory direction.*

The benefits are as follows:

1. Fundamental or root causes get addressed. Without the necessary direction, "bandage" problem fixing, based on restricted resources and limited understanding of the technological aspects of the special causes, is the likely result.
2. Individuals with the appropriate levels of authority, skills, responsibility, and knowledge are involved in the improvement work.
3. The larger systemic issues that contribute to the presence of special causes at the process level can be identified and resolved.
4. The data collection strategy allows for the formation of subgroups that delineate between the types of variation that are captured within and between the subgroups. Thus, cause and effect relationships are better understood.

It is management's responsibility to implement plans for systematic study of the process and to identify the appropriate individuals to be involved in the work. The types of problems identified by this work and the selection and implementation of effective solutions require departmental and functional representation in the selection of the individuals to be involved in the work.

The importance of management's involvement becomes evident when one considers the fact that what is identified as a special cause at a micro level in the process is actually a common cause in a larger, more complicated engineering system. For example, a special cause may be identified at the process level as a "maintenance" problem when, in fact, there is a policy in existence that limits correct and consistent maintenance practices. Also, study of a process may reveal that an unusually large number of nonconforming items are produced at times when an untrained operator is working on the process. This special cause can be removed by training of the operator. However, the fact that the operator started working on the process without adequate training indicates that a larger training and educational issue exists. Unless the issue is resolved by management, the same special cause (inadequately trained operators) will appear again and again.

> *Idea 4: p charts are most effective for process study if fraction nonconforming for only one specification, standard, or type of nonconformity is included on a chart rather than using one chart to report on an aggregate fraction rejected.*

The benefits are as follows:

1. Increasing or decreasing trends in the fraction nonconforming for a particular characteristic are not disguised or offset by contrary movements in other characteristics.
2. Values for fraction nonconforming indicating the presence of special causes are more easily understood or interpreted in terms of the process, because the value itself provides a reason for the rejection in that subgroup.
3. The form of the data presentation aids in problem identification and discloses opportunities to learn about underlying common and special causes.

A fundamental mistake often made in resource allocation occurs when managers or work teams attempt to report on numerous specifications or characteristics using one chart. In an attempt to make life easier, managers often direct the construction of a few charts with multiple characteristics on a single chart. In terms of time, effort, and other resources, data collection and basic charting is by far the least expensive. Problem identification, determination of causes, and testing and implementation of solutions are the more greedy absorbers of resources. When many types of nonconforming attributes are included on a single chart, the usefulness of the chart for gaining process knowledge is severely limited. Even if numerous common cause systems and various types of special causes are active, the chart will provide little information about these sources of variation. When multiple characteristics are included on one chart, time spent on inspection and data collection is usually wasted.

One exception to this suggestion exists. Before opportunities for improving a process or product can be identified, one must have an idea of what problems typically occur and with what relative frequency. A *p* chart that includes all types of nonconforming attributes in conjunction with a Pareto chart to communicate the larger problems or types of defects may be constructed to report on the current state of a process. Careful judgment must be used in setting the priorities based on this work. The severity of respective problems should be questioned according to their long-term impact on the customers' requirements and needs.

> *Idea 5: Where technically and economically feasible, inspection for the purposes of process and product improvement should be converted to a variables format.*

The benefits are as follows:

1. No information from the measurements is lost.
2. More information on the sources of variation and the types of problems is available.
3. More specific direction is provided for work on process improvement.

Sometimes process inspection consists of making a measurement, comparing it to specifications, and then recording "conforming" or "nonconforming" as the response. In such instances, valuable information is discarded. The *p* chart provides information about the ability of a process to conform to specifications. However, there is limited information about whether the problem is too much variation, an incorrect average, or both. Since improving process performance requires working on the sources of variation in the measurements, data collected in a variables format facilitates the work on process improvement.

3.8   PRACTICE PROBLEMS

3.8.1   At a food packaging plant, a recent increase in the monthly report on defective cans has prompted an investigation into the situation. Knowing that the sealed metal containers pass through a final inspection where they are checked for proper can height, label application, and vacuum and other surface characteristics, data on the proportion of defective cans found at final inspection are requested. The following data are reported.

Fraction of Defects in 22 Samples of 1000 Cans

Sample	Fraction Defective	Sample	Fraction Defective
1	.003	12	.003
2	.004	13	.004
3	.008	14	.001
4	.006	15	.002
5	.003	16	.001
6	.006	17	.002
7	.004	18	.000
8	.006	19	.001
9	.002	20	.001
10	.005	21	.000
11	.004	22	.001

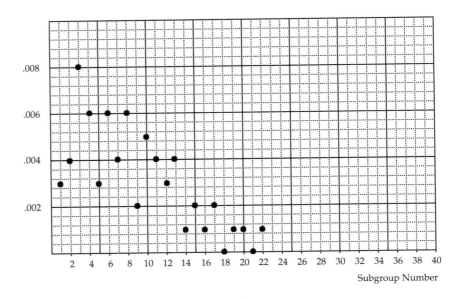

FIGURE 3.21 *Fraction Defective Cans in Samples of Size 1000*

a. Construct an appropriate control chart for these data. (The fraction defective have been plotted in Figure 3.21.)
b. Comment on the manner in which these data were obtained.
c. To control and/or improve this process, what future actions would you recommend?

3.8.2. In the manufacture of roll-on deodorant, deodorant is discharged into a container and a plastic (polypropylene) ball is forced into the top of the container. Following this operation, samples are collected to test product functionality. Each item in a sample is placed into a device with the ball pointed downward. An apparatus moves the ball into contact with a two-foot strip of vinyl. Pressure is applied until product is dispensed. If product is not dispensed when the pressure reaches 700 grams, the dispenser is considered nonconforming. Two hundred bottles are tested each day. Results for the past 25 days are shown in Table 3.6. A plot of the number nonconforming appears in Figure 3.22.

a. Calculate control limits for the chart in Figure 3.22.
b. Does the process appear to be operating in a consistent fashion?
c. If appropriate, estimate the fraction nonconforming bottles.
d. Would bottles that dispense deodorant too readily be a cause for nonconformity?
e. What sources of variation are captured within each subgroup? Between subgroups?

TABLE 3.6 *Roll-On Deodorant*

Day	Number of Nonconforming	Day	Number of Nonconforming
1	17	14	23
2	14	15	16
3	14	16	17
4	21	17	14
5	23	18	12
6	17	19	6
7	10	20	17
8	18	21	8
9	15	22	15
10	17	23	19
11	10	24	18
12	19	25	16
13	13		389

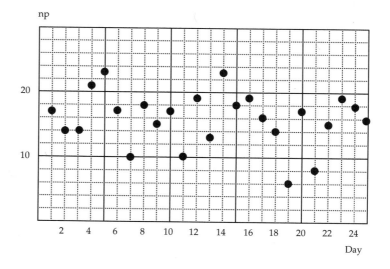

FIGURE 3.22 *Number of Nonconforming Deodorant Assemblies*

3.9   APPLICATION PROBLEM

Figure 2.2 in Chapter 2 contained a flowchart of a hornpad assembly process. A Pareto diagram of the counts of defects produced by the assembly process appeared in Figure 2.11. One of the largest categories on the Pareto diagram was for the number of defects due to excessive flash. This type of defect has become the focus of work efforts at the production site. A cause and effect diagram has been developed in order to capture possible causes of excessive flash. This diagram appears in Figure 3.23. From the cause and effect diagram it can be seen that the color of the hornpad may be one of the major determinants of the amount of excessive flash. It is believed that the different colors of plastic used to produce hornpads may have to be managed in different ways.

Data on excessive flash have been collected for each of the different color types. The data in Table 3.7 were the result of the investigation of the dark blue hornpads. One hundred consecutively manufactured hornpads have been examined every two hours for the past week. The fraction rejected because of excessive flash for each subgroup are recorded in Table 3.7 and plotted in the chart in Figure 3.24.

a. Is the process that produces dark blue hornpads in control? If so, what is the estimated proportion of dark blue hornpads produced by the process?

b. Other colors have been investigated for percentage nonconforming and found to behave differently with respect to fraction nonconforming.

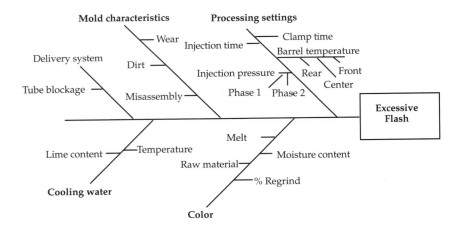

FIGURE 3.23 *Excessive Flash on Hornpad Assemblies*

TABLE 3.7 *Fraction Rejected of Dark Blue Hornpads (n=100)*

Sample Number	Fraction Rejected	Sample Number	Fraction Rejected
1	.04	11	.08
2	.03	12	.04
3	.01	13	.00
4	.03	14	.02
5	.06	15	.06
6	.05	16	.03
7	.07	17	.02
8	.02	18	.02
9	.02	19	.05
10	.05	20	.04
			.74

Further work on the excessive flash characteristic, in part, consisted of an examination of the effect of moisture content. Tighter controls on moisture content have been instituted; follow-up data on the dark blue hornpads are contained in Table 3.8. Again, 100 hornpads were selected every two hours after the changes to moisture management were made. What effect does tighter control on moisture content have on the proportion of dark blue hornpads with excessive flash?

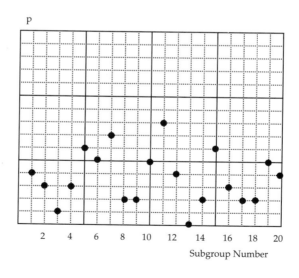

FIGURE 3.24 *Rejected Dark Blue Hornpads*

TABLE 3.8 *Fraction Rejected of Dark Blue Hornpads
(Tighter control on moisture maintained) (n=100)*

Sample Number	Fraction Rejected	Sample Number	Fraction Rejected
1	.03	11	.06
2	.02	12	.00
3	.05	13	.02
4	.07	14	.03
5	.02	15	.07
6	.02	16	.02
7	.06	17	.01
8	.01	18	.06
9	.03	19	.05
10	.02	20	.02

4 Control Charts for Attributes Data: *c* and *u* Charts

In Chapter 3, information on process behavior was gained from data collected by classifying each item inspected as conforming or nonconforming, according to a set of requirements. For many processes, a different kind of data collection will provide better information on the causes of variation acting on a process. The process study methods discussed in this chapter are useful in situations where a number of incidences, either defects or events, are likely to occur on an inspected item or throughout a given interval of time. In these situations, process managers and engineers will need to know how often these events occur, what causes contribute to their occurrence, and how these causes can be managed in order to reduce their occurrence. To address these needs, it is necessary to count the occurrences rather than to simply classify an item or time interval as acceptable or unacceptable. The behavior of these counts across time and conditions will provide insights into sources of variation that affect process performance.

The inspection of large rolls of paper used for printing newspapers illustrates the advantages of counting defects. During inspection, a process manager could choose to have each roll inspected for defects of a certain type, and if a defect is found, to classify the entire roll as nonconforming. Thus, once the first defect on a roll is identified, no further inspection is needed. Of course, the only information about paper rolls available from examining such data would be that there was at least one defect on the roll. Since a roll is either classified as conforming or nonconforming, little knowledge is gained concerning the timing and severity of that defect type over time. Consequently, the ability to relate the severity of defects to the techniques, materials, equipment, or other causes in the production process is also lacking. In this application, thin spots in the rolled paper, flecks of impurities, and wrinkles are defects that tend to cause paper tears and jams in the newsprinting process. These different types of defects also tend to be created by different parts of the paper-making process. Therefore, it is important that the behavior of the number of defects for each type occurring in the rolls be studied over time in order to begin relating the generation of these defects with process causes.

In addition to counting the number of defects or nonconformities in a given area of product, business environments commonly require a count of specific events in a period of time. For example, in most industrial settings, it is important to study accidents over time in order to provide the necessary OSHA safety information and, more importantly, to identify and remove the factors contributing to employee accidents. Although the proportion of employees that have one or more accidents each month could be recorded and studied by the use of p charts, information on the stability of particular types of accidents over time and the causes of these particular accident types will not be gained from the collected data. Such information can be obtained by taking counts of the particular types of accidents in a week or in a month. These counts of accident occurrences will exhibit variation over time, and conditions can be studied to provide guidance on possible actions to reduce the number of accidents occurring over time.

When count data for events in time or defects in an area of product are collected, the statistical tool useful for studying process behavior over time is a c or u chart. C and u charts provide information on the consistency of the occurrence of defects across time and conditions. These charts help those working on a process to determine the stability of the process behavior as well as to identify possibilities for process improvement. This chapter describes the types of situations where the use of the c chart or the u chart is appropriate, and it illustrates the construction and use of these charts through several examples and a case study.

4.1   DESCRIPTION OF ATTRIBUTES DATA REQUIRING THE USE OF c OR u CHARTS

A subgroup may consist of several jobs, parts, assemblies, or events. When each job or item is inspected against one or more requirements and the item is, in turn, classified as either conforming or not conforming to the requirements, then a p chart can be used for the statistical analysis of the fraction nonconforming over time. An np chart could also be used to study the number of nonconforming items. When each unit of inspection is compared against one or more specifications and the number of nonconformities or defects are counted, then statistical analysis from subgroup to subgroup can be performed by *c* or *u* charts as long as the criteria discussed in the following section are met. In Chapter 3, the following examples of situations in which attributes data are generated were discussed:

1. *Counts of the number of damaged cans in lots of 1000 cans.* When each can is judged as being either damaged or not damaged, a *p* chart can be used to analyze the data. However, counts of the number of dents or scratches occurring in the lots of 1000 cans would be analyzed with the use of a *c* chart.

2. *Counts of the number of shafts in samples of 200 whose diameters fail to meet stated specifications.* Each shaft either meets or fails to meet the specifications. Hence, a *p* or *np* chart is appropriate for data analysis.

3. *Daily counts from each day's production on the number of assembled motors that fail a voltage test.* Because each motor either passes or fails the voltage test, a *p* or *np* chart would be used to analyze the behavior of the data over time.

4. *Weekly counts of the number of invoices that contain at least one error.* If the proportion of invoices with errors is plotted over time, then a *p* chart is appropriate for studying the stability of the process data. However, more information can be gained about the process by counting the number of errors that occur in samples of, say, 50 invoices. Then, a *c* chart would be used to analyze the counts of errors in each sample of invoices over time.

In a plant that produces cut copper tubes for use in air conditioning and refrigeration units, the tubes are inspected for burrs prior to shipping. During inspection, samples of 10 cut copper tubes are taken and examined for burrs. As shown in Figure 4.1, a tube that has a "large" burr or that has two or more "small" burrs is rejected prior to shipment. A count of the *number of rejected tubes* (*np*) yields the numbers in the third column. In the example of Figure 4.1, 8 tubes are rejected in the three subgroups of 10 tubes inspected. To understand the causes of the burrs, more detailed information about the burrs is needed. Thus, another method for describing the behavior of the process is to count the *number of burrs*. These counts are recorded in the first column of Figure 4.1. In this example, the proportion of rejected tubes averages 0.2667. However, it is more useful to know that the average number of burrs is 0.7667. Studying the process using data on the number of burrs occurring provides more insights into the severity of the problems with burrs and therefore into the potential reasons for the occurrence of burrs.

Counts of occurrences of events in time or counts of nonconformities in a unit of inspected material are commonly analyzed by the use of *c* or *u* charts. Whether or not such an analysis is appropriate depends on the manner in

Any tube with a burr larger than allowable specifications (o) or with two or more "small" burrs (x) is nonconforming.

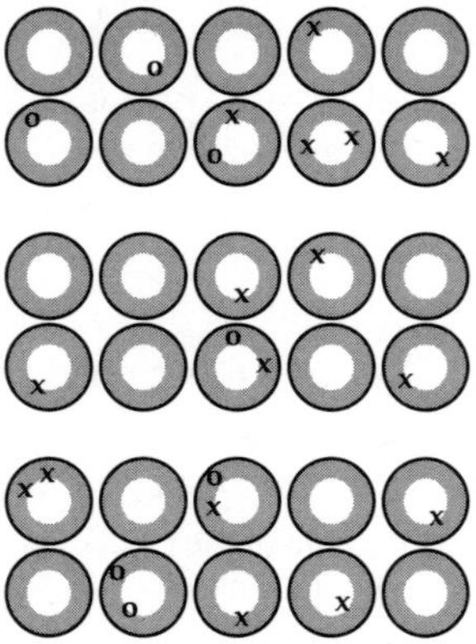

Subgroups of 10 Copper Tubes	Number of Burrs	Number of Rejected Tubes	Fraction Rejected
	8	4	0.40
	6	1	0.10
	9	3	0.30
	23	8	

Avg. Number of Burrs per Tube = 23/30 = .7667	Number Inspected = 30 Number Rejected = 8 $\bar{p} = 8/30 = .2667$

FIGURE 4.1   *Example of Inspection Yielding the Two Types of Attributes Data*

which such events might happen if only random or common cause sources of variation are present. Four criteria summarize the behavior of occurrences of events or nonconformities that should exist if only common cause sources of variation are present. The theoretical basis for these criteria, the Poisson probability model, is discussed in Appendix A.3. The use of a c or u chart to analyze the behavior of counts of occurrences is appropriate when the common cause sources of variation meet the four criteria shown in the box. The following two examples illustrate the practical application of the four criteria.

Example 1: Counts of Flaws on Sheets of Processed Aluminum

In the manufacturing of aluminum cans, it is critical that no holes be present in the aluminum prior to can manufacture. At one plant where aluminum cans are manufactured, past experience has shown that there are no large holes in the aluminum sheets received from the supplier, but tiny pinholes are occasionally found. The aluminum is therefore 100% inspected prior to use for can manufacture, and the number of pinholes found in a sheet of aluminum is counted.

Criteria for the Use of *c* and *u* Charts

Criterion 1

   The counts are independent of each other.

Criterion 2

   The number of possible occurrences is large.

Criterion 3

   The chance of an occurrence at any one time or place is small.

Criterion 4

   The expected number of occurrences is proportional to the amount of time or material that is included in an inspection unit.

If a *c* or *u* chart is to be used to understand the behavior of the process of making aluminum, it is necessary to consider the anticipated behavior of the pinhole occurrences if only common cause sources of variation are present in the process. If only common cause sources of variation are present, then one expects to see pinholes scattered in a random fashion over the sheet of aluminum. An active, special cause might have the impact of isolating the pinholes to only one area on the sheets of aluminum. If pinholes tend to occur in clusters, then the detection of a pinhole at one position suggests a high likelihood of other pinholes in the surrounding vicinity. When only common causes of variation are present, the counts of pinholes on a sheet of aluminum are independent of one another; this is the statement of criterion 1.

The second and third criteria for the appropriate use of *c* and *u* charts are easily visualized. When one considers a sheet of aluminum and the number of possible places a pinhole can occur, it is obvious that the number of possible occurrences of pinholes is large. Hence, criterion 2 is met. Now, if a sheet of aluminum is subdivided into many small pieces, each one of these pieces can possibly contain a pinhole. However, it is unreasonable to expect that every one of these many pieces contains one or more pinholes. Thus, criterion 3 is also met. The important concept in the second and third criteria is that there exists a large opportunity for pinholes, even though the anticipated number of actual pinholes is small when compared to the number that could possibly occur.

As stated in criterion 4, the number of anticipated or expected pinholes is proportional to the amount of aluminum that is inspected. If no special causes are acting and if a sheet of aluminum is divided in half, then the expected number of occurrences of pinholes on one half is the same as the expected number on the other half. When two sheets of aluminum are inspected, then about twice as many pinholes are expected than when only one sheet is inspected. Of course, the consistency of the number of anticipated pinholes from one sheet to the next describes the behavior one expects to see if the process is stable over time. Departures from this behavior, as in observing a very large number on one sheet, indicates the presence of a special cause. It is then useful to direct attention toward identifying the different conditions, materials, etc., that may have occurred to cause the increased number of pinholes.

Example 2: Counting the Number of Production Line Stops

Another situation where a *c* or *u* chart might be used is in counting the number of production line stops in, say, a shift. In the example of pinholes in aluminum, the unit of inspection was a sheet of aluminum. In the current example, the unit of inspection is a unit of time, in particular, one shift. The four criteria for using *c* and *u* charts must also hold for situations when occurrences of events in units of time are being counted. In a situation where the production line is shut down at planned intervals, maybe for maintenance or for a tool change, then the number of line stops is a quantity fixed by management and using a *c* or *u* chart to summarize this information is not appropriate or useful. Instead, the *c* or *u* chart might be used for studying a situation where randomly occurring events cause the production line to stop. For instance, in the production of paper, a number of different causes may be acting together or separately to result in breaks in the paper. The times at which breaks occur are random and cannot be predicted. The four criteria listed for the use of a *c* or *u* chart can be examined to determine if a *c* or *u* chart is appropriate for describing the stable behavior of the occurrence of production line stops.

Criterion 1 states that the counts of occurrences should be independent. In the present discussion, this implies that if one production stop occurs, it does not mean that one or more production stops are more or less likely to occur in the near future. The next production stop will occur at random sometime in the future, with the same chance of occurrence as if the previous one had not occurred. Now when considering the use of a *c* or *u* chart, a question about whether such an assumption is correct might be raised. "Is it possible that production stops tend to occur in clusters when certain conditions are present?" Believing that such might be the case does not mean that a *c* or *u* chart cannot be used. Rather, the *c* or *u* chart provides an objective method for studying the process to provide some guidance toward answering the posed question.

Careful consideration should be given to how data is collected from a process to see that this criterion 1 is reasonably well met. In collecting data on the number of line stops, it may be the case that once the production line stops, it is down for as long as an hour or two. In this case, it is not reasonable to have the inspection unit be an eight-hour shift. Suppose that there were a line stop toward the beginning of the shift; then in the remaining portion of the shift there would be less chance of a line stop occurring because the line is not running for a significant portion of the shift. Instead, a more appropriate unit of time in which line stops are counted might be one that corresponds to actual running time of the equipment. If a set number of parts is created in a fixed interval of running time, a unit of inspection could be the time to make some specified number of parts. In this fashion, the amount of actual running time of the equipment represented by a subgroup is fixed.

The applicability of criteria 2 and 3 is illustrated in much the same way as was done in the previous example. Suppose the unit of inspection for line stops is 10 hours of operation time. One could think of dividing the 10 hours into many small increments of time, even into seconds. A stop can occur in any one of these many small time increments. Thus, the possible number of occurrences is very large. However, the chance of seeing a line stop in a large number of these tiny time increments is very small. Most of the time increments will not contain a line stop. Hence, as stated by criteria 2 and 3, the number of possible line stops is large, while the chance of a line stop at any one small increment in time is very small.

Criterion 4 states that the expected number of occurrences is proportional to the amount of time that is included in an inspection unit. This criterion implies that from one unit of inspection to another, the expected number of line stops is the same if the process is subject only to common cause sources of variation. Departures from criterion 4 would not mean that a *c* or *u* chart would not be the appropriate analysis method. Instead, departures from this consistent behavior across time would be considered a special cause; it is this information about the process that could be gained by using a *c* or *u* chart.

Limits on a *c* or *u* chart indicate the number of occurrences that could be expected to occur if the process is behaving according to the given criteria. Figure 4.2 shows a *c* chart for production line stops, in which the upper control limit has a value of 6.45, and there is no lower control limit. If the process were stable over time, the expected number of line stops would be between 0 and 6.45. Since the chart has a point above the upper control limit, the chart indicates that nonrandom behavior is occurring.

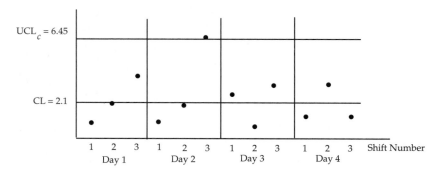

FIGURE 4.2 *Example of a c Chart for Production Line Stops*

4.2   CONSTRUCTION OF *c* CHARTS

Both c and u charts are used to examine the stability of a process over time and to identify sources of variation acting on the process when the information collected on the process is in terms of counts of occurrences. The c chart is used to study the counts of occurrences when the same size inspection unit is used over the course of the study. If counts of occurrences are made on varying sizes of inspection units, then a u chart is used. The data in Table 4.1 provide an example of a situation in which the amount of material inspected stays constant over the course of the study. The data in this table were obtained by counting the number of dye imperfections on 500-yard bolts of cloth. Each bolt was selected from a shipment received by a clothing manufacturer. The subgroup number corresponds to the order in which the shipments were received. For this example, each subgroup corresponds to one bolt of cloth. Since each bolt has the same yardage, the amount of material inspected for each subgroup is the same. The letter c is used to denote the number of nonconformities in each subgroup.

4.2.1   Calculating the Centerline and Control Limits
for a *c* Chart

As in the construction of a p chart, the construction of a c chart begins with plotting the data. With a c chart, the points plotted are the c's, or the counts of

---

Notation

k denotes the number of subgroups (For a c chart, k also refers to the number of inspection units.)

c denotes the number of nonconformities in each subgroup.

TABLE 4.1 *Number of Dye Imperfections in 500-Yard Bolts of Cloth*

Subgroup	Number of Imperfections	Subgroup	Number of Imperfections
1	22	11	15
2	29	12	10
3	25	13	33
4	17	14	23
5	20	15	27
6	16	16	17
7	34	17	33
8	11	18	19
9	31	19	22
10	29	20	27
			Total: 460

nonconformities. The data should be collected and plotted in time order. Figure 4.3 is a plot of the number of imperfections on each of the 20 bolts of cloth inspected. The horizontal axis corresponds to the subgroup number and the vertical axis to the count of nonconformities. For these data, the ordering of the subgroups corresponds to the time order in which the bolts of cloth were received at the plant. Note, however, that it is not known by the plant whether the timing at which the shipments were received correlates to the timing of manufacture.

The centerline on a c chart represents the average number of nonconformities per inspection unit. Because the total number of inspection units on a c chart is the same as the number of subgroups, the average, $\bar{c}$, is simply the total number of nonconformities found divided by the number of subgroups.

$$\bar{c} = \frac{\text{total number of nonconformities}}{\text{total number of units inspected}}$$

Table 4.1 lists 460 dye imperfections in the 20 subgroups, so:

$$\bar{c} = \frac{460}{20} = 23$$

This value of $\bar{c}$ is used to place the centerline on the c chart, as shown in Figure 4.3.

The upper and lower control limits for the *c* chart define the amount of variation that is expected in the recorded nonconformities if only common cause sources of variation are present. In other words, the control limits describe the amount of variation to be expected in the recorded *c*'s if the process is subject to the kind of random behavior described by the four criteria given in Section 4.1. The formulas used to calculate the upper and lower control limits for the *c* chart are, respectively:

$$\text{UCL}_c = \bar{c} + 3\sqrt{\bar{c}}$$

$$\text{LCL}_c = \bar{c} - 3\sqrt{\bar{c}}$$

where *c* is the average number of nonconformities.

For the data on dye imperfections, the upper control limit is computed as:

$$\text{UCL}_c = 23 + 3\sqrt{23} = 37.39$$

The calculation for the lower control limit results in:

$$\text{LCL}_c = 23 - 3\sqrt{23} = 8.61$$

Figure 4.4 displays the completed control chart. Identification of an unstable process when using a *c* or *u* chart is done using the same guidelines defined for situations requiring the use of *p* or *np* charts. Because there are no points on or outside the control limits and no runs of length seven or more in the plotted points, the completed chart shows no indication of a special cause operating. It appears that dye imperfections occur in a random fashion over

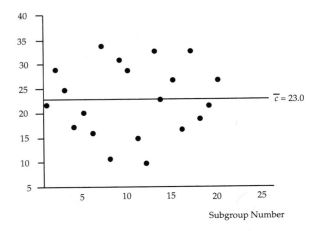

FIGURE 4.3 *Plot of Dye Imperfections*

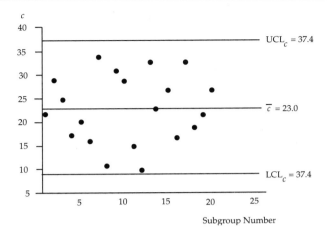

FIGURE 4.4 *Completed* c *Chart on Dye Imperfections*

time. One subgroup, or one bolt of cloth, does not have a significantly higher or lower number of imperfections to indicate that the process may have been behaving differently at the different times at which these bolts were made.

4.2.2   Summary

Construction of *c* **Charts**

k denotes the number of subgroups (k is the same as the number of inspection units) and c denotes the number of nonconformities in an inspection unit.

1. The number of nonconformities on each inspection unit (i.e., in each subgroup) is counted. The c values are plotted on the c chart.

2. Calculate the centerline on the chart as:

    $$\bar{c} = \frac{\text{total number of nonconformities}}{\text{total number of units inspected}}$$

3. Calculate the control limits as:

    $$\text{UCL}_c = \bar{c} + 3\sqrt{\bar{c}}$$

    $$\text{LCL}_c = \bar{c} - 3\sqrt{\bar{c}}$$

4.3   CONSTRUCTION OF *u* CHARTS

As with *c* charts, *u* charts are used to understand process behavior when the data collected on the process are counts of occurrences in a given unit of inspection. As seen in the previous section, *c* charts are used when the inspection unit or subgroup size stays the same across all subgroups. *u* charts, on the other hand, are used when the amount of material or the unit of time that forms a subgroup varies from one subgroup to the next. The following example illustrates the use and construction of *u* charts.

A manufacturer of steel framing systems supplies galvanized, pre-engineered studs and trusses to be used in the construction of commercial buildings. The company receives engineering specifications and architectural plans from the construction companies, converts them into engineering plans for the framing system, cuts the steel beams to length and width specifications, drills holes for the precision fit of the studs and trusses, and ships the completed system to the building site. Once the completed system is delivered to the construction site, no additional materials or cutting is needed to adapt the framing system to the architectural designs. The building crew simply erects the steel trusses and studs according to the component numbering system. Hence, each order is processed and manufactured to completion prior to the start of another order.

As part of an effort to work on improving the value provided to customers, one of the plant sites has begun work on improving its ability to supply precisely fitting systems within predictable delivery times. The plant has received several complaints from the building contractors that the received framing kits have not fit together precisely without additional drilling of holes. Due to the extraordinary amount of time involved with shipping the beams back to the manufacturer, having them recut or redrilled, and then shipping them back to the building site, many contractors choose to perform small alterations or redrills themselves. When this occurs, there are two major repercussions. First, the warranty on the framing system no longer holds if the system has been altered in any way by a building contractor. Second, additional cuts or holes in a steel beam weaken the strength of the frame and reduce its earthquake and wind loading ratings.

To begin studying the reasons for the complaints and to attempt to eliminate any adjustments performed by the building contractors, data were collected on the number of adjustments made by the building contractors per 10,000 square feet of framing. These data were obtained by the manufacturer's representatives who work with the building superintendents during the erection of the framing systems.

The data shown in Table 4.2 are in the time order in which the framing kits left the production line. The number of square feet differs from building to building. Subgroup 2 represents a framing system for a 74,000-square-foot building, whereas subgroup 10 consists of the number of adjustments to the steel frame for a building that has only 7000 square feet. Thus, the amount of material inspected changes over time. Because the data consist of counts of nonconformities on inspected material and the amount of material inspected changes over time, a *u* chart is the appropriate technique for studying the behavior of adjustments made over time to the pre-engineered framing systems.

The first step in constructing a *u* chart is deciding on the unit of inspection. The definition of an inspection unit is somewhat arbitrary; there is no one right way to choose the unit of inspection. Instead, the inspection unit is chosen by convention or for convenience. In the present situation, nonconformities are typically reported on a per 10,000-square-foot basis. Consequently, 10,000 square feet was adopted as the unit of inspection. The first subgroup was for a building with 12,000 square feet; the number of inspection units in

TABLE 4.2 *Data on Adjustments to Pre-Engineered Framing Systems*

Building	No. of Adjustments (c)	No. of Square Ft.	No. of Inspection Units (n)	$u = \frac{c}{n}$
1	10	12,000	1.20	8.333
2	21	74,000	7.40	2.838
3	14	15,500	1.55	9.032
4	12	9,000	0.90	13.333
5	13	17,000	1.70	7.647
6	9	14,500	1.45	6.207
7	18	36,000	3.60	5.000
8	9	21,000	2.10	4.286
9	16	48,500	4.85	3.299
10	11	7,000	0.70	15.714
11	14	13,000	1.30	10.769
12	12	19,500	1.95	6.154
13	8	16,000	1.60	5.000
	167		30.30	

the first subgroup is found by dividing 12,000 by 10,000. The number of inspection units in the first subgroup is recorded in the fourth column.

The data plotted on the u chart are the number of nonconformities per inspection unit. For the first subgroup, this value was found by dividing the number of adjustments, 10, by the number of inspection units, 1.20. This division results in a u value of 8.333.

Notation

c denotes the number of events or nonconformities in a subgroup.

n denotes the number of inspection units in a subgroup.

u denotes the number of events or nonconformities per inspection unit in a subgroup.

$$u = \frac{c}{n}$$

Other choices for an inspection unit are possible. For example, one square foot could have been chosen as an inspection unit. However, this choice would not have been as convenient, as can be seen by using one square foot as an inspection unit and calculating u for the first subgroup. With one square foot as an inspection unit, n for subgroup 1 would be 12,000 and u would be:

$$u = \frac{10}{12,000} = 0.000833$$

The small size of this and the remaining u's would make the plotting and calculations associated with these numbers tedious. For this reason, the choice of 10,000 square feet as an inspection unit is more convenient.

The values of n and u for the remaining subgroups are listed in the fourth and fifth columns of Table 4.2. The values of u have been plotted in time order on the chart in Figure 4.5. The centerline for this chart represents the average number of events or nonconformities per inspection unit. It is computed by dividing the total number of events by the total number of inspection units. The centerline is *not* computed by averaging the individual u values. If the subgroup size changes, averaging the u values provides an incorrect centerline value. The total number of events is found by summing

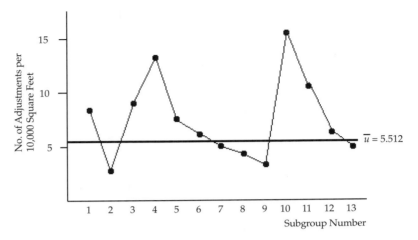

FIGURE 4.5   *Plot of Number of Adjustments per 10,000 Square Feet*

the number of nonconformities (number of adjustments) found in the second column. The total number of inspections units is found by summing the fourth column, the column which lists the number of inspection units. The centerline in this example is given by:

$$\bar{u} = \frac{167}{30.3} = 5.512$$

The formulas for the upper and lower control limits of a *u* chart are:

$$\text{UCL}_u = \bar{u} + 3\frac{\sqrt{\bar{u}}}{\sqrt{n}}$$

$$\text{LCL}_u = \bar{u} - 3\frac{\sqrt{\bar{u}}}{\sqrt{n}}$$

These formulas are a function of *n*, the number of inspection units in a subgroup, a value that changes from subgroup to subgroup. So the control limits for each of the subgroups must be calculated using the value of *n* associated with that particular subgroup. For example, the number of inspection units in the first subgroup was *n* = 1.20. The values for the upper and lower control limits for the first subgroup are calculated as:

$$\text{UCL}_u = 5.512 + 3\frac{\sqrt{5.512}}{\sqrt{1.2}} = 11.942$$

$$\text{LCL}_u = 5.512 - 3\frac{\sqrt{5.512}}{\sqrt{1.2}} = \text{none}$$

The upper and lower control limits for the other twelve subgroups are provided in Table 4.3. Figure 4.6 contains the completed *u* chart for the frame adjustment data.

Since there are two points above the upper control limit and one point below the lower control limit, the *u* chart indicates that the number of adjustments made to the steel frame systems at the building sites is not uniform across the buildings and the time studied. A closer look at this chart led the plant manager to begin considering the reasons for the out-of-control points. The plant manufactures frames for three different types of commercial buildings. Both of the subgroup values that fall above the upper control limit were for structures designed for commercial warehousing systems. To provide additional insights into the differences between building types, the data were rearranged by building type rather than in strict time order. This chart is shown in Figure 4.7. The first six points on the chart correspond to the framing systems for commercial warehouses. As in the previous chart, the same points are out of control. However, this chart does help to confirm that the number of adjustments made to the warehouse framing systems appears to occur at a higher average than for the other types of systems.

The group from the plant decided to focus their next efforts on identifying and eliminating the reasons for the adjustments to the warehouse framing

TABLE 4.3 *Control Limits for* u *Chart on Number of Adjustments to Steel Framing Systems*

Subgroup	*n*	*u*	UCL	LCL
1	1.2	8.333	11.942	none
2	7.4	2.838	8.101	2.923
3	1.55	9.032	11.169	none
4	0.90	13.333	12.936	none
5	1.70	7.64	10.914	none
6	1.45	6.207	11.361	none
7	3.60	5.000	9.224	1.800
8	2.10	4.286	10.372	0.652
9	4.85	3.299	8.710	2.314
10	0.70	15.714	13.930	none
11	1.30	10.769	11.689	none
12	1.95	6.154	10.556	0.468
13	1.60	5.000	11.080	none

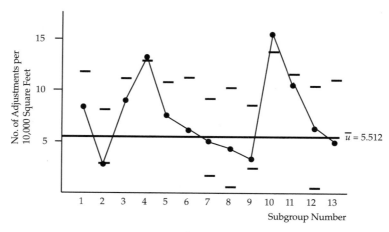

FIGURE 4.6   u *Chart for Framing Adjustments*

systems. They decided to continue collecting data on warehouse frame adjustments and to construct a separate u chart with a new centerline and limits for the warehouse frame data. Additional data points are needed prior to the construction of this chart. At present, only six subgroups correspond to warehouse frames. These six subgroups do not provide enough data for one to be confident about any conclusions reached from a separate control chart. The specific number of subgroups needed to reach meaningful conclusions depends on the nature of the process, the amount of time required to obtain a subgroup, and the information desired from the chart. However, as a "rule of thumb," at least 20 subgroups should be used to construct a control chart.

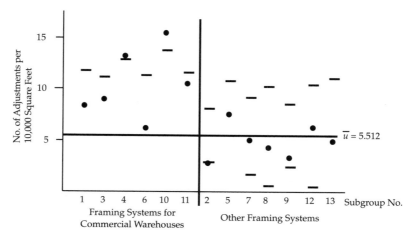

FIGURE 4.7   u *Chart on Adjustments (Plotted Points Rearranged by Building Type)*

The group also made plans to work with the contractors on identifying the actual reasons for the number of adjustments. Using a separate control chart for the commercial systems, they hope to obtain a better understanding of the reasons for inconsistent behavior by building type. With this information on causal factors, the plant will be in a better position to reduce the rework at the job sites and to eliminate the associated problems.

4.3.2   **Summary**

Construction of *u* Charts

c denotes the number of nonconformities in a subgroup; *n* denotes the number of inspection units in a subgroup; and *u* denotes the number of nonconformities per inspection unit in a subgroup.

1. Count the number of nonconformities on each inspection unit. Calculate *u*, the number of nonconformities per inspection unit, for each subgroup:

$$u = \frac{c}{n}$$

2. Plot the values of the *u*'s on the control chart.

3. Calculate the centerline of the control chart as:

$$\bar{u} = \frac{\text{total number of nonconformities}}{\text{total number of inspection units}}$$

4. Calculate the upper and lower control limits. The control limits for each plotted *u* will depend on the number of inspection units used to calculate the value of *u*. The formulas for the upper and lower control limits are given by:

$$\text{UCL}_u = \bar{u} + 3\frac{\sqrt{\bar{u}}}{\sqrt{n}}$$

$$\text{LCL}_u = \bar{u} - 3\frac{\sqrt{\bar{u}}}{\sqrt{n}}$$

4.4   # USING *c* AND *u* CHARTS IN ONGOING PROCESS IMPROVEMENT WORK: A CASE STUDY*

In a plant that fabricates insulators for use in large transformers, there are 40

---

* This case study was contributed by Dr. Ken Gilbert of the Management Science Department at the University of Tennessee, Knoxville.

unique product codes. Within each product code, there are several related product types. In all, about 360 different types of insulators are typically processed on the same equipment. To provide consistent, quality products to the plant's customers, the plant's management team has been focusing on the improvement of the critical systems within the plant. For this plant, on-time shipments are a critical customer value. Therefore, a major focus of the directed work groups is the improvement of schedule performance. Improved and consistent schedule performance will support marketing strategies that are related to customer inventory concerns. Based on this focus, one of the plant's key performance indicators is the percent of weekly orders shipped complete and on time.

Preliminary studies of schedule performance indicate that around 30% of all orders do not meet the criteria for on-time, complete shipment of customer orders. An in-depth look at shipments revealed that approximately 9% of the customers repeatedly receive unsatisfactory shipping service. Any one customer order typically includes several different product codes. In fact, 12% of the plant's products make up about 60% of the total volume of products shipped. With this information in mind, the management team began to realize that simply looking at the percent of unsatisfactory order shipments hides important information, such as the specific product codes or product types that have the most unsatisfactory shipping performance. Also, the managers questioned how the variation in such a performance indicator could be analyzed over time. Initially, they thought that a *p* chart was appropriate for tracking this indicator. However, when they attempted to chart the data using a *p* chart, they realized that the week-to-week shipment results are not independent of each other, because orders unfilled in one week are rescheduled for the following week. They also realized that weekly counts of the number of unsatisfactory shipments out of the total scheduled shipments provide no indication of the additional expenses and difficulties incurred when some orders are expedited to meet shipment schedules. Although the number of late or incomplete orders may provide a broad view of how well the plant is performing with regard to meeting schedules, additional information is necessary to identify ways to reduce this measure.

To obtain a better understanding of schedule performance, a work group constructed the cause and effect diagram shown in Figure 4.8. Late or incomplete order shipments was the effect. The team members agreed that downtime on the production line due to machine stops is an important contributor to the effect and that the major reason for the machine stops on the production line was machine blowups. Marketing and sales people who communicate regularly with the customers were asked to help the work group identify customers who are often affected by unsatisfactory shipments and the possible reasons for these occurrences. Four specific product codes that highly associated with poor shipping performance were identified. From this set of four product codes, the team chose one particular product type, 267C, within the product code 34FN, to use for a thorough study of machine stops as a significant contributor to late or incomplete shipments.

The study on machine stops during the production of insulator type 34FN-267C included:

- the development of a flowchart of the production process,
- identification of the possible causes of the machine stops, and
- the determination of an appropriate measure to provide information on the sources of variation contributing to the machine stops.

Figure 4.9 shows the macro flow diagram of the insulator production process that was developed. This diagram shows that the insulator blanks are initially formed in an injection molding procedure. Next, each insulator goes into 1 of 12 machines that tailors it to a specific product type. At this machine, numerically

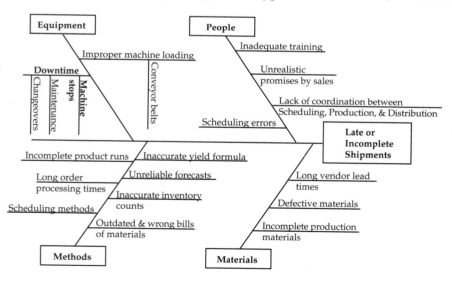

FIGURE 4.8 *Cause and Effect Diagram for Late or Incomplete Shipments*

controlled devices drill holes, cut wiring grooves, smooth all corners and edges, and form and attach mounting brackets. Each insulator blank passes through only 1 of these 12 parallel machines where, depending on machine setup, it is tailored for a particular application.

As previously stated, a suspected major cause for late or incomplete shipments is machine stops. In particular, blowups of the numerically controlled machines are a key concern. The term *blowup* refers to any unscheduled machine stoppage that occurs. One cause for a blowup is when an insulator blank gets free inside of one of these machines, jams the machine, and causes machine damage. A lot of time is required to replace damaged shear pins and bent tools and to perform other necessary repairs.

To study the occurrence of machine blowups over time, data were collected on the number of blowups that occurred during 20 consecutive shifts. These data are provided in Table 4.4. Currently, the production process only runs

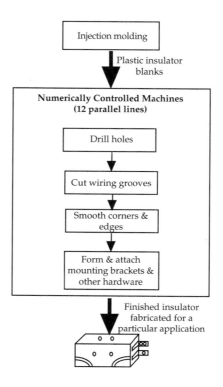

FIGURE 4.9 *Insulator Production Process*

one shift per day. So, each subgroup consists of the counts of machine blowups across all 12 lines during the same shift.

The number of blowups was plotted over time. Deciding that a c chart was most appropriate to chart these data, the centerline and control limits for the control chart were calculated to be:

$$\bar{c} = \frac{494}{20} = 24.7$$
$$\mathrm{UCL}_c = 24.7 + 3\sqrt{24.7} = 39.61$$
$$\mathrm{LCL}_c = 24.7 - 3\sqrt{24.7} = 9.79$$

The completed c chart is shown in Figure 4.10. The chart is in control; no points fall outside of the control limits and the longest run consists of four data points.

TABLE 4.4 *Machine Blowups by Day*

Day	Number of Machine Blowups
1	24
2	35
3	26
4	35
5	27
6	12
7	23
8	25
9	16
10	24
11	24
12	27
13	17
14	30
15	22
16	21
17	27
18	24
19	30
20	25
Total	494

The group working on the late shipment problem began to discuss the types of information provided on the machine blowups chart of Figure 4.10. They did learn that the process was averaging about 24 machine blowups per day. Given that there are 12 numerically controlled machines, this means that they were consistently averaging about two stoppages per machine per day. However, in discussing possible reasons for the machine blowups, such as machine differences, they realized that the collection of data was performed in a way in which all machines were included in a subgroup. This subgrouping of the data made it difficult to discover possible differences between the 12 machines. In fact, the only sources that this control chart is able to detect as special causes are those things that vary from day to day.

In addition to questioning the way in which the data were collected, they were also concerned about the appropriateness of the c chart for the data. One of the key assumptions for the use of a c chart is independence of the causes leading to multiple blowups. When a machine goes down, it often goes through two or three false starts prior to staying up. Therefore, the groups of several machine blowups due to the same cause should have been counted as one single event. All of these concerns about the collected data made the management team realize the importance of understanding process behavior and carefully defining the purposes of data collection prior to collecting and charting the data. Based on this information, they decided to go back and develop a detailed flowchart of the insulator process and construct a cause and effect diagram on the possible causes of machine blowups.

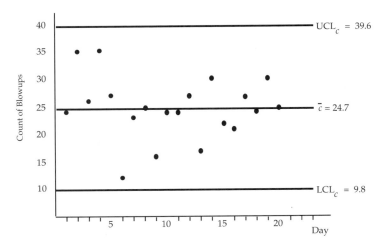

FIGURE 4.10 c *Chart of Machine Blowups by Day*

A more detailed flowchart of the process, provided in Figure 4.11, was developed. This flowchart shows that the insulator blanks are injection molded in an eight-cavity mold. After the molding stage, the insulator blanks are stored in a buffer area until they are loaded on a conveyor with 28 different hands that holds the insulators as they are carried by the machines. When an insulator blank passes an empty machine, the hand holding the blank loads it into the machine. The machine performs the drilling, finishing, forming, and the attachment of the mounting bracket. At the completion of machine operation, the insulator is dropped onto a conveyor belt that carries it into the packing area. Each insulator passes through one machine and each operator monitors three adjacent machines. The operator is responsible for appropriate machine adjustments as well as for loading the metal coils used to form the attachments.

After constructing the process flowchart, the operators were asked about the causes of the machine blowups. Some of their responses were:

> "I don't seem to have the problems that these new operators have. They don't know how to set up and adjust the machine so that it loads the insulator blanks straight."

> "Some of the machines are just hard to keep running. These three machines they've got me on now seem to have more problems than some of the other machines."

> "The hands on the conveyor all seem to load the machines differently."

> "It seems I always have a blowup in the afternoon when it gets really hot in here."

> "It's bad insulator blanks coming out of molding."

In addition to these comments, the supervisor in the molding area suggested that there is variation in insulator blanks formed in the different cavities. The supervisor also pointed out that each insulator can be identified by cavity since the cavity number is molded on the insulator blanks. These ideas and discussions about the possible causes of machine blowups were organized into a cause and effect diagram (Figure 4.12). Realizing that not all of the potential causes listed are necessarily valid, the group identified what they thought to be the most plausible reasons for machine blowups. The two most reasonable explanations for machine blowups were agreed to be differences in operators, including untrained operators, and bad insulator blanks from the molding process. To confirm these suspicions, plans were made for further data collection and analysis.

After much discussion, the decision was made to focus first on differences between operators as a causal factor for machine blowups. The data collection

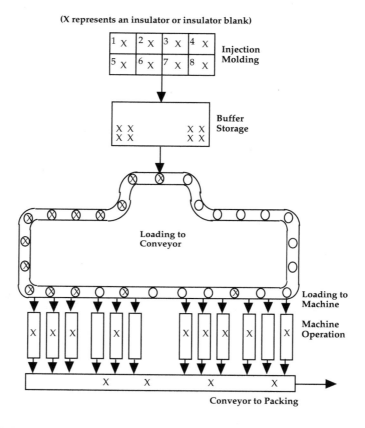

(X represents an insulator or insulator blank)

FIGURE 4.11 *Detailed Flowchart of the Insulator Process*

strategy was planned so that each subgroup contains data from only one operator. Counts of the number of machine blowups experienced by each of the operators over a period of four days were recorded. Five employees work as operators, so five subgroups of data were collected each day. These data are summarized in Table 4.5. To analyze these data appropriately, it is necessary to take into account the fact that the operators do not operate their respective machines for the same number of hours each day. Thus, the amount of time (or number of inspection units) differs for each operator. At first it was thought that an inspection unit could be determined by recording the number of hours worked by each operator. However, this count of the time worked would also include the time during which a machine was down, maybe because of a machine blowup. A different approach was adopted. When a machine is running, it produces insulators at a constant rate. So, the amount of time during which an operator can experience a machine blowup is proportional to the daily throughput of that operator. Instead of recording the time worked, the amount of throughput for each operator is used to determine

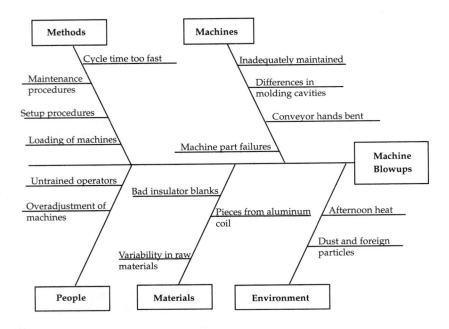

FIGURE 4.12   *Cause and Effect Diagram on Reasons for Machine Blowups*

the inspection units. Column three of Table 4.5 contains the throughput in hundreds of insulators for each operator who worked on each day. The numbers in column three are used to define the number of inspection units for each subgroup; an inspection unit is 100 insulators.

Since the amount of throughput varies from subgroup to subgroup, a *u* chart was constructed to analyze these data. The centerline for the chart was computed by dividing the total number of blowups, 99, by the total number of inspection units. Thus, the value for the centerline was computed to be 0.76. The control limits were computed for each of the three different values of *n* as shown below:

$$\text{For } n = 4, \text{ UCL}_u = 0.76 + 3\frac{\sqrt{0.76}}{\sqrt{4}} = 2.07$$

$$\text{LCL}_u = \text{none}$$

$$\text{For } n = 6, \text{ UCL}_u = 1.83 \text{ and LCL}_u = \text{none}$$

$$\text{For } n = 8, \text{ UCL}_u = 1.69 \text{ and LCL}_u = \text{none}$$

Two different charts were constructed from the data in Table 4.5. These charts
are shown in Figure 4.13. The first chart shows the data points plotted in the
same order as they are recorded in the table. The second chart has the points
for each operator grouped together with the time order maintained for each
operator's counts. Based on these charts, there is no evidence of any differences between the operators. Even though operator 5 is below the centerline
on all four days, there is not sufficient evidence to conclude that operator 5 is
significantly different from the other four operators. It is, of course, possible
that operator 5 does have on average a smaller level of machine blowups than
the other operators. Additional data could be collected to either confirm or
negate that operator 5 has a different average than do the other operators.
Based on the available data, the team members thought that any such difference is too small to be of practical significance. Thus, it was concluded that
each operator experienced about 0.75 blowups per 100 units of throughput.

TABLE 4.5 *Machine Blowups by Operator by Day*

Day	Operator	Throughput (in hundreds)	Number of Blowups	Blowups (per 100 units)
1	1	8	5	0.63
1	2	8	8	1.00
1	3	4	5	1.25
1	4	4	4	1.00
1	5	8	4	0.50
2	1	6	7	1.17
2	2	6	5	0.83
2	3	4	1	0.25
2	4	8	4	0.50
2	5	8	6	0.75
3	1	6	5	0.83
3	2	6	3	0.50
3	3	6	5	0.83
3	4	6	3	0.50
3	5	8	5	0.63
4	1	4	2	0.50
4	2	8	5	0.63
4	3	6	10	1.67
4	4	8	6	0.75
4	5	8	6	0.75
Totals		130	99	

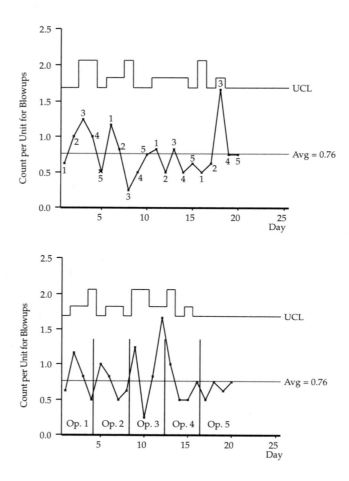

FIGURE 4.13 *Machine Blowups by Operator by Day*

There are two possible explanations for the charts not showing any differences between the operators. Either the operators are not a cause of the problem with machine blowups or, if they are associated with the problems, they are all consistent. For example, if all of the operators run the machines at a cycle rate that is too high, then this is a common cause of variation that is probably related to operator training or the incentive systems.

Before making additional investigations into operator practices, the focus of the study was turned to the effect of the molding process on machine blowups. The question that needed to be resolved is whether or not molds from some cavities are involved in more machine blowups than are molds from other cavities. Therefore, the data were grouped according to the cavity supplying the insulator involved in the machine blowups. Data were collected over a period of three days. These data are shown in Table 4.6. In this set of data, a subgroup consists of data obtained from only one cavity of the molding machine. For example, subgroup 1 represents a count of blowups during day 1 when the insulator in the machine came from cavity 1 of the molding machine. Since each cavity supplies the same number of insulator blanks for each shift (i.e., n is the same for each subgroup), a c chart is appropriate for the analysis of these data.

Figure 4.14 shows the completed control chart. The values for the number of blowups are plotted in the order of the cavity numbers associated with each subgroup. This c chart shows a point outside of the upper control limit. A careful study of the chart reveals that the corner cavities (1, 4, 5, and 8) seem to be involved in a disproportionate number of blowups. Therefore,

TABLE 4.6 *Blowups Subgrouped by Cavity*

Day	Cavity	No. of Blowups
1	1	5
1	2	2
1	3	1
1	4	3
1	5	6
1	6	0
1	7	1
1	8	3
2	1	3
2	2	0
2	3	2
2	4	6
2	5	3
2	6	0
2	7	2
2	8	8
3	1	6
3	2	2
3	3	0
3	4	10
3	5	7
3	6	0
3	7	0
3	8	8
Total		78

management decided to focus efforts on possible differences between the cavities in the injection molding stage. To be able to understand and remove the causes of the differences between cavities, the team planned to construct separate control charts — one for the corner cavities and one for the inside cavities. After constructing separate charts, work will begin on identifying those characteristics, such as wrong dimensions and excess plastic, that contribute to the differences between cavities and, therefore, to the machine blowups.

The study of the insulator process illustrates how combining process knowledge as captured on a process flowchart with speculations about sources of variation can be used to formulate subgrouping strategies to confirm or negate the effect of speculated sources of variation on process performance. The initial study of the insulator process showed that, when examined on a day-to-day basis, machine blowups were occurring at a consistent rate. On average, there were 24.7 blowups per day. Additional data collected on the process revealed that operator effects, if present, were consistent across operators. Finally, data were collected to verify the speculation that the molding process had an impact on machine blowups. The resulting chart (Figure 4.14) indicated that some of the cavities in the mold appeared to be involved in a higher number of machine blowups than others. The reasoning used to detect this cause of variation, cavity mold, should be clearly understood. Across the time of the three studies of machine blowups, no changes were made to the process, yet the first two charts of the process were in control and the third out of control. Cavity mold was revealed as a special cause on

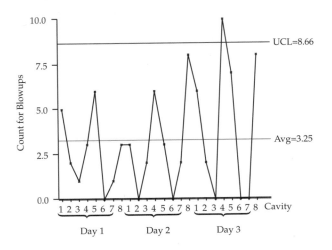

FIGURE 4.14   c *Chart for Machine Blowups by Cavity by Day*

the chart in Figure 4.14 because the effect of molded properties for, say, cavity 1, was present in some subgroups but not in others. In the first control chart constructed (Figure 4.10) on the number of machine blowups by day, the data were subgrouped by time only. Hence, causes acting within a day, such as differences due to cavity mold, are captured within each and every subgroup. Consequently, the first subgrouping scheme did not reveal these differences as special causes. Of course, none of the three subgrouping plans could be termed the "right" way to collect data on the insulator process. Subgrouping plans change as process knowledge is gained and as new questions concerning potential sources of variation arise.

4.5   SUPPORTING IDEAS FOR THE EFFECTIVE USE OF *c* AND *u* CHARTS

The ideas discussed in this section provide some additional thoughts on how *c* and *u* charts can be effectively used for process study. In the previous case study, even though the control chart of Figure 4.10 was in control, serious problems still remained in the performance of the process. When the aim of collecting and analyzing data from a process is to improve the efficiency and effectiveness of that process, getting the process "in control" is at best only a first step. Once the effects of special causes have been dealt with, process improvement is accomplished by identifying common cause sources of variation and then acting on these sources to reduce variation. It is with this intent in mind that the following ideas on using *c* and *u* charts are provided.

> ***Idea 1:*** *Process flow diagrams and cause and effect diagrams should be developed prior to data collection.*

As illustrated in the insulator case study, the use of *c* or *u* charts does not begin with the collection of data. Rather, the intent or purpose of the study needs to be well outlined. Current information about the process should be collected as a guide to what is currently known about process operation and what information is lacking. The collection of this information is aided by the construction of flowcharts and cause and effect diagrams. The flowchart is useful in describing how the process currently operates and in providing a useful reference for thinking about where critical process parameters may be impacting process outcomes. The cause and effect diagram is invaluable for describing what is currently known about causes of variation or for speculating about factors affecting process output. Armed with information about current process operation and possible sources of variation, data collection can be guided to provide further information on improvement opportunities.

> *Idea 2: Frequency, location, and number of subgroups are determined by the knowledge or speculations about critical sources of variation.*

In planning the data collection strategy, the goal of developing a subgrouping or sampling strategy should be to maximize the opportunity to learn about sources of variation. This goal is the first consideration in determining the frequency of subgroups, the sampling locations, and the number of subgroups. For example, subgroups should be collected with sufficient frequency and over a sufficient time to allow those items listed on the cause and effect diagram to change between subgroups. In addition, consideration should be given to collecting data from different process streams identified on a cause and effect diagram.

> *Idea 3: c or u charts are most useful for process study when only one type of nonconformity is counted.*

As plans are developed for collecting process data, the specific process nonconformities that need to be counted are chosen. People often attempt to record information about a number of different types of nonconformities on a single control chart hoping to attack multiple "problems" at once. In reality, counting several different types of nonconformities together makes it more difficult to attack any of the problems. An illustration of this kind of reasoning is provided by a firm that wants to study the number of temperature excursions that occur in processing a chemical product. The graph in Figure 4.15 illustrates a temperature profile. The horizontal axis is the processing time in a tank for one stage of a batch process. Temperature is graphed on the vertical axis. The protocol required a gradual warming and drop in temperature. To begin evaluating how well the protocol was being followed, specific descriptions of the manner in which these temperature changes should occur was required as well as a definition of what constituted a significant deviation from protocol. Figure 4.15 captures the type of limits drawn around a profile in order to define how close to the profile the temperature should be held. When the temperature falls above or below these limits, this phenomenon is labeled a temperature excursion.

It would have been possible to count other types of deviations from protocol at the same time. The number of deviations from the correct pH level, the number of deviations from the correct stirring procedure, etc., could have been included. However, this practice of including many different types of

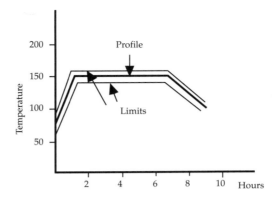

FIGURE 4.15 *Graphical Illustration of a Temperature Profile*

nonconformities should be avoided. Temperature excursions would likely be caused by the equipment that controls temperature; the ability to manage pH levels and stirring amounts will have sources of variation other than those affecting temperature. In general, a chart should report on nonconformities that have similar causal structures. Gathering all possible categories of nonconformities on one chart not only hides signals that would be visible if the categories were charted separately, but also hampers the determination of the reasons for any special causes that are active.

4.6 SUMMARY

c and u charts are used to provide information on the stability of a process when information on process behavior is gained by counting the occurrences of defects or nonconformities across time and conditions. The use of a c or u chart to analyze behavior of counts of occurrences over time is appropriate if the counts meet the four criteria provided in Section 4.1. c charts are used to study the counts of nonconformities when the inspection unit stays the same over the course of the study. If the counts of occurrences are made on varying sizes of inspection units, then a u chart is used to analyze the data. The upper and lower control limits for c and u charts define the amount of variation that is expected in the recorded nonconformities if only common causes of variation are present.

When the purpose for the use of c or u charts is to reduce the sources of variation acting on a process, the development of process flowcharts and the construction of cause and effect diagrams are critical. The development of process knowledge and information on causal factors allows for the appropriate determination of the type of nonconformities that need to be counted and charted. When the subgroups are formulated around a speculated source of variation, a c or u chart can negate or confirm the effect of that potential cause on process performance. The type of nonconformity counted will change as new process knowledge and questions develop during process study. Additionally, these questions and process knowledge will guide the determination of the frequency, location, and number of subgroups needed to learn about sources of variation.

4.7   PRACTICE PROBLEMS

4.7.1 A plant that manufactures mobile homes has noticed a recent increase in the number of discolorations in the wall boards. The plant receives glue from three different suppliers. It is believed that the discolorations are caused

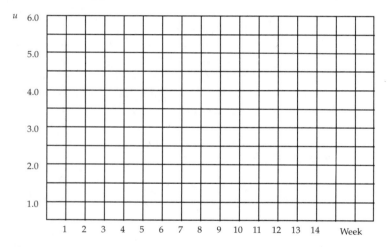

FIGURE 4.16

by an inconsistency in the glue used to attach the vinyl wall coverings to the boards. After construction is completed on a home, the home is inspected for any flaws in workmanship prior to shipment to the retail lots. A study of the process generating discolorations is begun; data are collected at final inspection on the number of wall board discolorations. Table 4.7 provides data on the total square footage of wall boards inspected and the number of discolorations identified in a week over a period of 12 weeks.

a. Letting a unit of inspection be 1000 square feet, construct an appropriate control chart for these data. (A blank chart is provided in Figure 4.16.)
b. Does the process appear to be behaving in a consistent fashion over time?

TABLE 4.7 *Number of Discolorations on Wall Boards*

Week	Square Ft. Inspected	No. of Discolorations
1	8,400	24
2	6,000	14
3	7,200	27
4	10,800	39
5	9,600	45
6	6,000	17
7	7,500	20
8	9,900	26
9	8,700	18
10	11,200	46
11	7,200	29
12	9,000	19

c. If appropriate, estimate the average number of discolorations expected per 1000 square feet of wall board.

d. Based on the believed cause of the discolorations, suggest an alternative data collection strategy that would provide more information on the causes of the discolorations in the wall boards.

4.7.2 At a metal-working machine shop, small metal parts are produced for assembly into pieces of heavy equipment. The parts are currently being run in lot sizes of 200 parts. After fabrication, any excess metal is machined off of the parts and the parts are deburred. To study the behavior of the process over time, two lots of parts are randomly selected and inspected at the end of each process for the existence of rough edges, excess metal, and burrs. The number of nonconformities found during the inspection of these parts has been recorded for the two shifts over a period of 10 days. The data are provided in Table 4.8, and a plot of the number of nonconformities is provided in Figure 4.17.

a. Calculate the upper and lower control limits for the chart in Figure 4.17.

b. Does the process appear to be operating in a stable and predictable manner over time?

c. Are shift-to-shift differences captured within subgroups or between subgroups? Do differences exist between the two shifts? How is this conclusion reached?

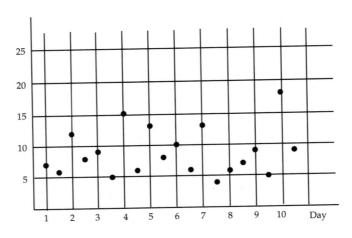

FIGURE 4.17   *Number of Nonconformities on Inspected Metal Parts*

TABLE 4.8   *Number of Nonconformities Found in Two Lots of Metal Parts*

Day	Shift	Number of Nonconformities
1	1	7
	2	6
2	1	12
	2	8
3	1	9
	2	5
4	1	15
	2	6
5	1	13
	2	8
6	1	10
	2	6
7	1	13
	2	4
8	1	6
	2	7
9	1	9
	2	5
10	1	18
	2	9
		266

4.8   APPLICATION PROBLEM

In the insulator case study of Section 4.4, data were collected and charted on the number of machine blowups by cavity by day. The resulting control chart (Figure 4.14) showed a point above the upper control limit. To determine differences between the cavities in the injection molding stage of the process, the team working on the process planned to construct two separate control charts for the data. Believing that the corner cavities (cavities 1, 4, 5, and 8) were involved in a higher number of machine blowups, data were collected for an additional two weeks and subgrouped by cavity location (i.e., by inside or corner cavity). The resulting data are provided in Table 4.9.

TABLE 4.9   *Blowups Subgrouped by Cavity Location*

Day	Cavity Location	No. of Blowups
1	Corner	17
	Inside	4
2	Corner	20
	Inside	4
3	Corner	31
	Inside	2
4	Corner	28
	Inside	4
5	Corner	26
	Inside	3
6	Corner	13
	Inside	5
7	Corner	20
	Inside	6
8	Corner	14
	Inside	5
9	Corner	23
	Inside	2
10	Corner	24
	Inside	5
11	Corner	10
	Inside	3
12	Corner	33
	Inside	3
13	Corner	23
	Inside	5

a. Construct two separate control charts for the data in Table 4.9 to help in identifying the causes of the differences between corner and inside cavities. Each cavity supplies the same number of blanks for each shift, and there are four corner and four inside cavities. Two blank charts are provided in Figure 4.18.

b. From the control charts constructed from the data in Table 4.9, if appropriate, estimate the difference in the average number of machine blowups between the inside and corner cavities.

c. To work on reducing the average number of blowups associated with insulator blanks from the corner cavities, the group working to improve the process decided to determine whether or not excess plastic on the blanks from the corner cavities was a major causal factor. During a period of six days, all blanks from the corner cavities were inspected for excess plastic. Any blanks with excess plastic were removed prior to being loaded onto the conveyor that feeds the machines. The number of machine blowups for the corner cavities were counted over this period of six days. These data follow:

Day	Cavity Location	No. of Blowups
1	Corner	8
2	Corner	13
3	Corner	20
4	Corner	9
5	Corner	14
6	Corner	17

Plot these data points on your previous control chart for machine blowups by corner cavity (Figure 4.18). Is there evidence to conclude that the elimination of excess plastic in the formation of insulator blanks reduces the average number of machine blowups?

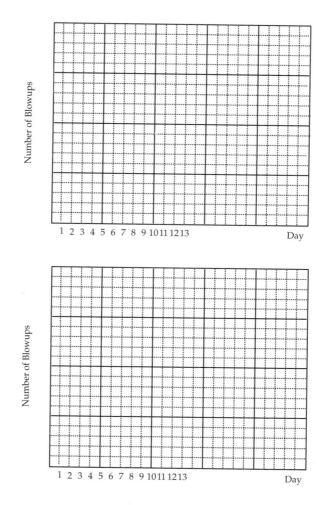

FIGURE 4.18 *Control Charts of Machine Blowups by Cavity Location*

5 Control Charts for Variables Data: Variability and Location

A wide variety of statistical techniques exists for analyzing data collected for process study. These techniques include run charts, control charts, scatter diagrams, analysis of variance, time-series analysis, and others. Selection of an appropriate technique depends primarily on the nature of the problem and on the questions and issues that motivate the data collection and on the type of data collected. Additionally, the background of the people involved in preparing, analyzing, and acting on the collected data impact the choice of an analysis technique. As previously discussed, measurements of process yield either quantitative or qualitative evaluations. Qualitative evaluations of goods and services typically result in (1) counts of the number of nonconforming items or in (2) counts of the number of errors, defects, or omissions. Suggested analytical treatments of these two types of attributes data have been discussed in Chapters 3 and 4. On the other hand, quantitative evaluations result from a determination of the degree to which a process condition, parameter, job or service possesses a characteristic. Examples of variables data that result from such quantitative measurements are measurements of density, pressure, temperature, resistance, force, hardness, dimensions, etc.

A critical part of process investigation is determining the effects that changes in material, methods, and equipment have on process outputs. The statistical techniques used for analysis, the sampling strategy, and the measurement process are critical determinants of the success of process investigation. To obtain accurate and relevant measurements, an adequate measurement process must be in place. In this chapter, it is assumed that a measurement process has been evaluated and found to be predictable and adequate for the current purpose. Consistent and adequate measurement processes are a prerequisite for successful process analysis. The methodology for evaluating the measurement process is discussed in Chapter 9.

Many of the issues regarding process performance are often initially addressed during product and process development when the methods, equipment, and design are proofed. Even when this is the case, problems still occur and effective, efficient process operation remains an issue. Necessary responsibilities of process management will include problem solving to achieve and maintain process control. In addition, the need for incremental and continual process improvement will not abate. To carry out these responsibilities, an evaluation of current process status is an imperative. This evaluation should include

- what is expected from the process,
- how work sequences, methods, and tasks contribute to the achievement of those expectations, and
- process performance measurements that allow us to compare what is actually produced with what is desired.

Obviously, processes undergo changes in materials, methods, equipment, and environmental conditions over time. Discovering when and under what circumstances these changes take place and verifying the effects of these changes on measured characteristics is an essential responsibility in process investigation. When organized and presented in an appropriate manner, process data provide the manager, engineer, and operator a medium through which changes taking place in the process can be seen and evaluated. This chapter provides an introduction to the use of control charts for variables data in order to support work directed toward process control and improvement.

5.1   DESCRIPTION OF TYPES OF VARIATION IN PROCESS DATA

The purpose of data collection is to provide information that supports the effective management of processes. Two critical components of process management are process control and process improvement. Data analysis must therefore contribute to identifying and evaluating sources of variation in process variables and outcomes. An insightful means of understanding the effects of sources of variation is to consider their effect on both "short-term" variation and the "level" of process data. Consequently, initial data analysis is directed toward measuring process variability and process level so that the predictability of these two process characterizations can be evaluated.

Suppose that measurements on a process output are made and plotted over an extended period of time. These plotted values might represent measurements of viscosity, length, density, flow rate, temperature, or thickness.

Data plots from five processes (A, B, C, D, and E) are shown in Figure 5.1. In each case, the vertical scale corresponds to a number line for measurements of a particular process characteristic. This characteristic might be a measurement of a process input, a process variable, or a process output. The horizontal scale corresponds to time. These examples collectively display patterns and characteristics of process behavior that demonstrate some of the central issues in process evaluation and data analysis. These issues are emphasized and discussed by comparing the time plots with respect to variability, average, and predictability. The analysis of the behavior represented by these five time plots is informal; the intent is to provide a conceptualization of the important characteristics of process behavior to be understood. Formal analysis of these characteristics of process behavior can be made by control charts for variables data. The construction of these charts is explained following the description of the information desired from such charts.

5.1.1   Process A: Consistent Short-term Variation and Consistent Level

The data plot representing measurements on a characteristic for process A of Figure 5.1 is first considered. For convenience, it is assumed that these measurements have been made on a process output; the same discussion could just as easily have taken place about a process variable. In common with all processes, process A does not provide exactly the same result each time. Process A obviously exhibits variability; all the data points are not the same in value. The magnitude and nature of this variability is of primary interest. How large is it? What causes it? Is it predictable? In the study of processes, it is useful to examine the behavior of variability over a long time period in order to understand process outputs or variables.

Variation exhibited in smaller time intervals, such as the shaded intervals in the plots of Figure 5.1, provides partial information about the total variation observed in process values. (The data values in each of the shaded areas might represent the values used to form a subgroup for evaluating the process.) For process A, the variations in the measured values in each subgroup appears to be about the same throughout the total, or "long," time period covered by these data. Because of this consistency, short-term variability is said to be predictable, stable, or "in statistical control." Short-term variation is defined to be that variability in process measurements which is due to the effect of sources active within a subgroup. A conclusion regarding stability in short-term variability marks a significant finding in process investigations.

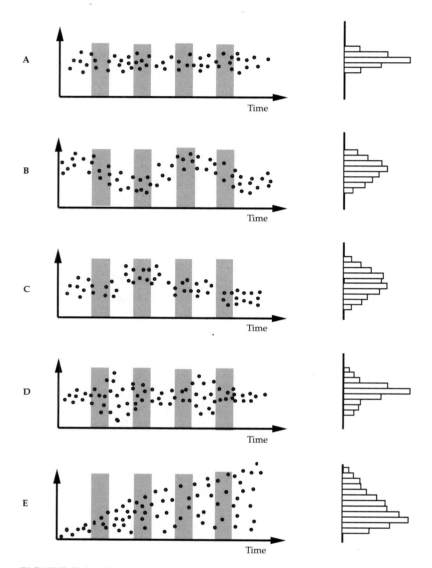

FIGURE 5.1   *Process Examples*

Long-term variation, as opposed to short-term variation, is defined as changes or shifts in process level or average. The variation seen in process measures over the long term may be due to changes over time in the sources contributing to the common cause structure of the short-term variation or to other sources that only become active during longer periods of time. Consequently, another important property of process measurements is the level around which individual values tend to fluctuate. Process A appears to operate at a nearly constant average value throughout the period for which data were collected. Because of this apparent constancy in average value, the process level is said to be predictable. Because process A is judged to have almost constant short-term variability and a nearly constant average, the process is said to be in statistical control.

5.1.2 Processes B and C: Consistent Short-term Variation and Changing Level

Given that process A is in a state of statistical control, we can compare observed behaviors of other processes to those of process A to gain additional insights into process information provided through data analysis. Consider the nature of the short-term variation in process B of Figure 5.1. For the measurements obtained within a short time span, the range appears to be of about the same magnitude regardless of the selected time interval. In other words, the short-term variability appears to be stable. This statement can also be made about the point-to-point variability displayed in the data plot for process C. For processes B and C, the point-to-point fluctuations, referred to as short-term variability, are about the same magnitude as that displayed by process A. However, the results for the three processes (A, B, and C), are obviously not equivalent. Specifically, neither process B nor process C operates at a consistent average over the long time period represented. Process B appears to undergo smooth, incremental changes in its average. From the given plot, it is not possible to judge whether this "cyclic" behavior replicates itself in a consistent fashion over longer periods of time. In contrast to the smoothness in changes displayed by the average for process B, process C exhibits more abrupt shifts in its average. Processes B and C do not have predictable averages.

5.1.3 Processes D and E: Inconsistent Short-term Variation

From the plot in Figure 5.1, it appears that process D is operating at a consistent average or level over time. However, process D differs from A, B, and C in an important way. Recall that process A maintains about the same amount of variation around its average. Process D, on the other hand, displays inconsistent short-term variability in individual measurements around its long-run average. The range in a subgroup from one short time period can be quite different from that of a different subgroup. Thus, the short-term variability in process D is unpredictable and unstable. The inconsistent variation of process D should be contrasted with the inconsistent behavior described for processes B and C. In process D, the short-term variation changes across time. In processes B and C, the short-term variation is consistent; it is the inconsistent average which is creating increased variation in these processes. Process E displays both a trending average and increasing short-term variation in individual measurements over time. Hence, the short-term variation and the level of the process outputs are unstable over time in process E.

5.1.4   Understanding the Impact of Process Variation

We emphasize that changes over time in an average as well as deviations in individual values around that average contribute to the total variation in process output. All values plotted for process A in Figure 5.1 have been used in constructing the histogram that appears to the right of the time plot. Histograms for processes B, C, D, and E are also shown on the right-hand side of Figure 5.1. It is easily seen from these histograms that the values from processes B and C are more variable than those from process A, even though the fluctuations during any short period of time appear to be of about the same magnitude for these three processes. This increased variability is seen by comparing the width of histograms B and C against that of process A. The changes in averages over a longer time period contribute significantly to the variability experienced by a customer. All of the values represented by the plotted points for the characteristic in question may be within specifications; that is not the issue. The central issue is that the patterns for processes B, C, D, and E, with their respective changes in short-term variability and in their average, contribute to the overall variation. Control charts for variables data are intended to identify the consistency of both the short-term variation and the level of a process, and the intent of this description of process variation is to support work directed at discovering the reasons for process variation so that it can be reduced.

5.2   IDENTIFYING VARIATION AND KNOWING ITS SOURCES

Processes A through E, described in the previous section, exhibit different types of variation over time. Subsequent understanding of these processes must include characterizing the sources of variation that act to create this variation. These sources of variation can be classified as common cause sources (those acting to create variation in all outputs) and special cause sources.

Special cause sources of variation are those that act to create inconsistent short-term variation or act to create inconsistencies in the process average.

Plotting data by production sources and production order helps in compiling a set of potential common and special causes of variation. Data plotted in time order also describe the manner in which the process output changes as the variable "time "changes. However, remember that time itself corresponds to numerous changes in processing circumstances that have the potential to affect either the short-term variation or the average of a process characteristic. For example, component parts and raw materials surely change over time. These changes, in turn, may generate increased variability around an average or may cause the average to shift up or down. As management, operating, and engineering personnel change over time, so may the methods or techniques for operating machines, assembling components, or mixing materials. These changes may affect the short-term variability, the average, or both. Environmental variables, such as temperature and humidity, can also change in a dramatic fashion over time. These changes may cause other changes, which then affect output characteristics. The operating efficiency of process and laboratory equipment changes with the passage of time, so, maintenance of equipment becomes a critical issue. If process output is to be consistent over time, processes cannot be affected by typical changes in process variables. It is management's job to ensure that either process output is robust to changes in process variables or that process variables are managed at specific averages with outcomes falling within given specifications.

As process analysis begins, information similar to that displayed in the time plots of Figure 5.1 is not typically available. The manager, engineer, or operator who begins to study the process may be said to be searching for a description of process behavior. In fact, the analysts are looking for the signs of inconsistency noted in processes B, C, D, or E. A process is judged to be consistent if there is no evidence of nonrandom behavior in the data collected over time. Process improvement work based on this judgment of the presence of nonrandom behavior typically occurs in the following order:

1. If a process is found to be inconsistent (i.e., out of control), then the first responsibility is to bring the process into control.
2. Once the process is in control, the ongoing responsibility is to maintain control.
3. Once a process is in control and the ability to maintain this control is demonstrated, then the variation of the characteristic in question is predictable. Those responsible for the process can begin to compare credibly the current output against what is required.

4. After control is achieved and maintained, responsibility moves toward working on the process to obtain improvement.

The use of control charts for variables data must serve these four needs.

5.3   CHARTS FOR VARIABLES DATA AND PROCESS ANALYSIS

As illustrated through the previous examples, the short-term variation and the average around which the process data fluctuate are two important characterizations of process behavior. Thus, statistical techniques for analyzing such data should provide both information on these two characteristics and insights into their causes. Control charts are excellent for providing this type of information because they are designed to aid in the discovery of the effects of changes in processes over time. Control charts signal when things are different. Combined with diligent data collection and a clear understanding of the measurement process and subgrouping strategy, the charts can assist managers and engineers in determining *why* things are different. The following example illustrates the basic computational and data analysis issues associated with using control charts for variables data to help in process investigation.

5.3.1   Practice Data Set

The data set printed in Table 5.1 is presented only for the purposes of illustrating the arithmetic necessary to use some basic statistical techniques for analyzing variables data. Twenty subgroups, each of size four, are shown. The numbers in Table 5.1 are meant to represent measurements of a single characteristic on a collection of parts. The data are organized in subgroups of four readings each. This implies that the four values were collected under similar conditions, or at the same point in time, or from the same source. There should be a rational basis for subgrouping the data. Discussions and practice in selecting appropriate subgrouping schemes are presented in Section 5.5 after some ideas about variables data control charts are in place.

Several kinds of information are desired from the data set of Table 5.1. For each subgroup, a measure of within-subgroup variation (such as the sample range) will be computed to describe how much the four parts differ among themselves. The average value is also required for each subgroup. All of the subgroup ranges will be used to provide information about the stability and magnitude of short-term variation. Subgroup averages and their time pattern will offer information about the process level and its predictability. For a controlled, predictable process, the individual observations can be formed into a histogram, which can be used to provide a "picture" of process output.

The symbol n is used to represent the number of measurements in a subgroup and the symbol k represents the number of subgroups. In this example, $n = 4$ because there are four measurements in each subgroup and $k = 20$ because there are 20 distinct subgroups.

TABLE 5.1 *A Practice Data Set*

| Subgroup | \multicolumn{4}{c}{Observations} | | |
	X_1	X_2	X_3	X_4	R	$\bar{X}$
1	36.13	32.85	34.05	38.04	5.19	35.2675
2	38.68	34.95	32.36	33.68	6.32	34.9175
3	34.34	35.69	35.06	29.72	5.97	33.7025
4	33.37	31.73	33.45	35.58	3.85	33.5325
5	32.42	35.58	34.35	35.79	3.37	34.5350
6	30.62	34.10	34.75	36.91	6.29	34.0950
7	31.76	33.29	37.41	31.50	5.91	33.4900
8	34.94	33.79	33.68	36.90	3.22	34.8275
9	34.66	32.02	36.34	33.50	4.32	34.1300
10	39.37	34.82	33.47	31.30	8.07	34.7400
11	33.06	38.97	35.88	36.07	5.91	35.9950
12	37.42	36.39	34.68	33.52	3.90	35.5025
13	37.18	34.43	36.34	33.88	3.30	35.4575
14	32.19	34.90	36.34	33.41	4.15	34.2100
15	33.36	34.36	33.38	33.68	1.00	33.6950
16	33.22	31.18	32.95	32.51	2.04	32.4650
17	34.22	33.01	36.63	35.10	3.62	34.7400
18	32.68	33.03	38.15	35.47	5.47	34.8325
19	31.49	35.84	31.00	35.47	4.84	33.4500
20	34.08	28.97	34.76	35.53	6.56	33.3350
Sums					93.30	686.5900

5.3.2   Measure of Location: The Subgroup Arithmetic Average

A subgroup value is represented by the symbol, X_i, where the i indicates the position of each number in a sequence of observations. The symbol for the average is – X, stated X-bar. The formula for the arithmetic mean or average is:

$$\overline{X} = \frac{\sum\limits_{i=1}^{n} X_i}{n}$$

where the symbol, Σ, pronounced "sigma," means "add" the values represented by the symbol X from $i = 1$ to $i = n$, with n being 4 in this example. For the first subgroup in Table 5.1, $\overline{X}$ is found to be:

$$\overline{X} = \frac{36.13 + 32.85 + 34.05 + 38.04}{4} = 35.2675 \cong 35.27$$

Each of the subgroup averages is computed in the same manner. These subgroup averages for all subgroups are shown with the original data in Table 5.1.

The arithmetic mean or average is the most frequently used measure of location. The arithmetic average is the balance point, or center of mass, for a collection of measurements. The average value need not be a number that occurs in the subgroup. For example, in the first subgroup, no observation has the value of 35.2675, the subgroup average. Also, the average value need not have an equal number of observations above and below it. Subgroup 3, as an example, has three values larger than the average and one smaller. If a data set has extreme measurements in one direction, this last property implies that an average value does not provide a "central" value of the data set.

5.3.3   Measure of Subgroup Variation: The Range

Numerous methods have been developed to measure the variation in a set of numbers. Two such measures used in this text are the range and the standard deviation. The range, indicated by the letter R, is defined to be the difference between the largest value and the smallest value in the set of numbers. The range for subgroup 1 is:

$$R = 38.04 - 32.85 = 5.19$$

The range is recognized as the simplest method for measuring the variation in a subgroup of size n. The value of the range depends only on the two extreme observations in the subgroup. The ranges for the remaining subgroups are also provided in Table 5.1.

5.3.4   Measures of Subgroup Variation: The Variance and Standard Deviation

The subgroup standard deviation, s, and its square, s^2, the variance, are also ways to describe the variation in a subgroup. Unlike the range, these two measures of variation use each of the data values in the subgroup. The variance of a subgroup of numbers, s^2, is calculated by the following formula:

$$s^2 = \frac{\sum\limits_{i=1}^{n}\left(X_i - \overline{X}\right)^2}{n-1}$$

An examination of the formula used to calculate the variance helps us to understand how the variance captures information about process variability. The number calculated is based on the squared deviations about the average. The farther away a data value is from the average, the larger will be the contribution to the variance.

For the first subgroup of Table 5.1 the variance is found to be:

$$s^2 = \frac{(36.13 - 35.2675)^2 + (32.85 - 35.2675)^2 + (34.05 - 35.2675)^2 + (38.04 - 35.2675)^2}{3}$$
$$= \frac{15.757275}{3}$$
$$= 5.252425$$

The standard deviation is defined to be the positive square root of the variance. For the first subgroup, the standard deviation is:

$$s = \sqrt{5.252425} \cong 2.29182$$

These computations are quite tedious without the use of a hand-held calculator. (Table 5.3, in Section 5.3.6, contains the same subgroups of data as in Table 5.1, but the standard deviation of each subgroup is included in the table instead of the range of each subgroup.)

The subgroup variance, standard deviation, and range are all used to measure the magnitude of subgroup variation. Any one can be used; however, the standard deviation or the range is preferred to the variance in control chart work. Since the standard deviation and range are both used in process study, comparisons of the two measures of variation are provided:

1. *For large subgroups or sets of data (i.e., n > 10), the standard deviation is considered to provide better information about variability.* Since a data set collected over time to study a process typically has subgroup sizes on the order of n = 2 to n = 6, this advantage of the standard deviation over the range does not matter.

2. *The range is considerably simpler to calculate.* With the current availability of hand-held calculators, this advantage of the range over the standard deviation is no longer critical.

3. *It is easier to conceptualize the information on variation that is captured by the subgroup range.* For purposes of communicating with different people with varying levels of mathematical skills, this advantage of the range can be important.

4. *The standard deviation is used extensively in studies of components of variation* (see Chapter 8). Studies of components of variation rely on combining information from numerous subgroups. Consequently, the standard deviation will provide a better measure of variation for components of variation studies. Additionally, it is the measure of variation typically reported by computer software packages that supports this type of analysis.

As these characteristics of ranges and standard deviations indicate, the choice of whether to use the range or standard deviation is typically a matter of convenience. For on-line study of the consistency of process behavior over time, ranges are most commonly used to summarize process behavior.

5.3.5   Construction of Range and Average Charts

In using control charts to evaluate process behavior with variables data, within-subgroup variation (i.e., short-term variability) is evaluated for stability first. In this section, the range chart is used to examine the magnitude and stability of within-subgroup variation. Values of the subgroup ranges in Table 5.1 are plotted in Figure 5.2. Then, the average value of these ranges is determined. The average range, $\bar{R}$, read as R-bar, is obtained from the ranges for all subgroups and, in this case, is calculated to be:

$$\overline{R} = \frac{\sum\limits_{j=1}^{k} R_i}{k} = \frac{93.30}{20} = 4.665 = 4.67$$

The formulas for the lower and upper control limits of the range chart are, respectively:

$$\text{LCL}_R = D_3 R \text{ and } \text{UCL}_R = D_4 R$$

where D_3 and D_4 are control chart constants that have values indexed according to the number of observations within a subgroup. Values of D_3 and D_4 are provided in Table 5.2. This table is also found in Appendix D. In this example, with $n=4$, $D_4 = 2.282$. Thus, the upper control limit for the range chart is computed as:

$$\text{UCL}_R = 2.282(4.665) = 10.646, \text{ or } 10.65$$

Note that the value used for $\overline{R}$, 4.665, has not been rounded. It is suggested that rounded values not be used in calculations to avoid rounding errors. In determining the lower control limit for the range chart, note that for subgroups of size six or less, no value for D_3 is reported in Table 5.2. There is no lower control limit for ranges from subgroups of size six or less. Since $n = 4$ in this example, there is no lower control limit for the range chart.

The range chart of Figure 5.2 provides a time profile of process performance with respect to short-term variation. An essential element in the reported profile is the ability to provide signals of changes in the variability of the measurement. Changes in the short-term variation would be recognized by

* one or more points on or outside of the control limits, or
* a run of seven or more consecutive points on the same side of the average line.

There are no indications of the presence of special causes in the R chart in Figure 5.2. There are no values of R that equal or exceed the value for the upper control limit. There is no pattern in the data points suggesting the presence of systematic influences. The conclusion is that the variation in the values of R is produced by common causes.

If the range chart contains signals of instability in within-subgroup variation, then the first order of business is to do the work necessary to stabilize the variability. If there is evidence or signals of the presence of special causes, then it is not theoretically appropriate to place control limits on the X-bar chart. If within-subgroup variability is not consistent, then the magnitude of

TABLE 5.2 *Factors for Use with $\bar{X}$ and Range Charts*

Number of Observations in Subgroup n	Factor for $\bar{X}$ Chart A_2	Factors for Range Charts		Factor for Estimating σ d_2
		LCL D_3	UCL D_4	
2	1.880		3.267	1.128
3	1.023		2.574	1.693
4	0.729		2.282	2.059
5	0.577		2.114	2.326
6	0.483		2.004	2.534
7	0.419	0.076	1.924	2.704
8	0.373	0.136	1.864	2.847
9	0.337	0.184	1.816	2.970
10	0.308	0.223	1.777	3.078
11	0.285	0.256	1.744	3.173
12	0.266	0.284	1.716	3.258
13	0.249	0.308	1.692	3.336
14	0.235	0.329	1.671	3.407
15	0.223	0.348	1.652	3.472
16	0.212	0.364	1.636	3.532
17	0.203	0.379	1.621	3.588
18	0.194	0.392	1.608	3.640
19	0.187	0.404	1.596	3.689
20	0.180	0.414	1.586	3.735
21	0.173	0.425	1.575	3.778
22	0.167	0.434	1.566	3.819
23	0.162	0.443	1.557	3.858
24	0.157	0.452	1.548	3.895
25	0.153	0.459	1.541	3.931

deviations from one observation to another is unpredictable, being larger at some times than at others. When this is occurring in a process, it is difficult to know if an "unusual" value of X-bar is due to a shift in the process average or due to the same reasons that resulted in an unpredictable spread in the individual observations. Consequently, it is not advisable to calculate limits for an X-bar chart when the range chart is out of control. Although the X-bar values should be plotted to look for gross patterns, control limits should be placed on the X-bar chart only after the short-term variation has been stabilized. Once that is done, then analysis of the stability of the process average can be conducted.

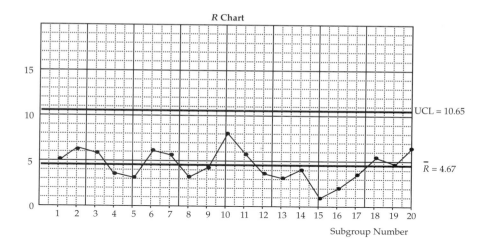

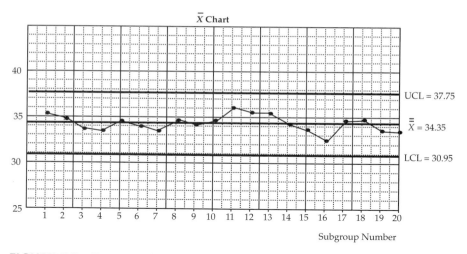

FIGURE 5.2 *Range and X-bar Charts for the Data of Table 5.1*

In our present example, the range chart indicated a stable process with respect to short-term variation. Therefore, it is possible to put limits on a chart of subgroup averages. As with other control charts, the X-bar chart requires a centerline and control limits. The centerline is the average of the subgroup averages. Formulas for the average of the averages and for the upper and lower control limits for the average chart are given next, along with the numerical results for this example problem. The average for the subgroup averages, represented by the symbol $\bar{\bar{X}}$ (read as "X double bar"), serves as the centerline for the X-bar chart:

$$\overline{\overline{X}} = \frac{\sum\limits_{i=1}^{k} \overline{X}_i}{k} = \frac{686.92}{20} = 34.3460$$

The formulas for the lower and upper control limits for the X-bar chart are, respectively:

$$LCL_{\overline{x}} = \overline{\overline{X}} - A_2 \overline{R}$$

$$UCL_{\overline{x}} = \overline{\overline{X}} + A_2 \overline{R}$$

where A_2 is a control chart constant whose value depends on the number of observations within a subgroup. The values of A_2 are also provided in Table 5.2. In this example, with $n = 4$, $A_2 = 0.729$. Thus, the upper and lower control limits for the X-bar chart are computed as:

$$LCL_{\overline{x}} = \overline{\overline{X}} - A_2 \overline{R} = 34.3460 - .729(4.665) = 30.945$$

$$UCL_{\overline{x}} = \overline{\overline{X}} - A_2 \overline{R} = 34.3460 + .729(4.665) = 37.747$$

The rules used to evaluate the range chart are also used for the X-bar chart. In this example, all subgroup averages are within the control limits and do not display any evidence of special causes in terms of unusually large or small values. The pattern of the data points must also be examined. There is no pattern in the subgroup average values to suggest the presence of special causes at work in the process.

As seen from the preceding formulas, the limits on X-bar charts depend on the average value of the range. When the range chart is in control, the numerical value of the average range (i.e., R-bar) captures the effect of variations acting during the time required to produce the material captured by a subgroup. Consequently, the process average is judged to be in or out of control according to the magnitude of these common cause variations captured within subgroups. The X-bar chart will be in control if the X-bars exhibit no more variation than that explained by the short-term variation estimated from the ranges. Points outside of the control limits on the X-bar chart indicate that the process is subject to additional variation across time which has not been captured within subgroups. This was the type of additional long-term variation exhibited by processes B and C of Figure 5.1.

Using the information provided by the charts in Figure 5.2, the process described by the data set in Table 5.1 is stable and predictable in terms of variability and average. It is to be immediately noted that stability, or "control," is an operational definition and says nothing about the utility of what the process is providing. A stable, "in-control" process means that the effect of material, equipment, or method changes is consistent. A stable process is not necessarily a satisfactory process. It could be that variation in the stable process is too large. Or, even though the process average remains consistent, it may be consistent at a level that is above or below the nominal, or target value, for the process.

5.3.6　　## Construction of Standard Deviation and Average Charts

The recommended procedure for constructing standard deviation and X-bar charts is similar to that used in constructing range and X-bar charts. The standard deviation, or s, chart is constructed first to check on the consistency of short-term variation. If this chart is in control, then the average of the standard deviations is used to calculate limits for the X-bar chart. The data set from Table 5.1 is used to demonstrate the construction of s and X-bar charts. This data set is repeated in Table 5.3. However, the column of ranges has been replaced with the standard deviations of the 20 subgroups. These standard deviations are plotted on the s chart in Figure 5.3. The centerline on this chart is $\bar{s}$, the average of the standard deviations from all subgroups. For the data set in Table 5.3, this value is calculated to be:

$$\bar{s} = \frac{\sum_{j=1}^{k} s_j}{k} = \frac{41.3949}{20} = 2.0697$$

Lower and upper control limits for the s chart are found by calculating, respectively,

$$LCL_s = B_3\bar{s} \text{ and } UCL_s = B_4\bar{s}$$

where B_3 and B_4 are constants that have values indexed according to the number of observations within a subgroup. Values of B_3 and B_4 are shown in Table 5.4. This table is also found in Appendix D. In this example, with $n = 4$, $B_4 = 2.266$, and the upper control limit for the s chart is found to be:

$$UCL_s = 2.266(2.0697) = 4.6899$$

For subgroups of size five or less, there is no lower control limit on the standard deviation chart. Hence, in Table 5.4, no values for B_3 are reported for subgroups of size five or less.

TABLE 5.3 *A Practice Data Set*

Subgroup	X_1	X_2	X_3	X_4	s	$\overline{X}$
			Observations			
1	36.13	32.85	34.05	38.04	2.2918	35.2675
2	38.68	34.95	32.36	33.68	2.7221	34.9175
3	34.34	35.69	35.06	29.72	2.7117	33.7025
4	33.37	31.73	33.45	35.58	1.5784	33.5325
5	32.42	35.58	34.35	35.79	1.5464	34.5350
6	30.62	34.10	34.75	36.91	2.6095	34.0950
7	31.76	33.29	37.41	31.50	2.7300	33.4900
8	34.94	33.79	33.68	36.90	1.4946	34.8275
9	34.66	32.02	36.34	33.50	1.8270	34.1300
10	39.37	34.82	33.47	31.30	3.4103	34.7400
11	33.06	38.97	35.88	36.07	2.4141	35.9950
12	37.42	36.39	34.68	33.52	1.7389	35.5025
13	37.18	34.43	36.34	33.88	1.5589	35.4575
14	32.19	34.90	36.34	33.41	1.8012	34.2100
15	33.36	34.36	33.38	33.68	0.4669	33.6950
16	33.22	31.18	32.95	32.51	0.9053	32.4650
17	34.22	33.01	36.63	35.10	1.5237	34.7400
18	32.68	33.03	38.15	35.47	2.5360	34.8325
19	31.49	35.84	31.00	35.47	2.5584	33.4500
20	34.08	28.97	34.76	35.53	2.9697	33.3350
Sums					41.3949	686.5900

The completed s chart in Figure 5.3 should be compared to the range chart in Figure 5.1. Since the s chart in Figure 5.3 and the range chart in Figure 5.1 both report on the consistency of within-subgroup variation, the information contained in the two charts should be the same. The comparison of the two different charts shows that this is indeed the case. In particular, note the following:

1. All points fall beneath the upper control limit on the s and range charts.
2. The pattern of points on the two charts is the same. This phenomenon can be explained by the fact that when there is large (or small) variability in the subgroup values, both the range and the standard deviation will be large (or small).

Since the s chart is in control, attention is now directed toward the subgroup averages. When within-subgroup variation is studied using standard deviations, control limits for the X-bar chart are calculated using the following formulas:

$$\text{LCL}_{\bar{x}} = \bar{\bar{X}} - A_3\bar{s} = 34.3460 - 1.628(2.0697) = 30.9765$$

$$\text{UCL}_{\bar{x}} = \bar{\bar{X}} + A_3\bar{s} = 34.3460 + 1.628(2.0697) = 37.7155$$

where A_3 is determined from Table 5.4 according to the subgroup size. A comparison of the upper and lower control limits found using $\bar{s}$ with those found using $\bar{R}$ shows that the two sets of limits differ by a very small amount. Whether using the R chart or the s chart to study within-subgroup variation, conclusions about out-of-control signals on the X-bar chart will generally be the same.

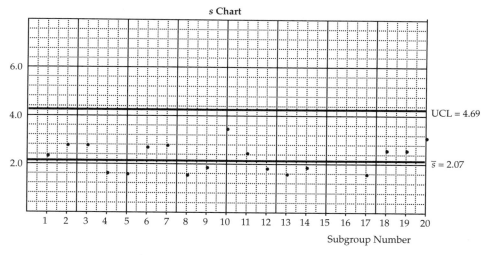

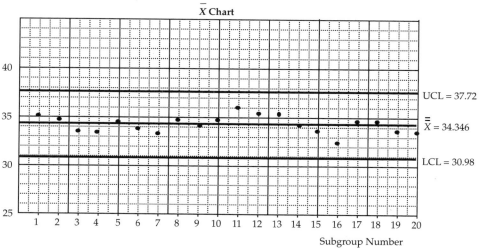

FIGURE 5.3 *Standard Deviation and X-bar Chart for the Data of Table 5.3*

TABLE 5.4 *Factors for Use with $\bar{X}$ and s Charts*

Number of Observations in Subgroup n	Factor for $\bar{X}$ Chart A_3	Factors for s Chart		Factor for Estimating σ c_4
		LCL B_3	UCL B_4	
2	2.659		3.267	.7979
3	1.954		2.568	.8862
4	1.628		2.266	.9213
5	1.427		2.089	.9400
6	1.287	0.030	1.970	.9515
7	1.182	0.118	1.882	.9594
8	1.099	0.185	1.815	.9650
9	1.032	0.239	1.761	.9693
10	0.975	0.284	1.716	.9727
11	0.927	0.321	1.679	.9754
12	0.886	0.354	1.646	.9776
13	0.850	0.382	1.618	.9794
14	0.817	0.406	1.594	.9810
15	0.789	0.428	1.572	.9823
16	0.763	0.448	1.552	.9835
17	0.739	0.466	1.534	.9845
18	0.718	0.482	1.518	.9854
19	0.698	0.497	1.503	.9862
20	0.680	0.510	1.490	.9869
21	0.663	0.523	1.477	.9876
22	0.647	0.534	1.466	.9882
23	0.633	0.545	1.455	.9887
24	0.619	0.555	1.445	.9892
25	0.606	0.565	1.435	.9896

5.3.7   Summary of Chart Construction

Construction of R and X-bar charts

k denotes the number of subgroups and n refers to the size of each subgroup.

1. Calculate R and $\bar{X}$ for each subgroup.

Construction of *R* and *X*-bar charts (*continued*)

R = largest measurement – smallest measurement

$$\overline{X} = \frac{\sum\limits_{j=1}^{n} X_i}{n}$$

2. Plot the R's and $\overline{X}$'s maintaining time order.

3. Calculate the centerline, $\overline{R}$, and the upper and lower control limits for the R chart.

$$\overline{R} = \frac{\sum\limits_{j=1}^{k} R_j}{k} \qquad LCL_R = D_3\,\overline{R} \qquad and \qquad UCL_R = D_4\,\overline{R}$$

4. If the R chart is in control, put limits on the X-bar chart.

$$\overline{\overline{X}} = \frac{\sum\limits_{i=1}^{k} \overline{X}_i}{k} \qquad LCL_{\overline{X}} = \overline{\overline{X}} - A_2\,\overline{R} \qquad UCL_{\overline{X}} = \overline{\overline{X}} + A_2\,\overline{R}$$

Construction of *s* and *X*-bar Charts

k denotes the number of subgroups and n denotes the size of each subgroup.

1. Calculate s and $\overline{X}$ for each subgroup.

$$s = \sqrt{\frac{\sum\limits_{j=1}^{n}\left(X_i - \overline{X}\right)^2}{n-1}} \text{ , where } \overline{X} = \frac{\sum\limits_{j=1}^{n} X_i}{n}$$

Construction of *s* and *X*-bar Charts (*continued*)

2. Plot the standard deviations and *X*-bars maintaining time order.

3. Calculate the centerline and upper and lower control limits for the *s* chart.

$$\overline{R} = \frac{\sum\limits_{j=1}^{k} R_j}{k} \quad \text{LCL}_R = D_3\,\overline{R} \quad \text{and} \quad \text{UCL}_R = D_4\,\overline{R}$$

4. If the *s* chart is in control, put limits on the *X*-bar chart.

$$\overline{\overline{X}} = \frac{\sum\limits_{i=1}^{k} \overline{X}_i}{k} \quad \text{LCL}_{\overline{X}} = \overline{\overline{X}} - A_3\overline{s} \quad \text{UCL}_{\overline{X}} = \overline{\overline{X}} + A_3\overline{s}$$

5.4   SUMMARIES OF PROCESS BEHAVIOR

Having judged that a process is stable with respect to both variability and average, estimates of those process properties can be reported. In addition, a histogram can be constructed from the individual measurements to provide a representation of the distribution of process measurements. These three descriptors, the process average, the process standard deviation, and the process distribution, not only provide useful ways of characterizing process performance, but also provide descriptors by which the current outcomes delivered by a process can be compared to what is required. The information provided by these three descriptors and the comparison of these properties to current specifications on a process are illustrated by a further examination of the previous data set.

5.4.1   Construction and Interpretation of Histograms

The 80 individual measurements in the practice data set from Table 5.1 have been used to construct the histogram shown in Figure 5.4. An examination of this histogram illustrates the important characteristics of a process, which often need to be summarized. The first of these characteristics is the shape of the distribution of measurements. Words typically used to describe the shape of the histogram in Figure 5.4 are "mound-shaped" or "bell-shaped." These words capture the idea that most of the observations appear to cluster around a center value and fall in a symmetric manner about that center. This mound behavior is similar to that described by a mathematical model of a distribution of measurements, referred to as the normal distribution. (Appendix A.1 discusses this normal model and its use as a theoretical basis for control charting work.) Thus, another description commonly used to characterize the data set in Table 5.1 is that it appears to be "normally distributed."

It is also useful to think about what information is *not* contained in the histogram of Figure 5.4. The time order in which the data were collected has been lost. For this reason, in process investigation or when representing process outcomes, it is recommended that histograms based on large data sets collected over long periods of time not be constructed until the process is shown, by the use of control charts, to be in control. In addition to the loss of important information about the process behavior over time, the use of a histogram to describe process behavior on an unstable process is highly suspect. As the variation or level in a process changes, so will the shape of the histogram. Without process stability, a histogram constructed from this week's data may look very different from what would be seen on a histogram completed from data collected next week.

The steps listed next provide a method for constructing a histogram. The use of this method is illustrated by showing how these steps were used to construct the histogram in Figure 5.4.

Step 1

> *Determine the range of the numbers in the data set.* Of the 80 numbers in Table 5.1, the smallest is 28.97 and the largest is 39.37. Consequently, the range is $39.37 - 28.97 = 10.4$.

Step 2

> *Choose the number of groups or classes into which to divide the data set.* Somewhere between 7 and 12 classes is recommended. Histogram shapes are affected, sometimes substantially, by the number of classes chosen. Statistical software often has default methods for selecting the number of classes; the number chosen may or may not be useful for a particular application. Given the differing behavior that may be observed by using different numbers of classes, the data analyst may choose to construct several different histograms from the same data set to see if different information is provided by the different histograms. For the histogram in Figure 5.4, the number of classes chosen was 7.

Step 3

Divide the range by the number of classes to obtain the class width. Dividing the range of 10.4 by 7 results in a class width of 1.486. Since it is more convenient to work with an integer value, a class width of 2 is chosen.

Step 4

Determine the class intervals for each of the classes. Seven classes each with a class width of 2 means that the range covered by the 7 classes will be 14. If the lowest class starts at 27 and the upper class ends at 41, the entire range of values in the data set is included. Table 5.5 lists the class limits for each class.

Step 5

Determine the frequency distribution by counting the number of measurements that fall within each class. Notice that the lower class limits in Table 5.5 have a plus sign after them. This indicates that a value must be larger than the lower limit to fall within a particular class. For example, the second class has a lower limit of 29_+. Consequently, a data value of 29.00 would fall in the first class but a data value of 29.01 would be placed in the second class. This convention means that a data value will belong in only one class.

The column labeled "Frequency Distribution" in Table 5.5 contains the number of data values in each class.

Step 6

Draw the histogram. The histogram is drawn by first marking off the class limits on the vertical axis and then drawing bars above these limits. The heights of the bars correspond to the number of measurements in each class. By convention, the bars drawn are contiguous.

As in Figure 5.4, instead of the class limits being marked off on the vertical scale, the class midpoints might be used instead. The midpoint of a class is simply the average of the upper and lower class limit. The class midpoints for the histogram in Figure 5.4 are also listed in Table 5.5.

TABLE 5.5 *Frequency Distribution for Measurements from Table 5.1*

Class Limits	Class Midpoint	Frequency Distribution
27^+–29	28	1
29^+–31	30	3
31^+–33	32	15
33^+–35	34	34
35^+–37	36	18
37^+–39	38	7
39^+–41	40	1

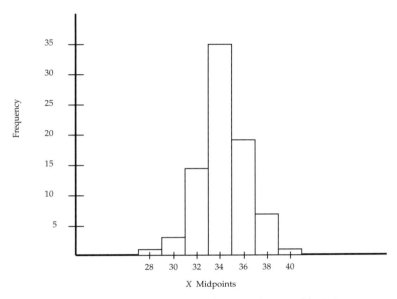

FIGURE 5.4 *Histogram of Measurements from Table 5.1*

The shape of a distribution of process measurements should be compared to the expected shape for that particular distribution of process measurements. The four histograms in Figure 5.5 illustrate some different shapes that occur in process work. Comparing these shapes with expected behavior provides insights into causes potentially affecting process operation. The importance of prior process knowledge in these comparisons should be noted. Growth in process knowledge occurs through a constant cycling between what is currently known or suspected about a process, collecting data to provide further information about the process, and then updating current knowledge in light of this information. An examination of the four histograms in Figure 5.5 illustrates the usefulness of a histogram as a tool for extending process knowledge.

Histogram A appears to have a large number of measurements right above the lower specification limit and right below the upper specification limit. One might expect the measurements to taper off on either side of the histogram in a more gradual fashion, since the variation pattern captured in a histogram rarely changes so abruptly as a result of inherent process variation. A possible reason for the observed behavior might be that any material near the upper or lower specification limit was reworked to ensure that it fell within specifications. Another possible reason for the observed behavior is that when measuring the material, values close to the upper or lower specification limits were remeasured until a reading was obtained that fell within specifications.

Histogram B appears to be chopped off at the upper and lower specification limits. As with histogram A, the explanation for this behavior must come from a closer examination of the process generating the results. One possible explanation is that the material being measured was sorted prior to the place at which the measurements were collected. The material that fell above or below the specifications was removed. If this supposed behavior is correct, it indicates that the process has more variation than can be tolerated and that the solution adopted for this problem is to sort out the nonconforming material.

Histogram C shows a pattern similar to that of B, but with a more dramatic drop-off at the lower specification limit. Again, the suspicion might be that there is more variation in the process than can be tolerated. Furthermore, one might imagine that the material measured prior to the sort had a normal distribution with the measured values clustering around a central value. Histogram C then suggests that this center does not fall in the middle of the specification limits, thus explaining the larger amount of material removed below the lower specification limit. Histogram D shows two distinct peaks. A number of reasons can exist for such behavior and additional investigation is

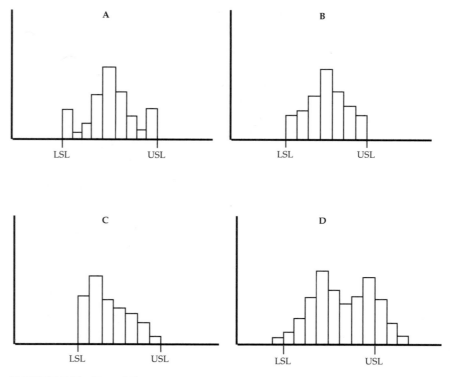

FIGURE 5.5 *Four Histograms*

needed. One speculation that might help such an investigation is that there are two different processes generating this set of data. For example, if two machines are creating the measured output, then it is possible that they are each creating material centered at different values.

One of the uses of the histogram is to provide insights into process operation. Another is the use of the histogram as a description of the measurements that will be produced by the process in the future. This second use is dependent on being able to use the histogram as a predictor of process behavior, which in turn is dependent on knowing that the process is operating in a consistent manner. The histograms of Figure 5.1 provide an illustration of this last statement. The histograms for process B and for process C look similar and their shapes might both be described as mound-shaped. However, since both histograms represent processes that are inconsistent over time, neither of the two histograms can be relied on to describe process outcomes. The histograms only capture the shape of process outcomes over the time frame investigated. Inconsistencies in the processes imply that for a different time period, the histograms will probably show a different description of process outcomes.

Shapes of histograms should not be confused with in-control and out-of-control processes. Histograms having shapes approximating that of a normal distribution may result from either a stable or an unstable process; the shape of the histogram by itself does not provide a description of process behavior over time. Data on dimensions of a part impacted by tool-wear may display a mound-shaped histogram over runs having a constantly increasing average. Data from a stable process that has consistent short-term variation and a stable average may produce a highly skewed histogram. Clearly, the shape of the histogram cannot be examined for clues as to process behavior over time.

5.4.2 Estimating Process Variation

Another look at the histogram of Figure 5.4 indicates that the variation in the measurements is another important property of the process that needs to be characterized. Examination of the histogram shows that the 80 measurements spread over a range from about 27 up to 41. The process standard deviation is a numerical measure whose intent is to capture this information on process variability. The process variation is the variability in all process results and can be reliably reported only for a controlled process. The process standard deviation, denoted by a Greek lowercase sigma, σ, represents this complete variability.

The concept of a process standard deviation is just that, a conceptual idea. An understanding of the process behavior that σ is thought to capture can be had by imagining that a very large number of measurements were available on a *stable* process. Calculating a subgroup standard deviation, *s*, from this very, very large subgroup would provide the type of information on process variability that σ is meant to represent. The process standard deviation describes the variation to be expected in measured values from a process. In other words, just as *s*, a subgroup standard deviation, will describe the variation that exists in the numbers in a subgroup, σ is meant to describe anticipated variation in measurements from a stable process.

If both the range and the X-bar chart are in control, an estimate of σ can be determined from the centerline on the range chart. Each of the ranges in Table 5.1 reports on the variation in one subgroup of four numbers; if subgroups of size five or six had been used to study a process, one would expect to see, on the average, larger values of *R* than for subgroups of size four. In addition, no one value of the ranges represents the complete spread seen in the range (10.4) of all 80 numbers. The formula for estimating σ is developed from the average range; the use of the divisor, d_2, frees the estimated standard deviation from its reliance on subgroup size. The formula for estimating σ and the resulting value are reported as:

$$\hat{\sigma} = \frac{\overline{R}}{d_2} = \frac{4.665}{2.059} = 2.27$$

The value for d_2 is found in Table 5.2. If an *s* chart had been used to study within-subgroup variation, then the process standard deviation could be estimated by:

$$\hat{\sigma} = \frac{\overline{s}}{c_4} = \frac{2.0697}{.9213} = 2.25$$

where c_4 is obtained from the last column of Table 5.4. The σ indicates the standard deviation while the " ^ " symbol indicates that the reported numerical value is an estimate of the process standard deviation. Because the range (or *s*) chart did not contain any signals as to abnormal behavior *and* because the $\overline{X}$ chart indicated a stable process average, σ is thought to be a reliable estimate of the process standard deviation.

The standard deviation of a process is a useful descriptor of process variation. It provides a measure by which the variation *delivered* by a process can be compared to the *required* level of variation. Such a comparison is often made by using the estimated process standard deviation to estimate the spread of measurements that will occur in the process. For normally distributed processes, almost all measurements will fall within a range of six standard deviations. For the data in Table 5.1, the standard deviation of the process was estimated to be 2.27. Thus, an estimate of the range over which measurements can be expected to be observed is $6 \times 2.27 = 13.62$. The histogram in Figure 5.4 shows that the measurements fall within a range from 27 to 41. This spread of $41 - 27 = 14$ is comparable to the value estimated by six standard deviations.

It is possible to compute a sample standard deviation for all the individual observations used to construct the R and X-bar charts. For the data in Table 5.1, the sample standard deviation calculated from all 80 measurements resulted in $s = 2.0993$. When the R and X-bar (or s and X-bar) charts are in control the resulting number will be in close agreement with $\bar{R}/d_2$ (or $\bar{s}/c_4$). However, when the range chart is in control but the X-bar chart is not, then $\bar{R}$ will understate the variation in the process. An out-of-control X-bar chart provides evidence that there is more variation in the process than that represented by the average range. It would be a mistake in reasoning to use $\bar{R}/d_2$ to represent process variation when the X-bar chart is out of control, since the long-term variation being evidenced on the X-bar chart is not reflected in the magnitude of $\bar{R}$. It is for this reason that both the range (or s) and X-bar charts must be in control before $\bar{R}/d_2$ (or $\bar{s}/c_4$) can be used to estimate process variation.

5.4.3 Estimating the Process Average

The histograms of Figures 5.4 and 5.5 were described as having distributions of measurements that clustered around a central value. The process average is a way of capturing information about this central value. The process average, denoted by a Greek lowercase mu, μ, is the numerical value around which process measurements can be expected to cluster. If an X-bar chart constructed from process data is out of control, then no such value can be reported. The best that can be said is that the average value of process measurements appears to be changing over time.

When signals as to abnormal behavior are absent from both the range and X-bar charts, $\bar{\bar{X}}$, which is the value of the centerline on the X-bar chart, can be taken as an estimate of the process average. For the data of Table 5.1, the process average would be reported as being 34.35, rounded from 34.346. The estimated process average is denoted by $\hat{\mu}$.

5.4.4   Process Capability

In its most general form, the term *process capability* refers to the capability of a process to deliver what is required. However, in response to the need to quantify "how capable" a process is, process capability is often more narrowly defined by whether the measured outcomes of a process exhibit small enough variation to fall within some set of specification limits. The engineering tolerance for a process is the difference between the upper specification limit and the lower specification limit. It is this more narrow definition of process capability that is discussed in this subsection. If the specifications stated that the measured dimension reflected by the data of Table 5.1 should fall between 30 and 40, then the engineering tolerance (ET) for the measurement would be:

$$ET = 40 - 30 = 10$$

The natural tolerance (NT) for the process refers to the range of measurements over which the process will produce material. Typically, the natural tolerance for the process is estimated by:

$$NT = 6\hat{\sigma}$$

The origin of this definition for the natural tolerance comes from the fact that almost all measurements (99.7%) from a normal distribution will fall within ±3 standard deviations of the mean. However, the definition of NT need not be restricted to processes with a normal distribution. For a wide variety of other distribution shapes, $6\hat{\sigma}$ provides a fairly conservative estimate of the range of possible measurements from a stable process. As discussed in Section 5.4.2, the natural tolerance for the process described by the data in Table 5.1 would be $6 \times 2.27 = 13.62$. A comparison of the NT with the ET allows one to decide whether the process spread is small enough that the specifications could be met if the process were properly centered. In the current example, the NT is larger than the ET so the process would be said to be "not capable."

Various capability indices have been defined as a method for reporting on the ability of a process to meet specifications. Two of the more common indices are C_p and C_{pk}. The C_p index is defined by the ratio:

$$C_p = \frac{ET}{NT}$$

The C_p index is an attempt to quantify "how capable" a process is. If this number has a value of 1 or greater, the process is said to be capable. Many organizations state that a preferred value for this number is 1.33 or more.

A drawback to using C_p to report on process capability is that the index does not capture information on the process average. A process could have a large C_p ratio and yet be producing many product items outside of specifications if the process is not targeted. The C_{pk} index is an attempt to report not only the effect of process variation but also the process average on the ability of the process to produce what is required. Index C_{pk} is defined to be the smaller of the numbers C_{pU} and C_{pL}, which are calculated from the formulas:

$$C_{pU} = \frac{USL - \mu}{3\sigma} \qquad C_{pL} = \frac{\mu - LSL}{3\sigma}$$

where USL refers to the upper specification limit and LSL the lower specification limit. From the data in Table 5.1, the following estimates of process properties have been obtained:

$$\overline{\overline{X}} = 35.35$$

$$\hat{\sigma} = 2.27$$

$$NT = 13.62$$

The upper and lower specification limits for the process were:

USL = 40

LSL = 30

From the preceding information, C_p would be:

$$C_p = \frac{10}{13.62} = 0.73$$

Of course, since C_p is less than 1, the process is not capable.

To calculate C_{pk}, C_{pU} and C_{pL} are first calculated to be:

$$C_{pU} = \frac{40 - 34.35}{3(2.27)} = 0.83 \qquad C_{pL} = \frac{34.35 - 30}{3(2.27)} = 0.64$$

Thus, C_{pk} would be 0.64, the smaller of these two numbers. The smaller value of C_{pk} as compared to C_p can be explained by comparing the process average to the center of the specification limits, or the nominal value. The nominal value in this instance would be 35; the process is not centered on this nominal value, but is targeted somewhat below 35 at a process average of 34.33.

The calculation of the two capability indices assumed that the process under study was operating in statistical control. One should be very skeptical about reported values of C_p and C_{pk}, because these indices may be reported even when not enough is known about the stability of a process. Too often C_p or C_{pk} is determined by a one-time application of a control chart or, even worse, by collecting, say, 30 consecutive readings from a process. If this were the case, then the reported process information, $\bar{\bar{X}}$ and $\hat{\sigma}$, cannot be relied on to summarize process behavior. These numbers are only useful if it is known that the process average and the process variation are stable over time. When these numbers are estimated from a one-time application of control charts, perhaps over a short time frame, then it is doubtful whether the estimated standard deviation will capture the variation in process outcomes that might occur from one run to the next, or that might occur as a result of incoming materials to the process changing, or that might result from changing crews, etc. Under these circumstances, reliance on C_p and C_{pk} to describe process capability is a doubtful, if not worthless, practice.

The interpretation of C_{pk} is dependent on the assumption that the distribution of measurements is approximately normal. This information on the process is only available by actually constructing a histogram of individual results to understand the shape of the distribution of measurements. Further, one would need to question closely an organizational practice that required a value of C_{pk} to be reported on all different characteristics of a process. Many measurements on a process could not reasonably be expected to behave according to a normal distribution. The amount of impurities and time to breakage are two examples where process measurements typically have a skewed distribution.

Capability indices provide summary measures of how current process performance compares to some stated specifications. However, their use as an aid to directing or prioritizing improvement efforts is limited. For example, when C_p or C_{pk} is reported as summary results of a process, the focus of study is often on the results of the process, rather than on the characteristics that need to be studied to improve the results. The sources of variation contributing to process results need to be worked and studied. Yet, implicit in the use of capability indices is that the correct characteristics of a process to be measured are known. This knowledge will only come from considerable process understanding. Even if sound reasons exist for investigating the characteristic under study, the behavior of this characteristic, in terms of the ability to target the characteristic at the correct average and maintain the characteristics at small levels of variability, is not well described by either of the indices.

It is, unfortunately, common practice in many organizations to state goals for C_p or C_{pk} for many, if not all, processes. In light of the discussed limitations and assumptions about the use of the indices, such practices do not provide the kind of information necessary to improve process behavior. Such goals for C_p or C_{pk} (like goals about 6σ) are just that, goals. They provide no direction on how or where to work to improve the value provided by a process.

5.4.5   ## Calculating Percent Outside of Specifications

When a stable process has been demonstrated and has a distribution that appears to be normal, tabled values for the normal distribution can be used to estimate the percent of product that falls outside of specifications. Appendix A.1 contains an explanation of how to find this kind of information for normally distributed data. Since the range and X-bar charts for the data of Table 5.1 were in control and since the histograms of these data appeared to have a normal distribution, the proportion of material outside of specifications can be determined by using the normal model. In particular, the normal curve of Figure 5.6 is labeled with the estimated process average, the upper and lower specification limits, and the z values associated with these limits.

From Table D.1, we find that the proportion of observations above a z value of 2.49 is .0064 and below a z value of −1.92 is .0274. Consequently, the proportion of material that is outside of specifications is .0064 + .0274 = .0338. Stated differently, the percentage of material predicted to be outside of the specification limits is 3.38%.

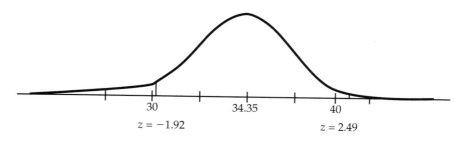

FIGURE 5.6 *Use of Natural Curve to Determine Percent Outside of Specifications*

5.5   EXAMPLES OF THE PROCESS USE
OF RANGE AND *X*-BAR CHARTS

5.5.1   Manufacturing of Particle Board

Particle board is an industrial product produced by manufacturing plants and sold to other industrial companies, such as furniture and mobile home manufacturers. These companies then process these boards and create products that go to other industrial users or to final consumers. Particle boards have numerous important features or attributes ranging from physical properties, such as internal bond or modulus of rupture, to surface or face characteristics such as flatness, smoothness, and moisture retention.

The manufacturing process takes wood chips and other materials and converts these into a finished product. A flow diagram of the process for manufacturing particle board is shown in Figure 5.7. The process begins with the specification and purchase of wood and other ingredients. Raw materials, wood chips, and sawdust are moved through milling and drying; through blending, where resin, wax, and urea are added, through the forming machine where multilayered mats are put down; and then taken to the press where the mats are pressed into a particle board. Boards are then allowed to cool, after which they are sawed, sanded, packaged, and moved into storage.

The flow diagram depicts the physical process flow; however, important aspects of the process are not represented or displayed on that diagram. How the work is to be performed is not represented; raw material requirements and specifications and equipment conditions are not described. Flow rates and volume parameters are not defined and neither are the required in-process characteristics as the material moves from one operation to another. Operational definitions have to be provided for process parameters, work methods, machine settings, and material characteristics. These comments are made merely to reinforce the statement that flow diagrams are an incomplete process description. Additional information is needed. It is this additional information that provides a basis for understanding the variation cause structure. In this process, other examples of additional information would include temperatures, pressures, time at the press, distribution of needed particle sizes, and amount and coverage of resin on the furnish. It is often within this additional information that explanations for special or assignable causes are found. Generally, process information of this type cannot be had on demand. It frequently is compiled only after managers, engineers, and operators begin work to improve process management and discover that such information is lacking.

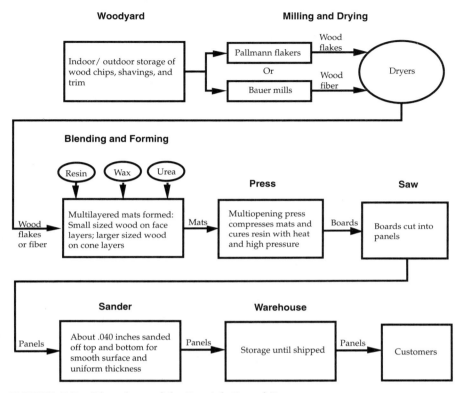

FIGURE 5.7 *Flowchart of the Particle Board Process*

In this brief application, the measurement to be studied is mat weight after the completed mat emerges from the former (a machine that "lays" down the furnish to create a mat to be pressed). For purposes of this discussion, the mats may be pictured as being several inches thick, several feet wide, and 20 feet long. The ability of the former to lay down the amount of material in a correct and uniform pattern is important for creating consistent final board properties after a correct press operation. Process operations upstream from the former affect the ability of the former to produce conforming mats correctly and consistently. Consistency of material properties, such as moisture and density, is important in achieving a nominal, product-specific, mat weight. Mat weight is a comprehensive number. It is a function of mat volume and mat density. Both of these mat parameters must be controlled; this control is accomplished by material characteristics going into the former and by equipment settings on the former that affect mat height and material distribution across and down the mat. Process management also requires information on density variations within and between mats. Variations within mats are created by certain process factors; variations between mats are created by all of these and by possibly other process factors.

A crucial point is that managers, engineers, and operators must immediately begin thinking about their process in terms of variations: within part, part to part, product source to product source, time to time, and in other useful ways. Decreased variations in mat weight, a specific characteristic for this example, provide a more uniform product, that can be worked with more ease and confidence by industrial customers, require less material, support timely schedule completion, and when joined with appropriate target values, provide product having the correct board and surface properties. Even with complete and comprehensive product design and manufacturing instructions, there will usually be further information and knowledge required to improve product and process and to be able to run to the future's tighter specifications. It is also correct that successful process management in today's environment requires not only product improvements expressed through design and manufacturing, but also improvements in *how* the work is done. This latter improvement requires a thorough understanding as to how work is currently done. Knowledge of manufacturing variations, their sources and their effects, are not only necessary for this work but in fact facilitate the exploration, discovery, and confirmation of new or revised process configurations. The concept of variations and process reasons for these variations are essential to process analysis. Knowledge of process reasons for variations is almost always incomplete; process improvement begins with this understanding and this incompleteness.

Measurements on mat weight for process evaluation are obtained by automatic measurement methods that read and store mat weight in pounds per cubic foot. For this example problem, one reading per mat is used to represent the mat weight. The subgrouping plan creates one subgroup of five mat weights ($n = 5$) each hour by taking one mat every 12 minutes. Weights for individual mats are recorded and the range and average of the subgroup of five each hour are computed and reported. Here, as with all uses of charts for process analysis, a distinction is made between factors affecting within-subgroup and those factors affecting variations between subgroups. In this application, within-subgroup variations, the deviations in five mats selected in the same hour, are created by the way in which process factors fluctuate or act within an hour's time. Between-subgroup variations are created both by those sources creating variation within an hour and by those process factors that tend to become active in usual or unusual ways from hour to hour.

In overly simple terms, points out of control on the range are caused by unusual variations or inconsistencies in those causes operating within subgroups. The average value of the effect of these sources, $\bar{R}$, is used to judge the stability of variations in the X-bars. An out-of-control condition on the X-bar chart would indicate that some source of variation is resulting in a changing process level across the time in which the subgroups were collected. Sources or factors of this general type or nature are described as driving long-term variations. A clear implication of these ideas is that the responsible manager should understand the sampling and measurement strategy relative to the factors or causes shown on cause and effect diagrams. A manager should understand which of the causes listed would drive the short-term or within-subgroup variation and which would be expected to affect long-term variation. Often it is the case that average values change over time because the process variables that are relatively constant while the subgroup is being produced take on different values at a later time. The effectiveness of chart applications is diminished by a lack of understanding regarding where the variations in particular causes are likely to be revealed. Subgrouping strategies and further discussion regarding subgrouping and its importance are discussed in Chapter 6. However, it is important to begin early to develop the ideas about the causal structure of within- and between-subgroup variation in order to learn about correct and productive use of charting techniques to support improved process management. Understanding the causal structure helps in the identification of special causes, but it also strengthens process knowledge about influential sources of variation, which supports informed decision making for achieving effective process improvement.

In the current example, where the sampling scheme selects five mats per hour, the within-subgroup variations are created by those process sources producing within-mat variations and mat-to-mat variations over the course of one hour. Hour after hour, the deviations in the weight of five mats yield range values for each subgroup. These values fluctuate according to causes operating consistently within an hour. A special cause may result in the deviations among five mats being unusually large or unusually small. Unusually large deviations, of course, are recognized by points on or above the control limit on the range chart. Unusually small deviations would be difficult to detect with subgroups of size five, since the range chart has no lower control limit. However, a production period that experiences sustained lower mat-to-mat variation would be indicated by a run of points below the average range. Within-mat variations are perhaps most influenced by process settings, maintenance status, and operational condition. Mat-to-mat variations occurring within one hour are affected most significantly on a common cause basis by short-term changes in material characteristics, in particular density and moisture changes. Changes in material characteristics are created, in turn, by the properties of the incoming wood materials, screening and milling equipment and practices, and the addition of resins and waxes. The magnitude of the average range value reflects the effect of these process conditions. An out-of-control condition on the range chart indicates that process factors acting on the process within an hour's span of time are not consistently practiced or achieved. Again, reasons for this inconsistency would be found in methods, equipment, and material variations and the inconsistent practice and operation of these factors. Common cause variations, measured by the average range for a controlled process, are those things that produce variations on the range chart within each and every hour. Obviously, the purpose of these somewhat detailed points is not to teach the reader how to make particle board; the purpose is to demonstrate ways of thought that are necessary for process control and improvement.

The X-bar chart, which contains values for hourly subgroup averages, measures both short- and long-term process effects. In this application, fluctuations of the averages for five mat weights are affected by what happens within any one hour and by those process inputs that tend to reveal their effects in periods of time that exceed one hour.

Data for demonstration and discussion of the above points are plotted in Figure 5.8. Results are plotted in two formats. On the left side of the figure, the sample data are plotted in production sequence, hour by hour. On the right side, the data are plotted by crew identification with correct production sequence maintained within data plot for a crew.

The range chart is plotted first. The range chart of the data in the natural production order indicates that the variations from mat to mat within an hour are consistent. That judgment is made because of the absence of any signals of assignable causes—there are no points on or outside the limits and there are no specific patterns in the plotted points. The average range is 0.04. Consistency, or stability, does not say that the average range is either small or large, nor does it imply that customer requirements are being met. The judgment is simply a statistical one that reports on the consistency of the effects of the observed variations typically occurring within one hour for mats separated by 12 minutes. The range chart indicates that the same within-hour deviations in mat weight are achieved and maintained regardless as to what time the data were taken or which crew was operating the process.

The X-bar chart with points plotted in natural order indicates the presence of special causes. There are numerous points outside of the control limits. The plotted points themselves contain no clue, however, as to why such large variations exist. Experience, careful logs and notes, other data, and a reorganization of the sample data according to other criteria are useful in identifying reasons for the larger variations. For example, different crews, using different interpretations of manufacturing instructions or having different run practices, might cause different average values for mat weight. To investigate this possible cause of the large variations, the original data need to be arranged in a different order so that differences among crews can be specifically examined.

In this case, the original data on mat weights are retained on an hourly basis, but have been plotted by crew. The range chart does not reveal any differences for within-hour deviations for the different crew operations; the ranges appear to be consistent across all crews. However, the *X*-bar chart clearly reveals that the different crews operate to different average values.

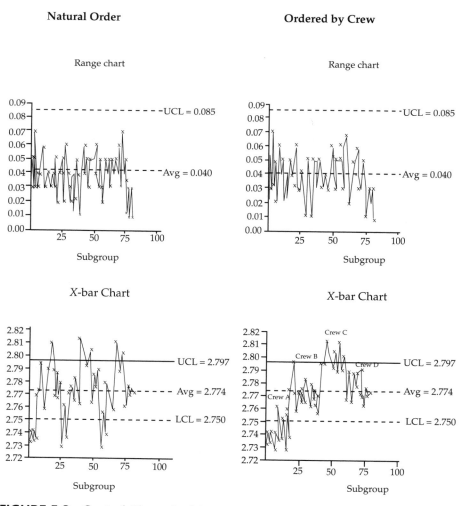

FIGURE 5.8 *Control Charts for Mat Weight*

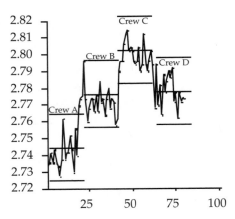

FIGURE 5.9 *X-bar Charts for Each of the Four Crews*

If an *X*-bar chart is constructed for each of the four crews (see Figure 5.9) using the common *R*-bar value of 0.04, each of these charts is in control. Consequently, it is concluded that, within themselves, the crews operate in a stable fashion. However, the different crews run or operate at different average values for mat weight.

At this point, the following information has been gained about the mat weights:

1. Within-hour deviations in mat weight are consistent, regardless of operating condition or crew.
2. The plant, as a production source, produces mats that are not in control with respect to the average mat weight; the presence of a special cause has been identified.
3. Crews produce mats to different average weights.
4. Each crew, with respect to itself, produces consistent mat weights.
5. The variability in mat weight is much larger than suggested by $\bar{R}/d_2 = .04/2.326 = 0.0172$. The fact that subgroup averages fall outside of the control limits indicates that other sources of variation, not captured by the within-subgroup ranges, are active in this process.

It is too easy simply to claim that the reason for the special cause noted on the *X*-bar chart is associated with crew practices. Although that may be the case, the reasons why the crews run in a consistent fashion to different averages have not yet been discovered. It may be due to inconsistent operating procedures; it may be due to differences in raw materials used by the crews; or it may be due to differences in equipment conditions that are beyond the control of the crew members. It is frequently the case that those working within the process do not have sufficient authority to carry out the necessary investigation or to address the removal of a special cause once its nature is understood. In this application, it may be that crews run to different averages for lack of policy and operating procedures that dictate the same setups, use of standard methods, and deployment of best current knowledge. No one line operator, crew, or supervisor has the authority to change or write policy and operating procedures. A person senior to all crews would have to decide on and implement consistent run policies and practices.

5.5.2   Availability of a Powdered Soap Product

The following case study is based on work with a company that manufactures consumer products. Senior management at this organization anticipate that their products will encounter increased competitive pressures over the next few years in terms of "customer value." This value will be interpreted by the customer as some combination of price, quality, availability, and consistency. Because these executives intend for their organization to remain a market leader by providing increased value to its customers, they have assumed responsibility for identifying critical systems and processes that must be improved in order to provide quality, consistency, and availability at selected prices.

One of the largest product lines in this organization is soap products. Consequently, it is decided to immediately focus process improvement efforts in this area. Investigation through marketing research in retail outlets and in the distribution chain reveals that product quality is consistent and meets current consumer expectation. However, it is discovered that the volume of product delivered in a container is often variable. This variability in container volumes is troublesome because it is known that consumers often make comparisons among containers regarding quantity. It is also found that the product is not always available in the quantities desired at retail outlets. Further investigation reveals that the distribution chain suffers from periodic stockouts of this and other products.

In light of this information on quantity variation and unreliable availability, attention is directed toward the organizational system that ties market forecasts into production needs, correlates purchasing with the production schedule, and relates manufacturing capability and run times with distribution and storage practices. An overview of the current operation of this sy stem is provided in Figure 5.10.

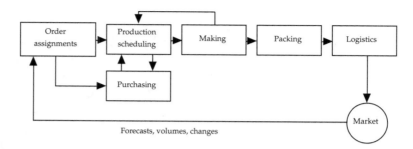

FIGURE 5.10   *Current Operation of Availability System for a Soap Manufacturer*

One of the performance measures developed for this system is the weekly deviation between forecasted demand and actual demand. Data on this measure have been collected for each of the major products of the business unit. As an example, weekly deviations for a powdered soap product have been recorded for the last six months of operation. The analysis of these data is performed in Chapter 7. However, the charts on the weekly deviations are in control; the estimated average weekly deviation is 43.16 cases and the estimated standard deviation is 31.66. These estimates imply that the range of weekly deviations to be delivered by this process can be predicted as follows:

$$NT = 6\hat{\sigma} \cong 190$$

With a process average of about 30, a natural tolerance of 190 means that in any week the forecast might underestimate demand by as much as 50 cases or overestimate it by as many as 140 cases. Similar numbers for other product lines reveal comparable levels of variation in weekly deviations. This large variation in weekly deviations is currently absorbed by the organization by maintaining large inventories of each product. Such a practice adds to the cost of the product. Also, as evidenced by the periodic stockouts at the distribution chain, the practice of maintaining large inventories is not entirely effective.

As a result of analyzing the given information, the senior group decides to change its approach and improve its product availability by manufacturing to demand. The ability to achieve this objective requires that purchasing, manufacturing, and packaging processes be managed in a consistent fashion with small levels of variation around the correct average. Data on the current performance of these processes regarding levels of variation and average are collected. Information on the packaging process is provided later. This information not only provides useful knowledge about the inconsistencies in product volume as reported by consumers but also addresses the issue of the consistency of levels of material usage.

The soap powder is packed by a machine that has four heads and typically runs across two shifts. There is a legal requirement that the fill of any container be no less than 123.5 ounces. In part, to verify that this requirement is being met, past practice has been for a quality control technician to select two containers from the line twice each shift and determine the net contents of each. At the end of each month, a histogram is developed in order to display the current operating ability of the process. The most recent histogram developed is shown in Figure 5.11.

The following pieces of information are contained in the histogram of Figure 5.11:

1. For the month summarized, all containers exceeded the minimum legal requirement of 123.5 ounces.
2. Some containers had as much as 131 ounces of soap.
3. The variation in fill weights for the data examined in *this* month was such that overpack was about, on average, 127.5 − 123.5 = 4.0 ounces.
4. The bimodal shape of the histogram suggests that there are different cause systems, maybe associated with the two different shifts, which are creating two distinct average fills.

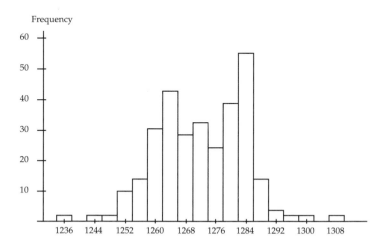

FIGURE 5.11 *Histogram of Weights of Soap Containers*

As discussed in Section 5.4.1, it is important to think about what information is *not* provided by the histogram in Figure 5.11. Because stability of the filling process has not yet been verified, there is no knowledge as to whether previous or future months will exhibit behavior similar to that captured by the histogram. Without knowledge that the fill process delivers consistent fill weights, the average overfill is not a credible number. It may or may not correctly reflect average overfill for the process. The histogram merely provides a picture of the data set that was used to construct it. It cannot be used to describe what is delivered day after day, month after month by the process.

Although it is apparent from the histogram that during the month studied there is considerable expense due to overpack, no information is provided concerning the reasons for the overpack. The "why," the causes of variation acting on the process, must be investigated. An understanding of these causes will be furthered by knowing how and when they impact the process. In other words, if the process provides consistent variation and averages day to day, then causes creating this large variation day after day are addressed. On the other hand, if the process is not stable day to day, then causes acting intermittently across days must be identified. Inconsistent within-day variation will lead to investigating one set of causes, whereas an inconsistent average from day to day will direct managers, engineers, and operators toward a different set of causal factors.

No knowledge is provided in the histogram about the procedures used to collect the data. The additional information that eight containers each day have been used (two taken at each of two times per shift) is not sufficient to address the issue of data reliability. The information provided by the histogram might be very different depending on the answers to the following questions:

- When were the containers selected? Were they selected when the process seemed to be running well or irrespective of current operating conditions?
- Are the same practices for selecting the containers used by personnel on the two different shifts? Is it possible that the practice on one or both shifts is to always select containers from the same filling head? If so, the data collected will not be representative of the entire machine.
- How well does the measurement method for determining container weight perform? Is it consistent from one day to the next and for both shifts?

Instead of providing a monthly histogram, many organizations provide monthly summaries of process behavior by reporting a monthly average and standard deviation or by reporting a monthly C_p or C_{pk} figure. Clearly, these monthly measures suffer from the same limitations as those described earlier for the histogram. These figures, C_p and C_{pk}, only provide information on the data set used to generate them. Without specific knowledge about where the numbers used to calculate the figures came from, they may not be descriptive of the entire month. Without an understanding of the stability of the process, they provide no basis for planning. But the most critical issue is that the ability to address the issue of overpack is not strengthened. It should be apparent from these deficiencies that attempting to lead process personnel to improve the capability of the filling process requires the manager to request and respond to more enlightening process information.

In this situation, management decided that additional process information was needed and that it must be reported in a different form. Since current fill applications are tested by randomly taking two subgroups of size two on each shift, these data are available for an assessment of process status regarding fill performance. The decision is made to use data collected by current procedures and to use time of selection as the basis for creating subgroups. The data set in Table 5.6 is a result of this request. This table contains recorded fill weights for the most recent eight days. Note that two subgroups are created for each shift. R and X-bar charts have been constructed from these data and are provided in Figure 5.12.

The R chart has several points above the upper control limit. (Since this chart is out of control, the X-bar chart has not been completed.) The points that fall above the upper control limit correspond to subgroups 4, 8, 24, and 28. All of these subgroups have measurements on containers selected from the second shift. Thus, a conclusion from these charts is that:

> *The numbers reported on the second shift display more variation than those reported by the first shift.*

This observation leads to questions concerning how the subgroups were formed. How were the containers to be measured selected? How reliable is the actual weighing process? Answers to these questions are necessary in order to interpret correctly the conclusion reached. For example, it cannot be concluded than containers filled on shift 2 have more variation that those filled on shift 1. It is possible that shift 1 always selects two consecutively filled containers, whereas shift 2 always chooses containers filled at least an hour apart. If this is the case, then the additional variation observed on the second shift would be due to changes in the filling process that occur from one hour to the next. These same changes might occur on the first shift; but

TABLE 5.6 *Fill Weight for Containers Selected Twice per Shift*

Day	Shift	Net Contents		$\bar{X}$	R
1	1	126.10	126.70	126.4	0.6
1	1	125.30	126.50	125.9	1.2
1	2	126.40	126.40	126.4	0.0
1	2	130.95	123.85	127.4	7.1
2	1	127.25	126.95	127.1	0.3
2	1	125.80	126.20	126.0	0.4
2	2	126.65	126.35	126.5	0.3
2	2	129.50	124.10	126.8	5.4
3	1	125.35	125.65	125.5	0.3
3	1	130.00	128.00	129.0	2.0
3	2	127.75	128.05	127.9	0.3
3	2	124.85	125.15	125.0	0.3
4	1	128.20	128.00	128.1	0.2
4	1	126.65	128.35	127.5	1.7
4	2	128.45	128.55	128.5	0.1
4	2	127.30	123.70	125.5	3.6
5	1	127.75	127.65	127.7	0.1
5	1	125.90	126.90	126.4	1.0
5	2	128.00	128.40	128.2	0.4
5	2	128.65	128.55	128.6	0.1
6	1	126.60	126.60	126.6	0.0
6	1	124.95	125.05	125.0	0.1
6	2	128.05	127.95	128.0	0.1
6	2	131.05	126.75	128.9	4.3
7	1	127.85	128.15	128.0	0.3
7	1	127.95	126.65	127.3	1.3
7	2	127.10	127.90	127.5	0.8
7	2	129.65	123.55	126.6	6.1
8	1	128.20	128.60	128.4	0.4
8	1	125.30	124.90	125.1	0.4
8	2	127.50	127.70	127.6	0.2
8	2	125.50	126.90	126.2	1.4
				4060.6	40.8

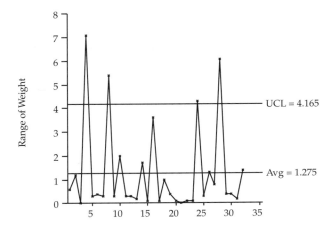

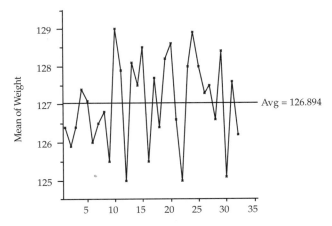

FIGURE 5.12 *Range and X-bar Charts of Soap Weights for Two Containers Randomly Selected Twice Each Shift*

because containers were selected consecutively, the opportunity to observe hour-to-hour changes in the first shift does not exist. Another possibility is that those who weigh the containers from the second shift are more erratic in their practices. Consider also these additional questions about the data:

- What is known about the consistency of the fill material across time? Is the material used by shift 1 consistent with respect to those properties, such as density, that affect fill volume? Is the material used by shift 2 consistent over time? How does the material used by shift 1 compare to that used by shift 2?

- The filling machine has four heads. Have both shifts selected containers so that all heads are represented by the data? It is possible that shift 1 only chooses containers from one head, while shift 2 includes weights from containers filled by all heads. This difference in sampling procedures may account for the observed differences between the two shifts.

In summary, the information provided to date on the filling machine has shown that fill weights have inconsistent within-shift variations when data are collected from both shifts. Because a poorly planned and poorly understood subgrouping method was used, there is little knowledge about why this variability exists. A follow-up study of the filling process is thus performed. Before identifying the subgrouping scheme to be used, the information desired from studying the filling process was identified. This desired information included

- determination that the measurement process provides reliable data on container weights,
- knowledge about the stability and magnitude of the within-head variation,
- comparison of within-head variations to determine whether or not within-head variability, container to container, is the same for all heads,
- knowledge about the stability and magnitude of within-head averages,
- comparison of variation among heads to see if the heads operate to the same average,
- information on how product changes in density, grain size, moisture, or other properties affect the variation in fill weights, and
- identification of how reduction in variation and/or an improved average will affect the plant's ability to achieve better material usage and improve consistency in volume.

Questions about the measurement process are not addressed in the present discussion. Methods for resolving these measurement issues are addressed in Chapter 9.

Based on the previous questions and to increase an understanding of the within- and between-head variations, further data are collected according to a subgrouping plan designed to answer these specific questions. The net contents of two consecutive containers from the same head are determined for each of the four heads. In recording the weights of these containers, head number and shift number are maintained. These data are reported in Table 5.7. Before examining the control charts for these data, it is useful to consider how this subgrouping plan will capture information on variation.

Since each subgroup contains weights for two consecutive containers, within-subgroup variation could be termed within-head variation. The magnitude of the ranges will be driven by those sources creating short-term variation (from one fill to the next) within a given head. It is thought that material properties, such as density, remain relatively constant during the time required to complete two cycles. On this basis, the deviation from container to container from the same head represents the head's ability to replicate itself, unconfounded by changes in material properties affecting weight. There are 10 subgroups for each of the four heads; the collection of the subgroups extends across both shifts and across five days. Consequently, an examination of the range chart will investigate the stability of the short-term, within-head variation across time and heads. The average for each subgroup serves as a measure of the level at which a particular head is filling at the time the two containers are selected. An examination of the X-bar chart will provide an understanding of whether inconsistencies exist in this average across time and heads.

Figure 5.13 contains the completed *R* chart. (Because this chart is out of control, no limits have been put on the X-bar chart.) The out-of-control points all correspond to the second shift. Additionally, they are all associated with head 4 of the filling machine. Attributing the out-of-control points to head 4 and to the second shift might easily have been overlooked in the charts of Figure 5.13 because of the way the data were plotted. Consequently, the ranges have been organized by shift and by head in Table 5.8. Figure 5.14 contains two *R* charts—one where the ranges are plotted by shift and a second one where they are plotted by head. In these two range charts the difference between the shifts and the difference between head 4 and the other three heads are more apparent.

At this stage of the investigation, follow-up work was done to determine why head 4 delivers fills with more short-term variation. Also, because concurrent work on the consistency of material being used by the two shifts revealed no marked differences in critical material properties, efforts were made to characterize the important differences in operating practices for the two shifts. The success of these endeavors can be investigated by referring to the application problem in Section 5.8 at the end of this chapter.

TABLE 5.7   *Fill Weights Subgrouped by Head and Shift*

Subgroup	Head	Shift	Net Contents		R	$\bar{X}$
1	1	1	126.72	126.12	0.6	126.42
2	2	1	125.42	125.82	0.4	125.62
3	3	1	126.43	126.53	0.1	126.48
4	4	1	124.38	126.78	2.4	125.58
5	1	2	126.88	125.88	1.0	126.38
6	2	2	125.99	125.59	0.4	125.79
7	3	2	126.17	127.07	0.9	126.62
8	4	2	128.14	122.74	5.4	125.44
9	1	1	126.28	126.58	0.3	126.43
10	2	1	124.75	125.45	0.7	125.10
11	3	1	126.27	126.57	0.3	126.42
12	4	1	125.89	125.79	0.1	125.84
13	1	2	126.02	126.72	0.7	126.37
14	2	2	126.49	124.79	1.7	125.64
15	3	2	126.28	126.68	0.4	126.48
16	4	2	127.33	123.73	3.6	125.53
17	1	1	126.29	126.39	0.1	126.34
18	2	1	125.71	126.01	0.3	125.86
19	3	1	126.64	126.24	0.4	126.44
20	4	1	125.85	125.85	0.0	125.85
21	1	2	126.34	126.44	0.1	126.39
22	2	2	125.22	125.32	0.1	125.27
23	3	2	126.62	126.42	0.2	126.52
24	4	2	123.23	127.53	4.3	125.38
25	1	1	126.43	126.73	0.3	126.58
26	2	1	125.92	125.52	0.4	125.72
27	3	1	126.01	126.81	0.8	126.41
28	4	1	124.84	126.84	2.0	125.84
29	1	2	126.00	127.20	1.2	126.60
30	2	2	126.12	125.72	0.4	125.92
31	3	2	126.95	126.45	0.5	126.70
32	4	2	126.24	124.84	1.4	125.54
33	1	1	126.06	127.06	1.0	126.56
34	2	1	125.62	125.82	0.2	125.72
35	3	1	126.95	126.35	0.6	126.65
36	4	1	125.42	125.82	0.4	125.62
37	1	2	124.77	127.77	3.0	126.27
38	2	2	125.50	125.00	0.5	125.25
39	3	2	125.81	126.61	0.8	126.21
40	4	2	124.11	127.01	2.9	125.56

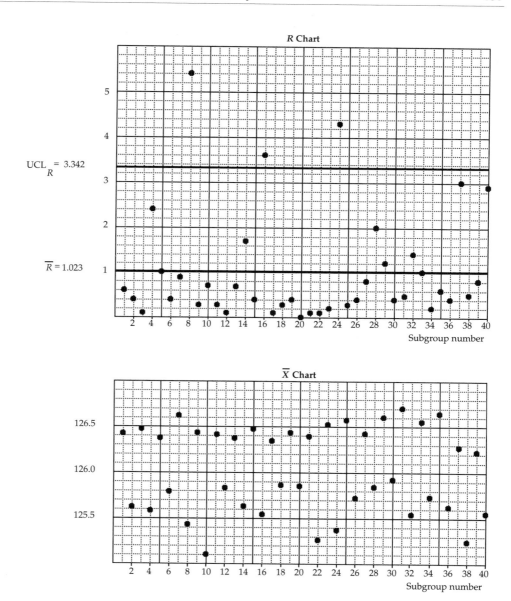

FIGURE 5.13 *Range Chart and Plotted X-bars for Fill Weight Data*

TABLE 5.8 *Subgroup Ranges Reorganized by Head and Shift*

	Shift 1	Shift 2	
Head 1	0.6	1.0	
	0.3	0.7	
	0.1	0.1	
	0.3	1.2	
	1.0	3.0	
			$\bar{R} = 0.83$
Head 2	0.4	0.4	
	0.7	1.7	
	0.3	0.1	
	0.4	0.4	
	0.2	0.5	
			$\bar{R} = 0.51$
Head 3	0.1	0.9	
	0.3	0.4	
	0.4	0.2	
	0.8	0.5	
	0.6	0.8	
			$\bar{R} = 0.50$
Head 4	2.4	5.4	
	0.1	3.6	
	0.0	4.3	
	2.0	1.4	
	0.4	2.9	
			$\bar{R} = 2.25$
	$\bar{R} = 0.57$	$\bar{R} = 1.475$	

Note: Values of range are recorded, and strict time ordering has been destroyed.

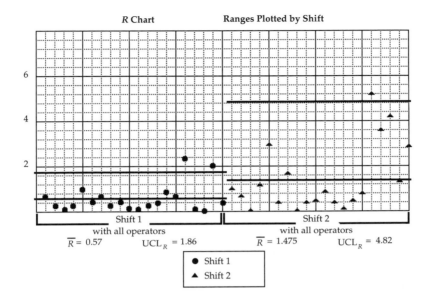

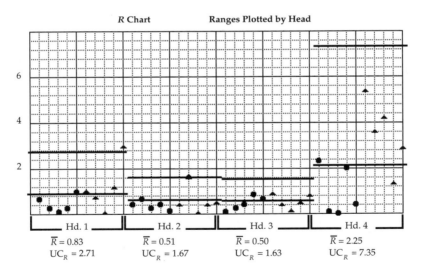

FIGURE 5.14 *Ranges Plotted by Shift and by Head*

5.6 SUPPORTING PRINCIPLES FOR THE EFFECTIVE USE OF *R* AND *X*-BAR CHARTS (OR *s* AND *X*-BAR CHARTS)

The effective use of the information provided by control charts for variables data is inescapably linked to the purposes for which these charts are used. The various purposes served by charting data are summarized (see box) by linking these purposes to the levels of process work that occur within an organization:

Level 4

Evaluation of the effectiveness of management practices

Level 3

Development of management and engineering knowledge for process redesign or improvement

Level 2

Establishment of process control

Level 1

Maintaining current levels of process performance.

Summary of Information Provided by Control Charts

1. The charts serve as a signaling system that can indicate the presence of special causes. The removal of special causes is necessary in order to maintain process control.

2. The charts are used to evaluate the effect of deliberate process changes. The plotted data provide information for comparing process outcomes before and after a process change. Charts provide evidence about the magnitude, direction, and stability of the effects of the process change.

3. After a process change is in place and its effect is verified, the charts provide ongoing confirmation that the change is maintained. Hence, the charts support holding or maintaining a gain achieved from a previous process change.

4. The charts provide a data representation by which operators, engineers, and managers can begin to discover, evaluate, and know the effect of various sources of variation on the process output. By understanding the sources of variation and their impact on the quality variable being studied, the sources of variation can be attacked in order to reduce or eliminate their effects on variation.

The information gained from control charts will seldom fall neatly into one of the categories given. When used as a means for confirming that a process change has been effective, these charts will also serve over time as a means for ensuring that the change is maintained. Then, these charts will be used as a means of maintaining control over the "new" process. For the knowledgeable manager, engineer, and operator, the charts will serve several purposes simultaneously. Therefore, these various purposes and their relationship to the levels of work within an organization deserve further discussion.

5.6.1 Level 1: Maintaining Current Levels of Process Performance

Obviously, one purpose of control charts is to assist an operator, engineer, or manager in maintaining effective control over a process. In this context, *control* is defined as the identification and removal of special causes. Processes, of course, do not naturally remain in control. They tend toward instability and chaos. Maintenance of stability is therefore an important responsibility for all managers and operators. Data collected and plotted in an ongoing manner support process control. The plotted points show the current process behavior and possible future behavior. They reveal the presence and timing of special causes according to the particular sampling or subgrouping strategy. The removal of these special causes is necessary in order to gain effective control over the process. Hence, when charts are used for control purposes, baseline data are collected, out-of-control conditions are identified, and work is begun with the purpose of preventing these special causes from occurring in the future.

The management and operation of a process so that process outcomes are predictable provides powerful advantages. Predictability enables accountable managers to make reasonable, objective evaluations of the effectiveness and efficiency of process performance in accomplishing the work of the organization. The effects of work methods, product and process designs, equipment operation and maintenance, work flow and supporting services can be evaluated with more certainty when processes display verified predictability. The effect of specific changes on any of the underlying process mechanisms can be judged and characterized for predictable processes. When processes are unstable (i.e., unpredictable), results regarding short-term variability and/or averages are erratic. When results cannot be anticipated, it is difficult to know and verify the effects of a process change. The abnormal behavior hides the intended results from process changes. Thus, it is strongly recommended that processes be continually evaluated over time. Small subgroups of measurements, collected and evaluated over time, combined with appropriate statistical analysis offer more power and insight regarding predictability than do large, infrequently collected samples. A data point by itself is difficult to interpret and understand; it is awkward to associate a single data point with particular process behavior or change. It is essential that the manager develop the idea of process history; process data should be examined in the light of past data. Ongoing processes should be examined by ongoing data collection.

> **Principle:** *Good practice requires that the root causes of out-of-control conditions be identified, verified, and eliminated.*

In practice, confusion often arises regarding the use of X-bar and R charts for control purposes. Engineers or operators may see a process go out of control, make an offsetting change in one or more process parameters, and then mistakenly consider that "control" has been regained. This is, in fact, a tactical and logical error both in process control and in improvement work. As an example, consider the production of alternator rectifiers. Data are collected on the inner diameter of pipe rivets in order to maintain good control. Five readings are taken during each shift and the range and average of these five readings are calculated. Given a data point above the upper control limit on the X-bar chart, unrelated to tool-wear or other well-understood causes, the operator simply adjusts the machine gauge to a lower setting in order to bring the process back into control without knowledge of the actual root cause. There is no assurance that the special cause will not reoccur in the future. Constant practices of this type build a culture that works against process control and improvement. Obviously, it may be necessary to make such offsetting changes, but it is more important for the accountable management, engineering, and operator groups to look for, verify, and eliminate the root causes of those out-of-control conditions.

> **Principle:** *When used for maintaining process control, control charts must be accompanied by knowledgeable selection of characteristics to chart, methods for responding to chart information, and means for adjusting charting techniques to take advantage of new knowledge.*

For purposes of establishing and maintaining process control, the dominant issue for management, engineering, and operations is "Are the essential causal factors known and are they managed correctly and consistently?" A clear understanding of the common cause system provides the base from which process maintenance effectively occurs. Too often operators are instructed to chart outcome measures of their various tasks with the mistaken belief that effective process management will then occur. Process management using only results data without knowledge of the common cause system provides for only reactive behavior and ends up impeding process improvement efforts. Management has the responsibility to put in place correct process management by ensuring that

- correct data are being taken,
- measurements are reliable, whether obtained by automatic equipment or a manual procedure,
- process personnel understand desirable behavior for taking process measurements and know what deviations from desired behavior to watch for,
- personnel have the capability to collect, report, and evaluate information on process performance, and
- appropriate supporting mechanisms are in place.

> **Principle:** *Control charts do* not *typically provide an adequate feedback control mechanism.*

With the previous principles in place, it is evident, as in the previous alternator rectifier example, why charts used simply as an engineering "feedback" mechanism are ineffective in preventing the reoccurrence of special causes. "Charting" is often used simply to represent the current process condition, that is, to identify when the process is out of control or is tending toward that condition. Results in this mode are usually disappointing. In general, the subgrouping methodology that provides the data for these charts is not useful as part of an engineering feedback control system. The limits, for example, say nothing about which process variables should be manipulated in order to maintain a desired level. Other required knowledge may also be missing. The statistical limits previously defined are not related to the time lag or other process dynamics, all of which are necessary in order to construct and operate a successful engineering feedback mechanism. Statistical control does not mean "keep it between the limits." Statistical control, again, means identifying special causes, often by changing subgrouping and measurement strategies, learning the root causes of those assignable events, and then removing or permanently nullifying the effects of those causes.

Many industries rely heavily on the use of automatic feedback mechanisms as a method for controlling process behavior. (Note that the word "controlling" in the previous sentence does not refer to statistical control but to an engineering mechanism used in process operations.) Most continuous and many batch-type processes rely on the use of such feedback mechanisms. Managers and engineers responsible for these kinds of processes would, justifiably, find the idea of using a statistical control chart as a feedback mechanism, instead of a computer-aided automatic feedback mechanism, a ridiculous suggestion. However, control charts are still useful in this environment because they can provide an analysis of the effectiveness of the continuous or batch process, which includes the engineering feedback mechanism. An example of this use of statistical control charts is provided in Chapter 7.

5.6.2   Level 2: Establishment of Process Control

Statistical process control of a manufacturing process is often taken to mean that key quality characteristics can be maintained at consistent levels with only random fluctuations. A broader definition of statistical process control is used in this book; this definition requires that we know the following: critical sources of variation that affect critical process characteristics; the way in which these causes act, whether continually or sporadically; and the effects that these causes have on process characteristics, which also need to be understood and managed. This definition requires knowledge of how variations in incoming material and changes in process parameters affect the average and variability of the output characteristics. Thus, control charts must be used as tools to identify sources of variation, rather than merely as feedback mechanisms for process monitoring.

> *Principle: Establishing process control requires verified knowledge of the effect that process inputs and variables have on critical process characteristics.*

Correct practice of process management requires an ongoing study of the process. Managers, with the assistance of engineers and operators, should first construct verified process flow diagrams indicating how, when, and where work is done. Cause and effect diagrams that contain verified information regarding relationships of critical input variables and the outcome of interest are constructed. The items listed on the cause and effect diagram call for specific work to describe the nature, type, and degree of the relationship between each item and the required effect. Effort is indicated and required when information regarding relationships or effects is missing or when suspicions about the actual existence of these relationships are unconfirmed. The cause and effect diagram becomes a means for compiling tested and verified causal structure, for indicating where work efforts must be focused, and for serving as a platform for generating ideas for improvement.

> *Principle: Standard operating practices must be developed, verified, and implemented.*

Charting work should confirm operating variations and levels of key input variables. Work methods are examined, tested, verified, standardized, and practiced in a consistent manner. Maintenance issues are addressed, corrected, and improved. It is reasonable to think that much of this information is documented during product or process introduction and released to manufacturing. However, in many businesses, this type of information is often not available. Lead managers must develop this information before process improvement efforts can be successful.

> *Principle:* The effective use of control charts for process study must be management led and directed.

Management has the responsibility to practice appropriate behavior in order to get processes in control and correctly manage these processes. Line supervisors, operators, and process engineers must know how to measure, sample, evaluate, and act and must know that these duties are expected of them. Verified process knowledge must be in place, deployed, and consistently practiced. For example, knowledge must be in place regarding past and current behavior of key input variables and work practices. The effects of variations in these variables or methods will have to be known in order to know which ones are critical for close tolerances and standard practices. Practical operating knowledge must be in place regarding potential or actual out-of-control conditions. Changes or adjustments in appropriate inputs or practices must be tested and verified. Management, not operators, is responsible for ensuring that such knowledge and practices are in place.

5.6.3   Level 3: Development of Management and Engineering Knowledge for Process Redesign or Improvement

Establishing process control is an inadequate process management goal. As previously emphasized, it is important to be able to predict process outcomes; however, improved efficiency and effectiveness result from moving the current process performance to improved levels of operation. The knowledge of statistical control is only a stepping stone to process improvement. The insights into process behavior gained from knowing the causal factors and how they impact process behavior lead to a consideration of how these causes should be managed differently. Observation, conjecture, and correlation of process parameters is essential for building process knowledge. (Further discussion of experimental techniques to verify causal relationships is contained in Chapter 10.)

> *Principle:* The relationship among process inputs, variables, and outputs will need to be understood and exploited to identify process improvements.

In evaluating current process operation, it is necessary to plot appropriate outcome measures in order to know process capability and to gather baseline information about current operations. Also, data on process inputs and process variables must be obtained in order to ensure consistent process operation. Knowledge of the relationship of these input and process variables with outcome measures is required. It may be that all process inputs and variables are maintained at recommended levels and that known important methods are practiced according to recognized standards. This information does not by itself provide information by which outcomes may be observed and judged. It is not possible to know the effects of process variables on outcome variables without understanding this relationship.

> ***Principle:*** *Control charts should be used to evaluate the effectiveness of process changes as well as to verify that observed process improvements are maintained.*

As previously stated, control charts serve as powerful indicators of the effectiveness of process change. However, process change must be supported by successful process control, because the ability to judge the sustained effects of a process change will be uncertain in the presence of an unpredictable process. Baseline data may indicate that special causes must be removed from the process before the effects of a process change can be studied. Once the process is in good control according to a defined sampling strategy, process changes may be recommended to affect sources of common cause variations. The intent of these changes may be to reduce short-term variation, to move the average to a more favorable value, or to remove long-term shifts in the average. The specific nature of the intended effect should be defined. Process changes, of course, are rooted in machine changes or revised machine parameter tolerances; changes in materials, material characteristics, or specifications; or changes in work methods or protocols.

Established charts on range and average will characterize the stable process. The previously established baseline data are then used for judging the effect of process changes. The range and average charts contain data that represent the process as it is prior to a specific process change. After the process change is made, data are collected and plotted with the same measurement and subgrouping strategy. The "new" data are plotted directly onto the established range and average charts. The effect of changes can be evaluated by comparing the new data against the established process data. The data plots following the change reveal the effect of the process change. Several outcomes are possible; the charts may reveal that the process change has had these effects:

1. *No effect on the process:* Evidence for this conclusion is provided by data points plotted on the baseline range or average charts that behave in a similar fashion to the data plotted prior to the process change. The new data will be consistent with the old and the charts will not reveal signs of a process change.

2. *The desired effect on either the short-term variation or on the average:* A decrease in short-term (within-subgroup) variation is revealed by data on subgroup ranges that plot at a lower average value than the average range on the R chart prior to the process change. The new data would suggest that the average value for the range has been decreased. A shift in the process average or mean value to a new level is revealed by plotting the subgroup averages for the new data on the baseline X-bar chart. A process that has moved to a new level will yield X-bars that have out-of-control signals when plotted on the previously established average chart. If the outcome of the process change has been determined to be beneficial, then it is necessary to ensure that the change is maintained. A successful process change is evidenced by a process that yields data that plot in control but centered about a new average value. After a process change is made, if the data plot consistently over time at a level different from the previous process level period of time, then the change has, in essence, resulted in a new process. Thus, new control limits and centerline can be computed.

3. *A deteriorating effect on the process:* It may be that the process change has not resulted in a process that is stable or predictable. In this case, it may be that an improved process would result if the new process can be brought into control by removing the "new" special causes that have appeared in connection with the changed process. On the other hand, the charts may reveal that the short-term variation has become larger or the average has shifted to an undesirable level.

After the effects of a process change have been determined, the responsibility of management is still that of ongoing process study. It is often the case that once a process change is made, and positive effects from the change are verified, then charts are used strictly to monitor outcome variables from that process and attention is diverted to other areas and processes. However, ongoing study of the process and its key input variables is necessary in order to hold the effects of the improvement and to make additional changes that have positive effects on process outcomes. It is only through continued process study that additional sources of variation can be identified and evaluated.

5.6.4   **Level 4: Evaluation of the Effectiveness
of Management Practices**

Charts on process outcomes provide information on process stability, magnitudes of variation, ability to maintain an average, and means for evaluating process capability. Consequently, control charts not only provide
information about current process performance, they also report on the
adequacy of the process architecture that management has put in place.
For example, the charts in Figure 5.12 revealed inconsistent short-term
variation in filled soap containers. At one level of thought, these charts
indicated needed work to correct this inconsistency. But at another level of
thought, these charts indicate that management has failed to put in place
consistent operating procedures, methods, or evaluation practices to
ensure consistent process operation.

> **Principle:** *Control charts should be used as a means for indicating needed
> improvements in management practices.*

The statistical techniques for evaluating process performance relative to variability, averages, and their consistency over time must be used by management to assess

- how effectively they have integrated raw material and component parts,
- the correctness of specifications and vendor selection,
- equipment selection and its operational effectiveness, efficiency, and
  maintenance,
- definition and selection of methods and practices, and
- selection of people and the adequacy of their training and performance.

If control charts are only viewed as a means of reporting on process operation and not on the management practices that established the operation,
then the opportunity to improve the establishment of the methods, materials,
practices, equipment, etc., will likely go unnoticed.

5.6.5   **Summary of Principles**

The principles discussed in the previous section are as important as the ability
to construct the appropriate control charts in learning how to effectively
manage processes. Therefore, these principles are restated here (see box).

Principles of the Effective Use of Control Charts for Variables Data

Principle 1

Good practice requires that the root causes of out-of-control conditions be identified, verified, and eliminated.

Principle 2

When used for maintaining process control, control charts must be accompanied by knowledgeable selection of characteristics to chart, methods for responding to chart information, and means for adjusting charting techniques to take advantage of new knowledge.

Principle 3

Control charts do *not* typically provide an adequate feedback control mechanism.

Principle 4

Establishing process control requires verified knowledge of the effect that process inputs and variables have on critical process characteristics.

Principle 5

Standard operating practices must be developed, verified, and implemented.

Principle 6

The effective use of control charts for process study must be management led and directed.

Principle 7

The relationship among process inputs, variables, and outputs will need to be understood to identify process improvements.

Principle 8

Control charts should be used to evaluate the effectiveness of process changes as well as to verify that observed process improvements are maintained.

Principle 9

Control charts should be used as a means for indicating needed improvements in management practices.

5.7   PRACTICE PROBLEMS

5.7.1 In the bearing manufacturing division of a company, one of the grinders, Machine 2325, is believed to be unstable with respect to the bearing sizes produced. To learn about the variability of the process, five successive measurements of Machine 2325 bearing diameters are taken twice daily for a week, as shown in the following table:

| | Measurements | | | | | | |
Day	1	2	3	4	5	$\bar{X}$	Range
1	2.763	2.583	2.401	2.614	2.469	2.566	.362
2	2.532	2.837	2.878	2.497	2.659	2.681	.381
3	2.538	3.114	3.122	2.893	2.836	2.901	.584
4	3.033	3.226	2.886	2.846	3.344	3.067	.498
5	2.909	2.819	2.808	2.637	2.989	2.832	.352
6	2.714	2.654	2.512	2.557	2.936	2.675	.424
7	2.829	2.697	2.363	2.754	2.437	2.616	.466
8	2.583	2.503	2.518	2.609	2.431	2.529	.178
9	2.582	2.918	2.809	2.734	2.798	2.768	.336
10	2.861	3.014	3.089	2.718	2.922	2.921	.371
11	3.293	3.147	2.859	2.962	3.177	3.088	.434
12	2.892	3.006	2.973	2.811	3.197	2.976	.386
13	2.542	2.780	2.573	2.767	2.455	2.623	.325
14	2.558	2.836	2.427	2.455	2.462	2.548	.409
						2.7707	.3933

a. The ranges and *X*-bars have been plotted in Figure 5.15. Calculate control limits for the range chart and, if appropriate, the *X*-bar chart.
b. Based on the subgrouping plan used, summarize the information available from these charts.

5.7.2 Capacity measurements for a certain type of pump are made on 10 successive pumps selected from production at three randomly selected times each day. Values of *s* and *X*-bar are computed for each subgroup of 10. After 27 subgroups, $\bar{\bar{X}}$ and $\bar{s} = 0.4582$.

a. What are the control limits for the *s* chart?
b. What are the control limits for the *X*-bar chart, assuming there is no evidence that the process is unstable with respect to variability?
c. Suppose that the blueprint specs for pump capacity are 8.5–11.5. Assuming that the process is stable, is it capable of meeting specs on pump capacity?

d. Assuming that pump capacities have an approximately normal distribution, what percentage of pumps is being produced with capacities that fail to meet the blueprint spec?
e. What are some relevant issues for the correct interpretation of these results?

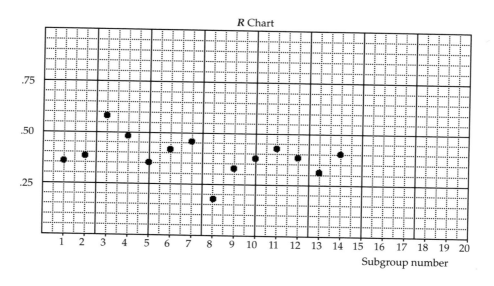

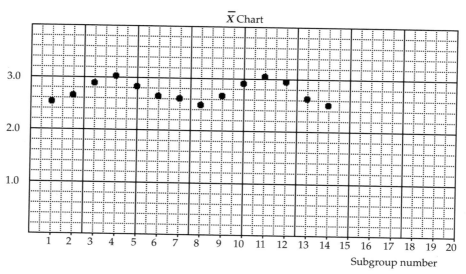

FIGURE 5.15 *Range and X-bar Charts for Problem 5.7.1*

5.7.3 Freeness measurements of kraft, a material created continuously as part of the process for paper production, are obtained six times an hour. Data for the last 25 hours of production have been collected. Ranges and averages for each of the 25 subgroups have been calculated and their values plotted on the control charts in Figure 5.16. The sum of the 25 ranges is 157.15 and the sum of the 25 X-bars is 487.03.

a. Complete the range chart in Figure 5.16 and, if appropriate, the X-bar chart.
b. If appropriate, estimate the process standard deviation.
c. If appropriate, estimate the process average.
d. Freeness measurements should fall between 18.0 ± 3.0. Is the process capable of meeting specifications? What percentage of measurements falls outside of the specifications?

5.8   APPLICATION PROBLEM

Section 5.5.2 contained a description of a filling machine used to fill a dry soap product. Investigation of the behavior of the four heads of the machine revealed that head 4 was more variable than the others. In addition, containers filled on shift 2 also showed more variation. After discovering this information, head 4 was found to have a specific malfunction and was overhauled. Investigation of the different levels of variation between the two shifts revealed that shift 2 personnel followed a different procedure in routing bins of material to the filling machine and more short-term variation in density was observed.

Maintenance completed work on head 4 and mixing procedures were reviewed on both shifts and standardized. Additional data are reported in Table 5.9, and the new data are plotted in Figure 5.17.

a. Is the process in control?
b. Does it appear that the process has changed as a result of the work on head 4 and on standardizing procedures? If so, in what respects?
c. Prepare any additional plots deemed necessary to complete the analysis of the current behavior of the filling machine.
d. What recommendations should be made for continuing operation of the filling machine?

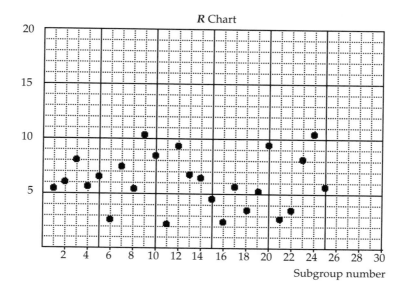

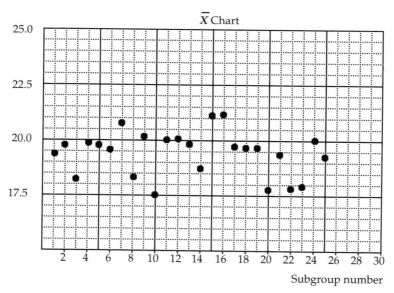

FIGURE 5.16   *Hourly Subgroups of Freeness Measurements*

TABLE 5.9 *Means and Ranges of Subgroups of Two Consecutive Containers from the Same Head*

Shift	Head	$\bar{X}$	R	Shift	Head	$\bar{X}$	R
1	1	126.4	0.5	2	1	126.2	0.4
1	2	125.4	0.2	2	2	125.6	0.5
1	3	126.8	0.0	2	3	126.3	0.7
1	4	125.7	0.2	2	4	125.4	0.2
2	1	126.5	0.1	1	1	126.5	0.0
2	2	125.8	0.3	1	2	125.5	0.4
2	3	126.7	0.1	1	3	126.4	0.7
2	4	125.4	0.4	1	4	125.5	0.2
1	1	126.4	0.3	2	1	126.6	0.0
1	2	125.7	0.3	2	2	125.4	0.6
1	3	126.8	0.7	2	3	126.8	0.1
1	4	125.7	0.8	2	4	125.2	0.6
2	1	126.4	0.0	1	1	126.1	0.2
2	2	125.8	0.4	1	2	125.8	0.5
2	3	126.4	0.2	1	3	126.5	0.8
2	4	125.5	0.3	1	4	126.0	0.3
1	1	126.8	0.4	2	1	126.4	0.1
1	2	125.2	0.2	2	2	125.7	0.7
1	3	126.4	0.6	2	3	126.4	0.6
1	4	125.8	0.2	2	4	125.6	0.4

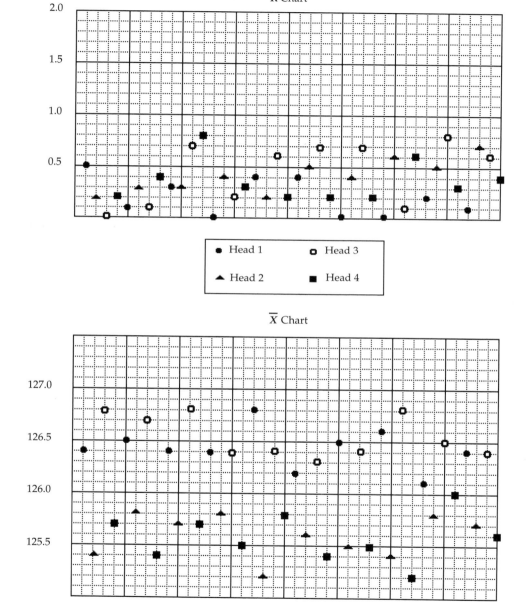

FIGURE 5.17 *Means and Ranges of Subgroups of Two Consecutive Containers from the Same Head*

6   Sampling and Subgrouping Principles

The purpose of this chapter is to provide further insights into collecting and converting process data into information that can be used to identify sources of process variation and to improve process operation. In the previous chapters, methods, concepts, and issues associated with the ongoing study of processes were introduced and discussed. Each of these ideas has been placed in the context of process investigation and improvement efforts that are guided by business strategy. The need for process control, as introduced in Chapter 3, follows from the fact that every process (and, therefore, process outputs) is impacted by sources of variation. Process control speaks to the need to be able to operate a process in the presence of these sources of variation in order to provide reliable process outputs of correct quality and quantity. To gain and maintain control of a process, process investigation is needed to

- understand the current status and capabilities of a process,
- understand how the stages of a process are interrelated, and
- identify the critical causes that influence the process outputs.

Without knowledge of process capabilities and critical causal factors, process improvement is almost impossible to gain and maintain over time. Therefore, a critical part of process improvement activities is the use of data to facilitate the identification of causes of variation acting on processes.

Many examples in the previous chapters emphasize the importance of planning the data collection activity in order to gain information on critical causal factors from the collected data. Data collection methods and data arrangement influence the amount and type of information that can be gained from charted data. The collection of a set of measurements in a given period of time is a *sample*. The way in which the samples of data are arranged, referred to as the *subgrouping structure*, is based on the purpose for data collection and the needed information. These three elements of process study, sampling, subgrouping, and purpose, are strongly interrelated. Sampling plans and subgroup formation should have an informed, meaningful, and articulated purpose. The subgrouping strategy will therefore need to be dynamic, changing according to the information needed during ongoing process investigation.

6.1   SAMPLING, SUBGROUPING, AND ELEMENTS OF PURPOSE

A standard need in all process investigations is the determination of the magnitude and behavior of both variability and level for critical process parameters. With this knowledge in place, the process can be assessed against requirements, resource use, and the need for process change. Sampling (or guiding data gathering) and subgrouping (or prescribing data arrangement) are particularly important in identifying and confirming effects and sources of variation. Both sampling and subgrouping should reflect current process understanding, the intention for further process study, and an understanding of questions about the process to be addressed. If process questions are vague or uncertain, then so will be the information in the data display. Some examples of benign, useless, or even detrimental sampling and subgrouping plans are provided later.

The following series of examples considers alternative sampling and subgrouping plans for a molding operation that supplies plastic switches to industrial customers for use in assembled control pads. The molding operation has eight machines; each machine has two molds; and each mold has 16 cavities. Process results over time indicate the following:

- Yields are erratic, producing an unpredictable fraction of acceptable parts.
- Parts are shipped that are out of specification.
- Shipment schedules are often incomplete.

Customers are seldom completely satisfied, sporadically returning parts for credit and occasionally complaining of assembly difficulties. Depending on intention, interest, and knowledge of the responsible person, operational directives might be stated in terms of increased throughput, using increased volume to attain sufficient quantity, despite inconsistent yields. However, another recommended approach would be to address yields, volume, schedules, and required quality by emphasizing process operation, articulated as follows:

1. How is the process currently operated? What are the operating procedures? Are they used? Can they be used? Always? What is in agreement with best knowledge and practice? What is questionable?
2. What is known about sources of variation and their effect on quality and schedule? Which of these effects have been verified? Which remain to be tested?
3. What are the actual results of this applied knowledge and current practice? What is the dimensional average? Is it sufficiently close to print nominal? How large are the part-to-part deviations? Is the operation capable of making parts to print? Can the operation reliably replicate itself over time?

Many sampling plans and subgroup formations will not provide useful information for answering these questions. For example, selecting one sample of molded parts from a certain time period will only reflect process conditions at that time and will not provide information on statistical control. This deficiency cannot be overcome with a larger sample. More parts taken from all molds and all cavities over, say, a two-hour period, even though it may be a more "representative" sample, remains fatally inadequate. One sample, no matter how large, does not permit process evaluation across time and conditions, the only realistic test for process stability.

Selecting many small samples across a significant time period does not assure useful data. Sample composition is also a determinant of information quality. Numerous smaller samples, each containing parts from multiple molds and cavities are not likely to be helpful in assessing process behavior. Variation in the results may be due to different machines, different molds, or different cavities—any of which may behave differently at different times and likely have different causal structures. When samples include parts from different molds, cavities, and several machine cycles, subgroups formed from these data contain these variations within subgroups, confounding possible deviations among molds and cavities with potential differences from cycle to cycle.

At least two detrimental consequences result from a subgrouping strategy that confounds sources of variation. One consequence is that with large and disparate effects occurring within subgroups, statistical evaluation through control charts is not likely to detect differences among these effects. However, the second consequence is more damaging—the amount of work required to learn about causes and to gain insight for correcting process effects is increased. The information and clues one hopes to gain from control charting subgroups over time are distorted and subject to misinterpretation. If statistical control charts signaled inconsistency in the range or average, the broad mix of possible causes acting within subgroups imposes a burden on evaluation, understanding, and correction. Different machines, molds, or cavities may contribute consistently or erratically to differences in ranges or averages; a similar statement could be made about the differing results from one cycle to the next. Inexact and ill-understood sampling and subgrouping detract from the effective interpretation and use of the process data. Well-drawn, regularly reviewed charts cannot overcome these deficiencies.

The previously listed process questions addressed the process means of attaining volume, schedules, and required quality for the molding operation; these questions should provide guidance on measurement, sampling, and subgrouping. In this application, a useful sampling plan and subgroup structure would be one that would allow us to assess specific, short-term variations by product source, by machine, by mold, and by cavity, as well as those short-term variations associated with different cycles. In addition, the sample plan and subgroup structure should permit statistical assessment of the average by the same recognized sources of variation. The data plots should reveal significant process differences, ideally indicating where and when they occur, and should provide data for supporting the correlation of causes with results for deviations among parts, changes in average, and stability characteristics. Useful information results from this intersection of purpose with sampling, subgrouping, and statistical and process chart capabilities.

6.2   CHART CHARACTERISTICS DIRECT SAMPLING AND SUBGROUP FORMATION

Sampling and subgrouping methods can limit the effectiveness of chart use for assessing process behavior and discovering process improvements. An understanding of the rationale used to construct control charts is therefore essential for developing sampling plans, designing subgroup structures, and constructing useful data displays. Because assessment of a measurable process variable focuses on variation, averages, stability characteristics, and the distribution of measurements, the basic tools for studying such a variable are variation and average charts. Note that the specification of sampling and subgrouping plans demands a sound understanding of the relationship between the variation and the average chart. The practical utility of these two charts will be directly proportional to the sampling plan and subgroup structure, whose development is determined by focused process questions.

An operational understanding of the relationship between $\bar{R}$ and $\bar{X}$ charts is clarified by examining control limit formulas for subgroup averages. Upper and lower control limits for the $\bar{X}$ chart are given by:

$$\text{UCL} = \bar{\bar{X}} + A_2\bar{R} \quad \text{and} \quad \text{LCL} = \bar{\bar{X}} - A_2\bar{R}$$

These formulas describe the functional role of the average range in determining control limits for $\bar{X}$. It is these limits that determine the $\bar{X}$ chart's capability for detecting differences in process behavior over time.

In a mechanical sense, the capability of the $\bar{X}$ chart to detect differences is governed by $\bar{R}$; by the subgroup size, n; and by the control chart constant, A_2. A process perspective on the ability of the $\bar{X}$ chart to detect differences considers the causes that influence the magnitude of $\bar{R}$. It is this process perspective on the role of $\bar{R}$ that provides the needed insights for developing sampling and subgrouping principles. The following section enlarges on the mechanics of chart construction and the process perspective that defines how these mechanics provide process information.

6.2.1   Defining Baseline Variation

The basic mechanics of chart construction dictate that the smaller the average range, the tighter the control limits on the average chart. The tighter the $\bar{X}$ chart control limits are, the more sensitive the chart will be for detecting differences of a given size acting between subgroups. Conversely, the larger the average range, the more widespread the $\bar{X}$ limits and, of course, wider limits will make the chart insensitive to variations of a given magnitude. The process implication of these statements about control limit width is that if effects of certain causes are consistently contained within subgroups and these causes consistently act in about the same way, then those effects will not result in detectable differences on the $\bar{X}$ chart. This property of $\bar{X}$ charts is a reflection of the statement "the average range defines baseline variation." Common cause variation is defined to be that variation present in every data point entering the sample, subgroup, and chart plot. Special causes are those that create variation which incrementally "adds to" the common cause variations. As a practical matter, the special cause effects need not be larger than the average range to be detectable; they need only add to common cause variation by a sufficient magnitude to be detectable.

Evidence of special causes may occur on the range chart or on the average chart. The illustrations in Figures 6.1 and 6.2 provide pictorial illustrations of subgrouped data that would provide evidence of (1) special causes on range charts and (2) special causes on average charts. Individual measurements from subgroups of size four are plotted in Figure 6.1. The spread of the four data points in the vertical direction on the plot of individual measurements represents within-subgroup variation. The first six subgroups in Figure 6.1 exhibit typical within-subgroup variations for a controlled process. The individual data plots for subgroups 7 and 9 show the effect of special cause(s) acting *within* subgroups. The information on the range chart is that some, but not all, of the material or parts within a subgroup are affected by some cause acting to increase dispersion within a subgroup. The ability to determine what the cause or causes might be rests on the specificity of the sampling and subgrouping plan. Knowing what potential sources of variation may be active within subgroups aids in speculating which of these may behave erratically or what other sources might be present at some times but not others during the time in which a sample was collected.

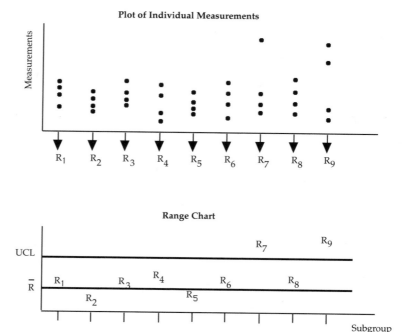

FIGURE 6.1 *Evidence of Special Causes on Range Chart*

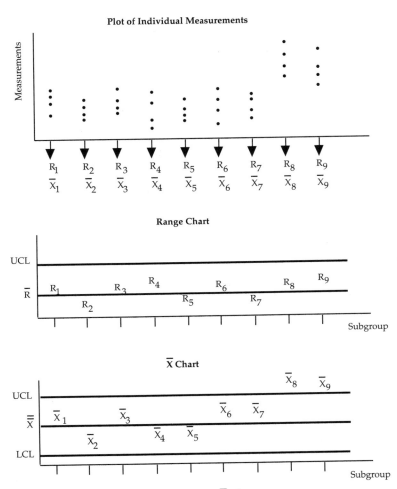

FIGURE 6.2 *Evidence of Special Causes on* $\overline{X}$ *chart*

Figure 6.2 is a graphical portrayal of between-subgroup events. The first seven subgroups plotted represent typical subgroups from a controlled process; these subgroups provide a baseline image of both variation and average. Subgroups 8 and 9 reveal the effects of a special cause occurring *between* subgroups. All material or parts within a subgroup are affected by process sources of variation in about the same way. The information on the accompanying range chart is consistent with that conclusion. The cause or causes that make the average change significantly leave the subgroup range unaffected. The average chart signals that either causes captured within subgroups are behaving differently over time *or* events are occurring that are not typical of those found within subgroups. Process results shown on the plot of individual values indicate that one or more causes has acted on all parts, leaving the within-subgroup variability consistent with previous experience, and increasing the subgroup averages by a magnitude larger than that pre-judged by $\bar{R}$.

The cause or causes of the effect evidenced by an increased average may be one that was not previously experienced in the process, or it may be one whose effects were captured within subgroups but which is acting differently to create a different effect. Simultaneous effects may happen in rather subtle ways; for example, the interaction between two variables may result in effects not recognized previously. There are, of course, numerous other possibilities. But a similar statement concerning the identification of these causes can be made here as was made concerning the range chart. Knowing what potential sources of variation may be active *between* subgroups aids in speculating about which of these may behave erratically or may have caused a possibly sustained change in the average. As with the range chart, the ability to determine what these causes might be rests on the specificity of the sampling and subgrouping plan.

6.2.2   The Effect of Subgrouping Strategies on Chart Information and Interpretation

In addition to providing evidence of erratic behavior in process causes of variation, range and average charts can meet other needs for process information. For example, an improved process may require the need to know how to run the process with fewer shifts in the long-term average, or run to a different average, or run with less variation. Control charts such as those in Figures 6.1 and 6.2 can be used to verify the effects of deliberate changes made to the process in an attempt to achieve the defined improvement. Evidence that intended process changes had proceeded as planned would be revealed on the appropriate chart, range, or average. As another example, a causal effect could be deliberately left out of within-subgroup causes; its impact on the process would then be observed by between-subgroup activity. The graphs in Figures 6.3 and 6.4 provide a simple, pictorial illustration of this tactic. In Figure 6.3, subgroups have been formed by selecting parts from each of two machines. Consequently, sources of variation affecting within-subgroup variation will include differences between machines. Both charts are in control.

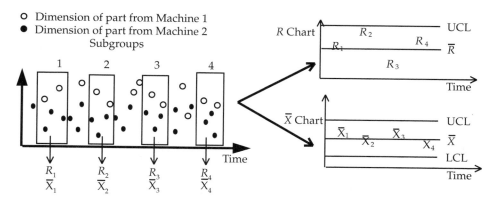

FIGURE 6.3 *Subgroups Consisting of Parts from Different Machines*

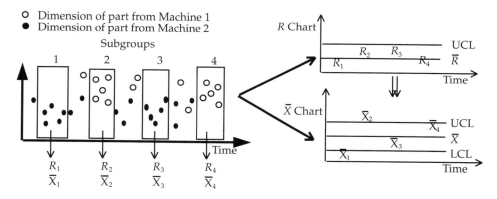

FIGURE 6.4 *Subgroups Consisting of Parts from the Same Machine*

A different subgrouping scheme is illustrated in Figure 6.4. Subgroups 1 and 3 contain parts only from machine 2. Subgroups 2 and 4 only have parts produced by machine 1. The X-bar chart formed from these data is out of control, because the variation reflected by the differences between machines is captured between subgroups, but not within. This tactic is an example of linking process knowledge of causal activity with an understanding of chart capability and how causal effects can be revealed by control charts.

6.2.3   Choosing Within-Subgroup Variation

Those individuals who use charts for process study deliberately design a sampling plan and construct subgroups that determine how and where the effects of sources of variation show up, either between or within subgroups.

Statistical decisions are based on the magnitude of $\bar{R}$ relative to other variations evidenced on either the range or average chart. The average range is not just a number; it must be interpreted as a collection of causes, since the magnitude of the average range determines process judgments. For example, it would generally not be practical to have a subgroup consist of multiple measurements on the same part at the same location. If this subgrouping structure were used, then within-subgroup variation would only reflect variation in the measurement process. If $\bar{R}$ is used to put limits on the $\bar{X}$ chart, then the $\bar{X}$ chart cannot be used to evaluate stability or consistency of effects of manufacturing causes. If measurement variation is small compared to process variation, then numerous $\bar{X}$'s will be out of control on the $\bar{X}$ chart. The out-of-control condition would only indicate that process variation is large compared to measurement variation, but would provide little information about the underlying causes of manufacturing variation. The reason is that measurement variation is not the correct variation for assessing the magnitude and stability of manufacturing variation.

Other sampling and subgrouping plans used in practice require judgment about cause activity to assess their effectiveness in gaining process knowledge. For example, some manufacturing processes have very good short-term capabilities, providing small variations in several consecutive parts. A plan that forms subgroups from a small number of consecutively manufactured parts may not be of practical use for assessing the stability of various manufacturing effects. The nature of the causes represented within subgroups would be one portrayal of common cause sources affecting manufacturing. Many $\bar{X}$'s may be out of limits; the information provided by these out-of-control points would be overwhelming to those attempting to use the chart to identify the reasons for these out-of-control conditions. It might be useful in these instances to increase the time over which the parts are taken in order to increase the causes that affect within-subgroup variation and, consequently decrease the number of signals on the X-bar chart indicating the presence of special causes. Process judgment should be exercised in the determination of a useful and practical within-subgroup variation.

Another example of an inappropriate sampling and subgrouping plan would be one that inflates the value of $\bar{R}$. Such an inflation could occur by the deliberate inclusion of numerous long-term effects or process differences within subgroups. This practice would tend to result in wide limits on the average chart, hampering the discovery of opportunities for reduced variation.

6.2.4   General Advice for Sampling and Subgrouping

The use and understanding of the variation occurring within subgroups and that occurring between subgroups is critical for providing a numerical, objective response to structured process questions. Variations occurring within and between subgroups are effects due to process causes, broadly listed under material, equipment, people, environment, and method. Sampling and subgrouping determine where these effects are revealed. By knowing the causes creating variations and where these are likely to be portrayed by data results, the user group is strengthened regarding objective judgment of variation, its practical meaning for process interpretation, and identification of likely causes for evaluation and correction or reduction. The source and revealed impact of these variations should not be dependent on chance. Sampling and subgrouping are to be guided by purpose, linked to some knowledge of process causes, and supported by an expanding understanding of the relationship between cause activity and process behavior.

The concepts just discussed can be used to begin establishing principles for sampling and subgrouping. The functional role of the average range in X-bar and range charts provides the basis for two elements of traditional advice regarding sampling and subgrouping.

1. *The first element is that sampling and subgrouping should be done in order to "let the $\bar{X}$ chart do the work."* The operational translation of "let the $\bar{X}$ chart do the work" is to devise sampling tactics and a subgrouping structure so that abnormal variations or evidence of detectable process differences are disclosed by average values outside limits on the $\bar{X}$ chart or by unusual, nonrandom patterns in the data points. Data collection is designed so that the causes of these events act between rather than within subgroups. An illustration of this concept can be had by recalling the data displays of Figure 6.4. The effect of the differences in machines was noted on the X-bar chart because the effect was not included in the within-subgroup variation.

    Successful practice of the "let the $\bar{X}$ chart do the work" principle presumes that factors contributing to the range chart have been made stable and predictable. For an initial sample plan and subgroup structure, specific and challenging work may be required to stabilize the range chart.

    Part of the reasoning governing this first principle rests on statistical properties of ranges and averages. The $\bar{X}$ chart is more sensitive to detecting departures from typical variations than is the range chart and, consequently, is more likely to discover differences of a given magnitude.

    One of the reasons why this principle has proven so valuable is process based. In many processes, causes are more likely to remain consistent in their activity and effect during a short time interval; and it is on this basis that samples are drawn from within a short interval of time. For processes that behave in this fashion, sample plan and subgroup structures are constructed to emphasize that behavior.

When the $\overline{X}$ chart is the primary vehicle for defining the nature of the effects of sources of variation, the range chart contains variations due to causes that are not the subject of focused work. Sampling and subgroups have been constructed so that effects due to special causes or ones deliberately left out of the range chart will occur between subgroups. Within-subgroup effects are accepted as an inherent part of the process. There is then the danger that this level of variation becomes "acceptable." However, if process variation is determined to be too large, a critical evaluation of those sources affecting the range must be done. Some of these sources may only affect short-term variations within any reasonable span of time; others have the potential to change the average significantly; and yet others may do both or neither. It may be necessary to differentiate between causes of within variation and between variation. Work of this type may require the use of designed experiments to identify and sort the respective causes as to their effects.

2. *A second element of traditional sampling and subgrouping advice is that the sample plan and subgroup structure should provide for minimal variation within subgroups and maximum opportunity for variations between subgroups.* The motive for including minimal variation within subgroups and maximum opportunity for variations between subgroups is easily understood in terms of the role of the average range in the interpretation of the X-bar chart. The smaller $\overline{R}$ is, the more sensitive the $\overline{X}$ chart; whereas if $\overline{R}$ is large, the less sensitive the $\overline{X}$ chart. Requirements of minimal variation within subgroups and maximum opportunity for variations between subgroups place an obvious demand on the sampling and subgrouping. Realistically, the recommended practice of minimal within-subgroup variation must support practical process work. Sampling and subgrouping should be designed so that variations on the average chart exceeding limits imposed by the "minimal" range will be due to causes that are to be investigated for reduction or elimination. As always, process knowledge must play a key role in assessing the information to be provided by specific sampling and subgrouping plans.

Advice on practice imbedded in "let the $\overline{X}$ chart do the work," combined with a stable, and passively viewed, range chart is entirely consistent with and supported by advice on sample plan and subgroup structure drawn from "make within-subgroup variation small" and "maximize the opportunity for subgroups to differ in their averages." Sampling to make within-subgroup variations small tends to move the effects of causes to occur between subgroups, utilizing the theme of letting the $\overline{X}$ chart do the work.

6.3   THREE SAMPLING AND SUBGROUPING PLANS IN A DIMENSIONAL STUDY

To illustrate the previously discussed ideas regarding sampling and subgrouping plans, three different plans for the study of a dimensional characteristic are explored. These examples serve two purposes. One is to elaborate on the concepts that support principles of subgrouping. The second is to emphasize the need for knowledge and awareness of process causes, the value of developing hypotheses regarding their activity, and the sampling and subgrouping plans derived from these activities.

Example 1: Judging Process Variations in Subgroup Averages Using Inappropriate Within-Subgroup Variation

Within one production run, k parts are sampled from a process. For each selected part, four independent measurements are made at the same part orientation by one person using the same instrumentation. The measurements for each part are treated as a subgroup, resulting in k separate subgroups. The range for each subgroup represents measurement variation since the same part has been measured repeatedly at the same location. Each subgroup average represents a part. Figure 6.5 contains a graphic displaying how the measurements were made, how the subgroups were formed, and symbols for range and average. Consider these three questions about the sampling and subgrouping:

1. How are sources of variation reflected in the values of R and $\bar{X}$?
2. What information is then provided by R and $\bar{X}$ charts?
3. Is the information provided consistent with the process information desired?

Causes within subgroups primarily reflect the measurement process. For any part, the four determinations differ because measurement is not exactly replicable; the same value is not obtained each time the part is measured. Each average represents a part. Subgroup averages differ because parts differ. Manufacturing causes create differences in parts; and it is these manufacturing causes that have the potential to act between subgroups. Limits on the $\bar{X}$ chart are based only on measurement variation and so would not provide useful baseline variation for assessing the stability of the manufacturing process. It is inappropriate to use measurement variation to evaluate consistency of manufacturing variations. If the purpose of the study is to assess manufacturing stability, the within-subgroup variation must include variations creating part-to-part differences in the short term.

The sampling and subgrouping of Figure 6.5 are useful for other purposes. For example, as part of a measurement study, a determination of the ability of the measurement process to discriminate between parts may be needed. In this case, the average range *would* provide an appropriate baseline variation against which to judge part averages. Values of $\bar{X}$'s out of limits would suggest that part-to-part variations are larger than variation predicted by measurement, indicating that the measurement process has the ability to discriminate among parts. Chapter 9 contains an extended discussion of this situation.

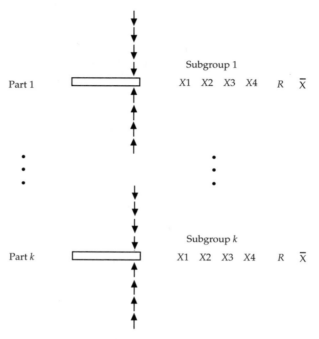

FIGURE 6.5 *Four Measurements on the Same Part at the Same Location*

Example 2: Judging Part-to-Part Variation Against Within-Part Variation

In this second example, k parts are again sampled, but in this instance measurements are made at four randomly selected locations rather than at the same location. Figure 6.6 provides an illustration of this sampling and subgrouping plan. Again, each part is the basis for a subgroup. However, in this instance, the numbers within each subgroup differ because of measurement error and within-part variation. Variation within subgroups, reflected by the subgroup ranges, is due to measurement causes and those manufacturing causes accounting for lack of uniformity within parts. Neither of these causes can account for variations from part to part (or average to average.) The process causes creating part-to-part differences are likely to be quite different from those causes responsible for within-part variation or measurement error. If the purpose were to evaluate consistency of the manufacturing process, it would not be appropriate to use the resulting average range to compute control limits for the average chart. Subgroup, or part, averages, vary for reasons not contained within subgroups or represented by the average range. Consequently, points outside of limits on the $\overline{X}$ chart would either indicate the *presence* of causes acting to create part-to-part differences or suggest that measurement or within-part causes are inconsistent. No information on the *consistency* of between-part differences can be gleaned.

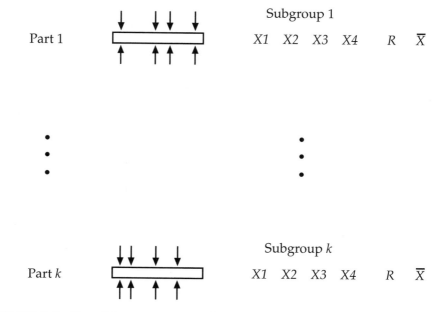

FIGURE 6.6 *Four Measurements on the Same Part at Different Locations*

The range in this application is a measured characteristic of a part; it measures lack of uniformity, taper, or some other characteristic. Although it is often useful to have such a measure of a part or material characteristic, that use should not be confused with its intended role in charting applications as a measure of variation within subgroups.

Example 3: Capturing Within- and Between-Part Variations in Within-Subgroup Variation

As a third example, consider the situation where four parts are sampled within a time span of a specific, designated length. This sampling is repeated at k distinct time intervals throughout a lengthy production run. There are four parts to a subgroup, each measured once, with k subgroups being formed, one subgroup per time period. A graphic is presented in Figure 6.7. With one measurement per part, four parts per subgroup, measurement and manufacturing causes for part-to-part deviations are contained within subgroups. If measurements are made at random locations on the part, then within-subgroup variations would also contain the effects of causes creating variations within parts. If the parts are measured at a fixed location, then the within-subgroup variation would be less affected by within-part variation. A controlled range chart reports that these effects are consistent. A subgroup average represents four parts. The common causes of variation in $\bar{X}$ have been accounted for by the manufacturing variations represented by $\bar{R}$. Control limits for the average chart represent variation in process causes against which process control for the average can be reasonably and legitimately judged.

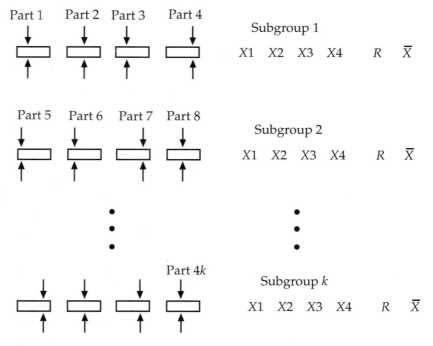

FIGURE 6.7 *Four Measurements on Each of Four Different Parts*

Summary of the Three Plans

The three examples of this section provide three different sampling and subgrouping plans for evaluating the same process. Neither of the three can be identified as "the correct" plan. The choice between the plans must be based on the purpose for which the data are to be collected. If the intention is to evaluate the stability of the manufacturing process, then the third plan provides for this assessment.

If the intention is to compare within-part variation to between-part variation, then the second plan might be productively used. And if the intention is to evaluate the ability of the measurement process to discriminate between parts, the first plan would provide valuable information about that issue. The key points to be gained from these examples are that knowledge of which causes of variation occur within subgroups and which occur between subgroups should be deliberately chosen to achieve process study purposes, and that correct interpretation of $\overline{X}$ and R charts depends on the choices made.

6.4   SUBGROUPING PLANS FOR PROCESSES WITH POSSIBLE FIXED DIFFERENCES

In the fifth section of Chapter 5, a study of a filling machine was discussed. At issue in the study of this machine was the amount of variation in fill weights from bottles filled by the machine. This same physical situation can be used to provide a framework for an examination of sampling and subgrouping plans in the presence of fixed differences. The filling machine under investigation has four heads. To understand how fill machine characteristics impart variation to filled weights, a consideration of the sources causing variation was undertaken. It was decided that knowledge of variations within and between shifts and within and between heads must be had. Data that captured these sources of variation were captured by subgrouping across shifts and heads.

Other sampling and subgrouping plans for a filling process could have been proposed. Three different plans are considered here in order to discuss the manner in which the possibility of fixed differences in a process characteristic should inform the choice of a sampling and subgrouping plan. The fixed differences to be studied in this example are those that could exist between four heads of a filling machine. Before collecting data on such a machine, the possibility that the four heads are filling to different averages must be considered and this consideration must inform the manner in which the data are taken. Of course, prior to collecting process data, there is no knowledge of how the fill weights for the respective heads are actually performing; but in order to provide a reference for considering the effects of different subgrouping plans it will be speculated that the average fill weights for each of the four heads are in control and can be described by the histograms in Figure 6.8.

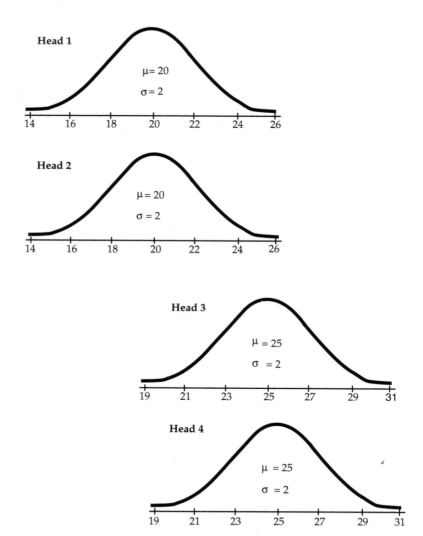

FIGURE 6.8 *Fill Weights for a Four-Headed Filling Machine*

Each of the heads in Figure 6.8 results in fill weights with similar amounts of variation ($\sigma = 2$) but heads 3 and 4 deliver higher average fill weights. The type of behavior displayed in Figure 6.8 is not, of course, confined to differences that exist between heads on the same machine. Other process situations that could result in just such fixed differences would be ones that had different machines performing the same tasks, or different lines, different shifts, etc. The discussion of the sampling and subgrouping plans for the four heads would be applicable to these situations as well.

Three different subgrouping plans should be evaluated in connection with the filling machine:

Subgroup Plan 1

> Four consecutively filled containers are taken from the first head. The weights of these containers form the first subgroup. Then four consecutively filled containers are taken from the second head, then the third and then the fourth. These form the second, third, and fourth subgroups, respectively. At later times four additional subgroups are formed by the same sampling and subgrouping process until there is a total of 20 subgroups.

Subgroup Plan 2

> Four containers are selected, at random, from the containers filled by the machine. Consequently, any and all of the heads might or might not be represented in the subgroup formed from these four containers. This sampling and subgrouping is repeated until there is a total of 20 subgroups.

Subgroup Plan 3

> Four containers are taken from the filling machine, one from each of the four heads. The weights of these containers form one subgroup. This same sampling and subgrouping procedure is repeated at later times until there is a total of 20 subgroups.

The intention of the following discussion is to explore how the fixed differences will affect the information about the process provided by the $\bar{X}$ and R charts formed from the subgroups described in the three plans. The expected behavior on these respective charts will be described assuming that the filling machine is behaving as described in Figure 6.8. Of course, in an actual physical situation, one would not know that the filling machine exhibited this kind of behavior. But speculating on how fixed differences would be captured by each subgrouping plan will provide guidance for choosing the most appropriate plan.

Subgrouping Plan 1

The subgroups in this plan contain fill weights from only one head. Subgroup 1 would have fill weights from head 1, subgroup 2 would have weights from head 2, and so on. Consequently, ranges calculated for these subgroups would only capture sources of variation that create variation in weights within a head. This variation might be described as those causes that result in the standard deviation, σ, of two displayed in Figure 6.8. If each head produces about this same level of variation in fill weights, as the graphs in Figure 6.8 indicate, then all 20 ranges would be of a similar size. In other words, a range chart for these 20 subgroups could be expected to be in control. Using $\bar{R}$ to put limits on the $\bar{X}$ chart implies that the control limits on the $\bar{X}$ chart will only reflect within-head variation. But the graphs in Figure 6.8 indicate that an additional source of variation impacting the process is the fixed difference in the average fill weights for the four heads. This source of variation would result in values of $\bar{X}$ falling outside of the control limits. Figure 6.9 provides a set of $\bar{X}$ and R charts which typifies the kind of chart behavior that would occur with subgrouping plan 1.

This first subgrouping plan is similar to the one used in the study of the filling machine in Chapter 5. This plan was found to be useful because the difference in averages of heads 1 and 2 versus heads 2 and 4 is immediately apparent on the $\bar{X}$ chart. In practice, this subgrouping plan would also be useful for determining whether the heads have consistent and approximately equal levels of variation. This information is, of course, provided by the range chart.

Subgrouping Plan 2

In the second subgrouping plan, subgroups are to be formed from fill weights of containers selected from the output of the filling machine without regard to which head filled the container. Variation captured within subgroups can be considered with the aid of Figure 6.10. A histogram of weights of containers from the combined output of the four heads might look like the one shown at the bottom of this figure. It is from this combined output that the samples of four containers have been selected. The histogram of the combined output has a much wider spread than those of the individual heads. This additional spread is due to the differences between average fill weights. Variation within subgroups will be due to sources of variation affecting both within-head and between-head variations.

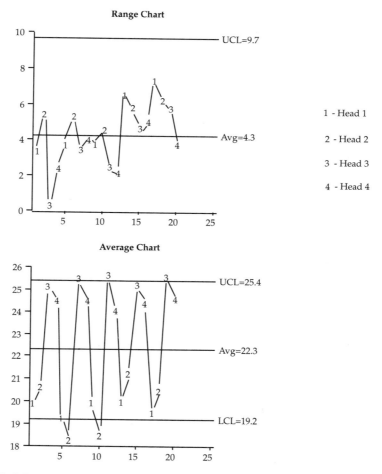

FIGURE 6.9 *Range and Average Charts Expected from Subgrouping Plan 1*

Figure 6.11 contains $\overline{X}$ and R charts that would be anticipated if subgrouping plan 2 were used to collect data on a process described by the graphs in Figure 6.10. Of particular interest on these charts is the following:

1. *The range chart is in control with a larger $\overline{R}$ than was observed using subgrouping plan 1.* Since each subgroup captures both within- and between-head differences, each of the 20 ranges is impacted by sources of variation acting to create both within- and between-head differences. In particular, the differences in the head averages influence the magnitude of the ranges.

2. *Unlike subgrouping plan 1, the $\bar{X}$ chart for plan 2 is in control.* Because $\bar{R}$ is influenced by additional sources of variation, the control limits on the $\bar{X}$ chart are wider. A different way of considering this phenomenon is that the ranges reflect short-term variations affecting the filling machine. These short-term variations include sources that act to create both within-head and between-head variation. Consequently, special causes that would result in an out-of-control condition on the $\bar{X}$ chart are causes that appear erratically or systematically across time.

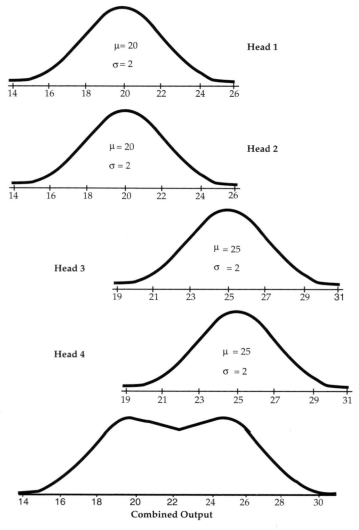

FIGURE 6.10 *Combined Output from a Four-Headed Machine*

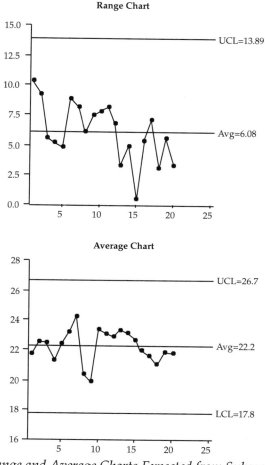

FIGURE 6.11 *Range and Average Charts Expected from Subgrouping Plan 2*

Because both the range and average charts are in control, $\bar{R}/d_2$ can be used to estimate the variation of filled weights of containers from this process. This statement is consistent with the earlier observation that $\bar{R}$ captures those sources of variation affecting both within- and between-head variation and no additional variation is evidenced by the $\bar{X}$ chart.

A comparison of subgrouping plans 1 and 2 should note the fact that subgrouping plan 2 captures variations on the range chart that might be termed inherent process variation. It is this variation that is present in the combined output of the four heads that is observed by the customer who purchases or uses the containers. However, the second subgrouping plan does not indicate the fixed differences in average fills of the heads which was apparent on the $\bar{X}$

chart of subgrouping plan 1. In other words, a subgrouping plan that does not include a possible fixed difference within the range can provide information about the existence of this source of process variation.

Subgrouping Plan 3

The third plan formed subgroups by choosing one container from each of the four heads. This type of a subgrouping plan is, unfortunately, too often used since it seems reasonable to obtain a subgroup that is "representative" of all four heads. An examination of how fixed differences in the heads would affect the interpretation of data generated from such a plan makes it clear why this approach should be avoided.

By selecting a container from each head, this subgrouping plan ensures that fixed differences in the averages of the four heads will affect each and every subgroup range. Consequently, every range will be large and the resulting $\bar{R}$ will also be large. What is distressing about this plan is that the fixed differences in the four heads will have little if any impact on the subgroup averages. This conclusion is clear by considering the fact that one container from each head is always obtained. So when the four fill weight values are averaged, this source of variation is averaged out. Consequently, for subgrouping plan 3 a source of variation affecting the ranges (deviations among heads) does not affect the averages. Since limits for the $\bar{X}$ chart are based on $\bar{R}$, the width of these limits will be wider than the variation in the subgroup averages. Figure 6.12 contains typical $\bar{X}$ and R charts that would result from using subgrouping plan 3 on a process described by the graphs of Figure 6.10.

The fact that not all sources of variation that affect the ranges are affecting the averages is evidenced on the $\bar{X}$ chart of Figure 6.12. The points on this chart appear to hug the centerline. This behavior on the $\bar{X}$ chart is often referred to as "too much white space," since most of the area on the chart between the upper and lower control limits does not contain any plotted points. Caution must be exercised when interpreting a chart that has too much white space. The control limits define the amount of variation that should be observed in the plotted points. The fact that considerably less variation is observed does *not* mean that the process is operating with less variation than is expected. "Too much white space" indicates some deficiency in the subgrouping plan; in the present instance this deficiency is that fixed differences between the heads are captured by the ranges but not the averages.

It is instructive to consider how misleading the use of subgroup plan 3 would be in evaluating process stability. The process on which the charts of Figure 6.12 are based was one for which each head operated consistently, with heads 1 and 2 filling to different averages than heads 3 and 4. Suppose sub-grouping plan 3 were used on a process that had not only fixed head differences, but also exhibited fluctuations in these averages across time, maybe due to changes in incoming materials. The width of the limits on the $\bar{X}$ chart make it insensitive to time-to-time fluctuations. Individuals attempting to use the charts to detect such fluctuations may not have any indication from the $\bar{X}$ chart that they are occurring.

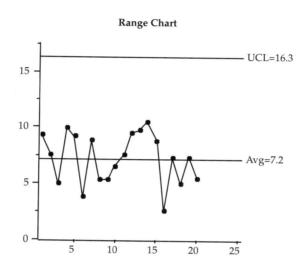

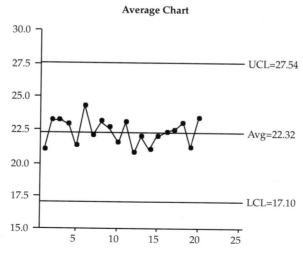

FIGURE 6.12 *Range and Average Charts Expected from Subgrouping Plan 3*

6.5   SUBGROUPING PLANS FOR STUDYING COILS, WEBS, OR ROLLS

In the paper, steel, aluminum, glass, polyfilm, and other industries, product is created in coils, sheets, rolls, or ribbons. Measurements on thickness, density, chemistry, moisture, and other product properties are made across the coil, sheet, roll, or ribbon. The graphic in Figure 6.13 represents one of these processes. The sketch illustrates that three measurements are made in the cross direction and sets of three are made in the machine direction. In practice, there may be numerous measurements in the cross direction, each in a "zone." However, considering a situation where only three measurements are made in the cross direction will provide for a discussion of the study of this processing situation. The learnings can easily be generalized to a situation where more measurements are made in the cross direction.

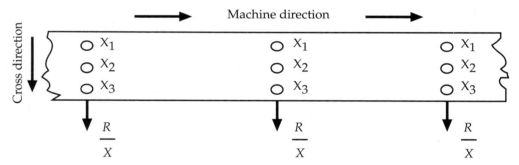

FIGURE 6.13   *Measurements in the Cross Direction and in the Machine Direction*

Every set of three measurements in the cross direction could be viewed as a subgroup. Alternatively, each of the three cross-directional zones could be studied separately using moving range and individual charts as the analytical tool. (These procedures are discussed in Chapter 7.) Although, the second analysis will generally be the more advantageous one, there are numerous process possibilities in these various industries and one view or description will not serve for all. Additionally, particular process characteristics afford useful discussion for different sampling and subgrouping plans. Several frequently encountered situations are discussed in the following material.

In industries that build and handle material in a roll, coils, or sheets, it is often the case that a "fixed" difference is designed and built into product characteristics in the cross-direction. For example, steel and aluminum coils typically have a distinct difference in caliper across the coil, called a "crown," which is required for forming the product. Similarly, in making glass, there is a definite difference in edge thickness compared to center thickness. In these cases, distinct differences in the cross-direction serve a process or product requirement. The existence of fixed differences indicates that individual measurements made in the machine direction will be subject to different process conditions, resulting in different averages and possibly different variation. Variation in the cross-direction will include the fixed difference. Of course, other processing conditions, such as variation in equipment, will also create variation in the cross-direction; this element of variation is often referred to as a *random component*. Two issues must be considered in designing sample collection, subgroup arrangement, and data display. The presence of fixed differences is one. The other is that the causes of variation in the cross-direction are generally not the same as causes in the machine direction.

The similarity between the formation of cross-direction subgroups, as illustrated in Figure 6.13, and subgroup plan 3 of Section 6.4 should be noted. Both of these sampling and subgroup methods include possible fixed differences within a subgroup. The analysis of this approach when studying a filling machine with fixed differences in average head fill can be applied to the current situation. In particular, two consequences of such a subgrouping approach should be considered in the present situation:

1. By including fixed differences in a subgroup, the average range reflects that variation and it tends to be inflated. There is a real possibility, then, that abnormal variations in subgroup averages will go undetected.
2. Unusual behavior on the range and average chart may result from this subgrouping plan. Since the average range tends to be too large to correctly describe variation in individual measurements for the respective conditions occurring within subgroups, two behaviors on the range and average chart often occur:
   a. The centerline on the range chart is "too large" so the plotted ranges tend to fall near the centerline.
   b. Subgroup averages are also affected by the inclusion of fixed differences within subgroups, causing them to vary less than random variation would require. The averages tend to plot too closely or narrowly about their centerline with, in extreme circumstances, all data points falling within one standard deviation for averages on either side of the chart average.

In light of these observations about the effect fixed differences in the cross-direction may have on interpreting process data, the process engineer or manager will need to assess how data collected as described in Figure 6.13

can be used to understand process sources of variation. The appropriateness of subgrouping in the cross-direction must be determined in light of the fact that subgroup averages on or outside limits indicate that machine direction causes create detectable differences compared to average cross-direction variation. Is this information helpful? Can it be used to suggest cause and focus work? The answer to this question is, of course, process dependent. In answering this question, the investigator will be deciding which of the following three scenarios most aptly describes the process being studied:

1. Variations in the cross-direction represent relevant process causes and provide an appropriate baseline variation against which to judge process stability in the machine direction. In this instance, the proposed subgrouping scheme would provide useful information for studying those sources of variation. This approach is not generally useful. The cross-direction range is scarcely affected by time-to-time causes; these variations occur in the machine direction. The following subgrouping plan must be used to study these time-to-time variations.
2. Some critical sources of variation act primarily in the machine direction and do not display strong effects in the cross-direction with the consequence that cross-direction variations understate variations to be expected in subgroup averages. If this is the case, then the investigator should consider a subgrouping scheme that includes machine direction variations within subgroups. A possibility would be to calculate moving ranges of data collected in the machine direction. This topic is discussed further in Chapter 7.
3. Fixed zone differences may exist in the cross-direction and, combined with random variations in the cross-direction, provide an inappropriate average range against which to judge process consistency in the machine direction. However, a range chart that captures these fixed differences can be used to examine the consistency of this fixed difference across time and conditions.

6.6   PRINCIPLES FOR RATIONAL SUBGROUPING

A rational subgroup is one that is homogeneous with respect to influential sources of variation. Operationally, a rational subgroup is one for which parts or material contained within that subgroup are subjected to a common set of inherent process causes and these causes affect the deviation among the parts or material in a consistent way. Furthermore, factors that may create changes in subgroup averages are thought not to be active during the processing period from which the sample was taken. The following brief examples will help describe what is meant by a rational subgroup.

Example 1

Suppose that the effect of changes in material lots on a part characteristic is to be investigated. A rational subgrouping plan requires that all parts in the same subgroup be manufactured from the same material lot. All parts in a subgroup are considered to have been made from material having essentially the same, or "constant," value for the relevant material properties since they were all made from the same lot of material. (Note that the "relevant material properties" may or may not be definitively characterized.) Differences among parts made from the same material are not considered as being due to material variations; material from the same lot is thought to be consistent. This is not the same thing as stating that the level of a material characteristic for a given lot of material has no effect on the average variation in parts created by that lot. A determination of whether the level of a lot or material affects the variation in parts made from that lot can be made using a range chart. Then, the effect of material variations, lot to lot, can be made by examining subgroup averages. Lot-to-lot differences that act to significantly change the average would show up as out-of-control conditions on the $\bar{X}$ chart.

Example 2

Different manufacturing fixtures may contribute to part-to-part differences. In this instance a rational subgroup consists of parts made on the same fixture. Several distinct subgroups from the same fixture will provide information on variation and average for parts made on that fixture. To evaluate consistency across different fixtures, subgroups will, of course, need to be had from the different fixtures. Differences between fixtures will be revealed in differences between averages for fixture subgroups.

In process investigations, subgroups can rarely be so neatly defined with respect to process sources of variation, such as different material lots or different fixtures, as is implied by the preceding examples. Consequently, a rational subgroup is often formed by selecting a subgroup in a short period of time during which process causes of variation do not tend to change in ways that create different averages during that time period. Rational subgrouping implies that the effects of variation within subgroups are "random" (as opposed to "fixed") in their effect on the unit, part, or material property being measured and that similar sources of variation are contained within each subgroup.

As illustrated by the numerous examples throughout this chapter, the development of sampling and subgrouping plans cannot be divorced from the purpose of process study or the particular process characteristics of the process to be investigated. However, several principles of rational subgrouping can be developed from the previous discussions to serve as guidelines for developing sampling and subgrouping plans.

Principles of Rational Subgrouping

Principle 1

> *Know the process factors.* Useful sampling and subgrouping plans are those that specifically address which sources of variation contribute to variations within subgroups and which contribute to variation between subgroups. Knowledge and intuition of what are or might be critical sources of process variation are imperative for developing useful sampling and subgrouping plans.

Principle 2

> *Sample from homogeneous production conditions.* Significant process factors should remain relatively constant during the time in which a sample is collected, and it should be possible to form at least one subgroup from the sample. The sampling should include more variation than measurement. Sampling should be specific with respect to source; in other words, defining and recording what, when and how different lines, machines, runs, heads, shifts, etc., are to be sampled must be done. Different subgroups can then be formed within the different sources.

Principle 3

> *Sample so as to obtain small within-subgroup variations that are consistent with defined purposes.* Sampling within homogeneous conditions tends to minimize the variation within a subgroup. The rationale behind this third principle is to avoid including so many sources of variation within a subgroup that the resulting $\bar{X}$ chart is not sensitive to significant process changes.

Principle 4

> *Sample intervals should be separated by a sufficient amount of time so that process causes have the opportunity to become active between subgroups.* The timing of sampling activity is obviously process dependent. If too much time passes between sampling, then a profusion of process causes may have been active, resulting in difficulties in identifying and confirming causes of variation.

Principle 5

> *Never knowingly create subgroups so as to include fixed differences within a subgroup.* As discussed in connection with subgrouping plan 3 of Section 6.4, including fixed differences within a subgroup means that common cause sources of variation affecting ranges may not be the same as the common causes affecting subgroup averages. The practical implication of

this phenomena is that the $\bar{X}$ chart cannot be used to study common cause sources of variation that affect the process average.

Principle 6

Sources of variation affecting within-subgroup variation should be consistent with sources that affect differences in the subgroup averages. Causes whose effects are contained within subgroups determine the magnitude of $\bar{R}$ and define the judgment made about variation expected in values of $\bar{X}$. If the subgroup structure is such that the primary causes of variation in $\bar{X}$'s are not also included within subgroups, then realistic assessment of consistency in values of $\bar{X}$ cannot be made. The result will be many signals on the $\bar{X}$ chart of the presence of special causes with little possibility of determining which of the myriad possible causes resulted in the out-of-control condition.

6.7   AN APPLICATION OF THE PRINCIPLES OF RATIONAL SUBGROUPING

Processes that have multiple spindles, dies, heads, molds, cavities, and/or workstations provide an exemplary framework for demonstrating the principles of rational subgroups. Process study in these situations will require that a broad set of pragmatic issues be examined. Decisions about sampling procedures must address where, when, how often, and how many. Criteria for subgroup formation must be decided: time, process condition, material source, and work method are possible criteria. In addition, the sample design and subgroup structure must be evaluated against both study purpose and process causes. These sampling and subgrouping issues will be discussed by considering the manufacture of a generic metal or plastic part from a multiple-source process. A description of the actual characteristic to be measured on the part is not provided, since any critical part dimension or property could serve equally well as a basis for discussion.

The graph in Figure 6.14 provides on overview of the process to be examined. The parts to be studied are manufactured at four different positions. Four parts are produced at each cycle of the machine. One measurement is to be made on each part. Because multiple measurements are not made on the same part, within-part variations cannot not be examined; they are not a focus in the following discussion. Observed variations in numbers reported on the process will include measurement variation. Because parts may be measured by any operator, perhaps on any one of several devices, at any time, typical measurement variation would include deviations within and among all technician-instrument combinations and time-to-time differences, all contributing to observed variability. However, in this investigation, measurements are made by one individual using one device following a measurement procedure that has been qualified and serves correctly for the selected characteristic.

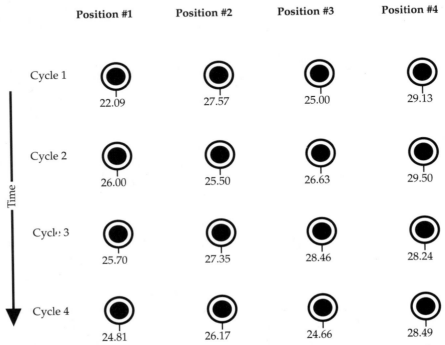

FIGURE 6.14 *Illustration of Part Production with Numerous Sources of Process Variation*

6.7.1   Defining the Primary Objectives

A primary requirement of any process investigation is a statement of major process management objectives prior to specifying a sampling and subgrouping plan. As illustrated later, particular process and product characteristics must be referenced and translated into guidelines by these statements. Sample and subgroup considerations are shaped by the basic questions posed from the consideration of these statements. The following questions were developed relative to the study of the process typified in Figure 6.14 are:

1. What is the short-term variation within each position? Is it predictable? What is its magnitude? Is it the same for all positions? What are the causes of this short-term variation?

2. What are the long-term variation behaviors for each position? Is each position average stable? What is the average value for each position? Do averages differ by position? What are the more important causes?
3. What is the distribution of part measurements for the process?
4. Does the process conform to requirements? Is the process capable? Does the process have the correct average?

These questions are very specific. To provide specific answers, parts must be taken from each position over time and arranged in subgroups. Other questions may develop as work proceeds.

6.7.2   Developing the Sampling Plan

Several basic sampling and subgrouping principles immediately apply to the current situation. One principle is that sampling should be from homogeneous conditions, resulting in small, random variations within subgroups. For this multiple-stream process, this principle suggests several criteria for sampling:

- Sampling should be done for each position since position averages may be different and those fixed differences should not be included in a subgroup.
- The sampling base time period should be of "short" duration to minimize the time-to-time variation captured within subgroups.
- The number of sampling time periods should be numerous.
- The total time covered by the sampling intervals should include opportunities for major, long-term sources of variation to become active.

Maintaining homogeneous conditions within the time period that a sample is selected will be aided by not changing tools or other fixtures during base time periods in which parts are taken, by not making machine adjustments during any of the time periods, and by not changing material lots during the sampling periods. All of these required activities can be done between sampling time periods. The effect of any of these changes will be noted in the behavior of the subgroup averages rather than the subgroup ranges. Only "random" effects should be included within subgroups; sampling by product source, sampling in short time intervals, and subgrouping by position operationally helps to ensure that only random effects are captured within subgroups.

The general principles we have discussed for sampling are made operational by the instructions developed for selecting parts and by ensuring that these instructions are implemented correctly. The following instructions are provided for sampling the process: Twice a shift a part is taken at four equally spaced subintervals of time from each of the four positions, with sampling extending over five days. Each base sampling period yields 16 parts, one part from each of the four selected cycles for each of the four positions. There are two sampling periods each day. Measurements made on these parts are recorded in Table 6.1. Consistent with good practice, data are coded to facilitate display, analysis, evaluation, interpretation, and re-arrangement. Individual observations are listed by day, 1 through 5; by sample time, 1 or 2; position, P1, P2, P3, or P4; and by cycle number, 1 through 4. The cycle number indicates part sequence, by time. Parts are sampled at equal intervals across the fixed time period. Although the length of time across which the parts are taken remains fixed, the sampling times themselves, two per day, are to be randomly selected. Sampling over five days represents the operational effort to evaluate the possible effects of major long-term sources of variation. Multiple sampling periods within each shift permit an assessment of within-shift effects and consistency of intermediate-term causes that might come and go in the hours covered by one shift.

TABLE 6.1   *Measurements of Parts Produced at Different Positions*

Day	Time	Cycle	Position 1	2	3	4
1	1	1	22.09	27.57	25.00	29.13
1	1	2	26.00	25.50	26.63	29.50
1	1	3	25.70	27.35	28.46	28.24
1	1	4	24.81	26.17	24.66	28.49
1	2	1	24.46	27.02	25.32	26.13
1	2	2	24.24	25.77	24.45	27.78
1	2	3	24.09	26.72	25.48	26.98
1	2	4	24.87	26.23	25.60	26.62
2	1	1	23.95	27.41	24.75	26.48
2	1	2	24.10	26.35	26.72	27.63
2	1	3	22.77	26.06	25.48	27.15
2	1	4	23.81	27.27	26.09	25.15
2	2	1	25.55	27.04	26.58	28.01
2	2	2	25.28	25.80	25.18	26.86
2	2	3	25.40	25.45	25.30	27.67
2	2	4	25.73	25.47	26.23	28.43
3	1	1	25.08	24.83	25.07	24.81
3	1	2	23.64	24.99	23.54	24.53
3	1	3	22.17	25.73	24.54	25.56
3	1	4	22.24	24.76	22.48	25.63

TABLE 6.1 *Measurements of Parts Produced at Different Positions (continued)*

			Position			
Day	Time	Cycle	1	2	3	4
3	2	1	25.01	25.75	26.57	27.34
3	2	2	24.14	26.46	27.66	26.64
3	2	3	25.35	26.48	26.14	27.16
3	2	4	23.66	27.04	25.47	26.74
4	1	1	26.40	26.49	26.49	30.03
4	1	2	24.01	27.05	25.96	28.34
4	1	3	24.39	26.12	26.98	29.54
4	1	4	24.85	28.59	25.47	27.55
4	2	1	23.20	25.20	24.27	25.82
4	2	2	24.13	27.01	25.76	24.88
4	2	3	23.10	26.36	25.62	27.15
4	2	4	23.39	26.08	23.94	26.80
5	1	1	25.74	27.81	27.16	26.91
5	1	2	24.64	27.30	24.47	28.86
5	1	3	24.53	27.36	27.60	30.19
5	1	4	26.80	27.99	27.53	27.86
5	2	1	24.71	26.31	25.94	27.35
5	2	2	24.37	28.43	27.45	27.00
5	2	3	24.71	26.11	28.79	28.74
5	2	4	24.95	27.08	27.32	29.33

6.7.3 Identifying Major Sources of Variation

Process factors contributing to the observed variability in this application might be classified into four categories:

1. measurement causes
2. cycle-to-cycle causes
3. position-to-position causes, and
4. time-to-time causes.

These categories, with specific sources of variation listed under each, can be used to consider where effects will be "seen," within or between subgroups for the proposed sampling plan and different subgroup structures. Developing categories of process causes and identifying how particular processing characteristics should be classified according to this list of categories serves to strengthen cause identification. By forming possible causes responsible for certain types of process behavior, more specific process knowledge can be gained, and sample and subgroup alternatives can be discussed and described in terms of these causes.

A cursory scan of Table 6.1 reveals differences in cycle-to-cycle results for any position. Variability from cycle to cycle, exceeding measurement variability, indicates distinctive causal effects, described as cycle-to-cycle causes. Effects due to these causes may or may not be stable over time and they may or may not have approximately the same effects, either for variability or for average, for the respective positions. Specific process factors cataloged as cycle-to-cycle causes may be dependent on the length of time during which the four parts are taken. If parts were taken consecutively, for example, there may be a select group of very high frequency causes responsible for part-to-part deviations. If the four parts were taken one hour apart, all of the previous causes would still be at work and, for many processes, the effect of others would be experienced. If the four parts sampled during one time period of some fixed length were treated as a subgroup, effect size would typically be smaller or larger depending on the length of time covered by the sampling plan. Cycle-to-cycle causes and realized effects play out over some time period. These causes are not independent of those listed as time-to-time causes, which may be considered as those acting longer term.

Position-to-position variation defines another set of possible sources of variation. Process causes resulting in differences among positions may not be well known. Positions may differ in the magnitude of part-to-part variability or by running to different averages or by both. Only data examined over time for variation, average, and stability will provide confirmed results. If positions tend to have about the same within position variability, attention would turn toward possible differences in averages. Differences in position averages may be considered in several respects. Differences may be random or fixed, consistent or inconsistent, large or small, and process significant or not. Definitive statements will not be possible until the data are examined. If differences are purely random, consistent, and satisfactorily small, there is no obvious need to maintain separate data by position. However, where conjecture suggests that distinctly different averages may be created by circumstance or condition, initial sampling and subgroup structure must respect that possibility. The different circumstance or condition should not be included within the same subgroup. If position differences in averages are confirmed and consistent, then deviations are considered to be "fixed," in cause and magnitude. These differences may or may not contribute significantly to total process variation. If the effects are significant, then an identification of the reasons for these differences must follow. The cause or causes of the differences are probably different than those creating cycle-to-cycle variations.

Time-to-time variations refer to effects of causes that become operative across time periods longer than that making up the basic time span. As discussed earlier, the distinction between cycle-to-cycle and time-to-time causes is partially dependent on the time frame over which the effects of cycle-to-cycle causes are observed. Typically, as the time frame is lengthened, more opportunity for various cause activity is available. Time-to-time effects, when influential and larger than within-subgroup variations, would occur between subgroups. It is likely that these time-to-time changes will be evidenced by erratic behavior of subgroup averages or by sustained shifts in the process averages.

6.7.4   Possible Subgroup Structures

At least three different structures for forming subgroups are possible in the current situation. These structures differ in several important ways:

1. Each structure distributes the respective classes of causes differently, within and between subgroups.
2. Because different sources of variation are assigned to within subgroups, the structures may have significantly different capability for detecting abnormalities or differences.
3. The structures have different capabilities for providing answers to various questions.
4. In all structures, *measurement* causes occur within subgroups, contributing to that observed variation.

Subgroup Structure 1: Capture Cycle-to-Cycle Variations Within Subgroups

One possible means of forming subgroups is to capture cycle-to-cycle differences (as well as measurement effects) within subgroups. Different subgroups would be formed for each different position. Each position would have its own range and average chart. The four range charts permit assessment of the consistency of cycle-to-cycle effects for the respective positions. Variation in subgroup averages will reflect the long-term variation, if it exists, at each position. How effects of the other two classes of causes will affect ranges and averages must be examined for this subgroup structure. Because a range and average charts are used for each position, position-to-position causes account for observed differences in the centerlines of the respective charts for the different positions. Widely divergent values for average ranges would suggest that positions differed in within-position variability, which for this subgrouping structure is created by process factors that cause cycle-to-cycle differences among parts. Similarly, deviations among the centerlines for the subgroup average charts would indicate that positions differ in average values. However, long-term sources of variation for each position would typically be active between subgroups and the presence of these effects would be seen in subgroup averages outside control limits or by nonrandom patterns in the data plot. As seen in the next section, this subgrouping structure, which has cycle-to-cycle differences within a subgroup for each position, is preferred over the other two discussed.

Subgroup Structure 2: Capture Position-to-Position Differences Within a Subgroup

A second means of subgrouping the data would be to form a subgroup at each cycle. Each cycle delivers one part from each of the four positions so the subgroups would contain four measurements. Only one range chart and one average chart would be used. In this structure, position-to-position and measurement variations occur within subgroups along with effects due to within-position variations. The range chart permits evaluation of the consistency of these effects within each cycle. The impact of these within-cycle variations is reflected in the value of $\bar{R}$, which is used to determine control limits on the average chart. One subgroup average represents the cycle result across all four positions. The occurrence of significant effects of the remaining two categories of causes remains to be identified.

For this subgroup structure, cycle-to-cycle and time-to-time variations tend to reveal their effects between subgroups. Consequently, differences over time in effects due to those two classes of causes are revealed on the subgroup average chart by one or more points on or outside the limits. A first level explanation for reasons would be in the process factors listed in those two categories. There would be some uncertainty as to which category is more likely responsible; that is, results would be confounded as to causes and this could make the charts difficult to use in practice. There is another potential difficulty in the use of this structure. Position-to-position effects are contained within subgroups. By itself, this is not a difficulty. But there is an issue if differences among positions are fixed rather than random. If there are fixed position-to-position differences, these may dominate within-subgroup effects, creating a large average range. A consequence is that this structure may result in a chart for averages having limits that are not sensitive to effects or differences rooted in cycle-to-cycle or time-to-time variations. This structure should not be used when position averages are expected to differ because of this potential insensitivity of the $\bar{X}$ chart. Control charts are presented later in this Section to demonstrate this phenomenon.

Subgroup Structure 3: Capture Both Between-Cycle and Between-Position Differences in a Subgroup

Subgroups could be based on the sampling period; in other words the 16 parts measured after each four cycles form a subgroup. This third structure clearly has variations due to measurement, cycle-to-cycle, and position-to-position causes within subgroups. Within-subgroup effects may be due to both random and fixed effects, reflecting any fixed differences among position averages as well as random variation between cycles and positions. Subgroup averages measure results for sampling time intervals. Even with the large subgroup size of 16, fixed differences among positions may tend to make the structure insensitive to the third major category of causes, those acting long-term, time-to-time, which are captured by variations between subgroups. Unusual effects due to time-to-time causes would be expected to create variations on the average chart larger than that predicted by the within-subgroup average range, resulting in points outside limits. But, an inflated average range, reflecting fixed differences among position averages, would tend to hide time-to-time effects of a certain magnitude, with the possibility that the presence of those causes in the process would go undiscovered. The structure suffers from the possibility of large, fixed differences among positions and should be considered questionable for that reason.

6.7.5 Forming Subgroups

Forming subgroups is an early step in extracting information from collected data. Data should be subgrouped initially according to the fundamental principles for forming rational subgroups, influenced by process judgment as to causal structure and by process questions to be addressed. These sampling and subgrouping principles leave little doubt in defining the best approach in the current application. For the data obtained by sampling this process, subgroups should be formed using subgroup structure 1. Because individual subgroups have parts taken from four selected cycles across a fixed time span, the subgroup averages provide capability for assessing the effect and stability over time of long-term, time-to-time variations. Separate, charted results should be posted for each of the four positions. Table 6.1 has been revised and the subgroup ranges and averages for the proposed subgrouping plan have been included in Table 6.2. Each row of these data contains a distinct subgroup. The data from the four positions have been separated and the time order in which the measurements were collected has been maintained. Measurement and cycle-to-cycle variations are both captured within subgroups, although it is expected that cycle-to-cycle variations will dominate measurement variation. There are 10 subgroups for each position, the number being limited to 10 for convenience of reference and presentation.

Subgroup ranges and averages are calculated for each subgroup and reported subgroup by subgroup, line by line, position by position. Control charts for range, R, and average, $\bar{X}$, are constructed for each position and displayed, by position, in Figure 6.15.

6.7.6   Process Knowledge Gained from Subgroup Structure 1

The four R charts in Figure 6.15 do not display any signals suggesting a lack of control because there are no points on or outside the limits and no unusual patterns in the data plots. Although the judgment that variation captured by the ranges is in control must be tentative because of having only 10 observations, predictability in the R charts implies that the effects of process factors contributing to cycle-to-cycle causes are consistent over time. The charts for subgroup averages are not all in control; this observation has implications for knowledge gained about the behavior of process variation, possible sources of this observed variation, and work to be done to address these sources.

TABLE 6.2   *Measurements of Parts with Cycle–to–Cycle Variations Within Subgroups*

| Day | Time | Position | Cycle | | | | Range | Average |
			1	2	3	4		
1	1	P1	22.09	26.00	25.70	24.81	3.91	24.6500
1	2	P1	24.46	24.24	24.09	24.87	0.78	24.4150
2	1	P1	23.95	24.10	22.77	23.81	1.33	23.6575
2	2	P1	25.55	25.28	25.40	25.73	0.45	25.4900
3	1	P1	25.08	23.64	22.17	22.24	2.91	23.2825
3	2	P1	25.01	24.14	25.35	23.66	1.69	24.5400
4	1	P1	26.40	24.01	24.39	24.85	2.39	24.9125
4	2	P1	23.20	24.13	23.10	23.39	1.03	23.4550
5	1	P1	25.74	24.64	24.53	26.80	2.27	25.4275
5	2	P1	24.71	24.37	24.71	24.95	0.58	24.6850
1	1	P2	27.57	25.50	27.35	26.17	2.07	26.6475
1	2	P2	27.02	25.77	26.72	26.23	1.25	26.4350
2	1	P2	27.41	26.35	26.06	27.27	1.35	26.7725
2	2	P2	27.04	25.80	25.45	25.47	1.59	25.9400
3	1	P2	24.83	24.99	25.73	24.76	0.97	25.0775
3	2	P2	25.75	26.46	26.48	27.04	1.29	26.4325
4	1	P2	26.49	27.05	26.12	28.59	2.47	27.0625
4	2	P2	25.20	27.01	26.36	26.08	1.81	26.1625
5	1	P2	27.81	27.30	27.36	27.99	0.69	27.6150

TABLE 6.2   *Measurements of Parts with Cycle–to–Cycle Variations Within Subgroups (continued)*

			Cycle					
Day	Time	Position	1	2	3	4	Range	Average
5	2	P2	26.31	28.43	26.11	27.08	2.32	26.9825
1	1	P3	25.00	26.63	28.46	24.66	3.80	26.1875
1	2	P3	25.32	24.45	25.48	25.60	1.15	25.2125
2	1	P3	24.75	26.72	25.48	26.09	1.97	25.7600
2	2	P3	26.58	25.18	25.30	26.23	1.40	25.8225
3	1	P3	25.07	23.54	24.54	22.48	2.59	23.9075
3	2	P3	26.57	27.66	26.14	25.47	2.19	26.4600
4	1	P3	26.49	25.96	26.98	25.47	1.51	26.2250
4	2	P3	24.27	25.76	25.62	23.94	1.82	24.8975
5	1	P3	27.16	24.47	27.60	27.53	3.13	26.6900
5	2	P3	25.94	27.45	28.79	27.32	2.85	27.3750
1	1	P4	29.13	29.50	28.24	28.49	1.26	28.8400
1	2	P4	26.13	27.78	26.98	26.62	1.65	26.8775
2	1	P4	26.48	27.63	27.15	25.15	2.48	26.6025
2	2	P4	28.01	26.86	27.67	28.43	1.57	27.7425
3	1	P4	24.81	24.53	25.56	25.63	1.10	25.1325
3	2	P4	27.34	26.64	27.16	26.74	0.70	26.9700
4	1	P4	30.03	28.34	29.54	27.55	2.48	28.8650
4	2	P4	25.82	24.88	27.15	26.80	2.27	26.1625
5	1	P4	26.91	28.86	30.19	27.86	3.28	28.4550
5	2	P4	27.35	27.00	28.74	29.33	2.33	28.1050

The $\bar{X}$ charts for positions 2, 3, and 4 have points on or outside limits, indicating that variations in $\bar{X}$'s are demonstrably larger than the average effect of cycle-to-cycle causes as measured by $\bar{R}$. Process interpretation of these results is provided in the following four conclusions below.

Conclusion 1

The lack of statistical control for positions 2, 3, and 4 indicates inconsistency in the respective position averages. A number that describes the average dimension observed at positions 2, 3, or 4 cannot be provided, because the $\bar{X}$ charts indicate that this average changes over time. Consequently, statements about the process average value are suspect. The total variation in the dimension of parts from this process is unknown because position averages tend to either drift or move by unknown amounts at unknown times.

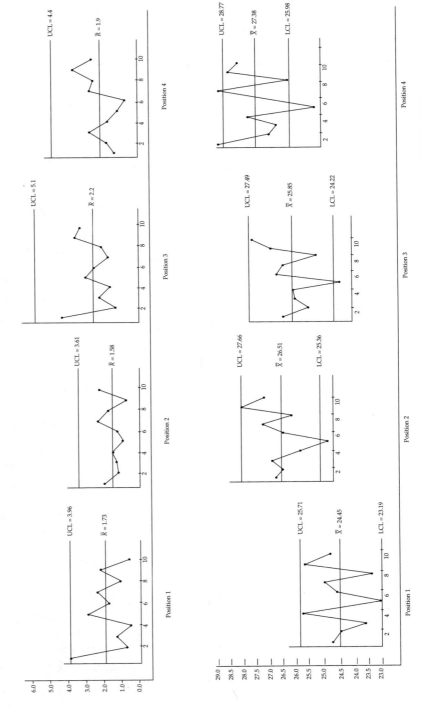

FIGURE 6.15   *Range and Average Charts by Position, Subgroup Size of Four, Cycle-to-Cycle is Within Subgroups*

Conclusion 2

Questions about what causes create the inconsistencies in position averages and the work to be done on those causes must now be addressed. Causes for the inconsistent behavior are most likely to be found among those sources that have the opportunity to occur between subgroups. Because subgroup averages display more variability than predicted by cycle-to-cycle causes, high on the list of suspected reasons for the inconsistent averages would be those causes described as time-to-time variations. Several possible reasons follow:

1. One or more of the causes active within subgroups (cycle-to-cycle factors) may have changed; this change resulted in an increase in dimension of all parts in the subgroup, which changed the average but left the deviations among the four parts essentially unchanged.
2. Factors either typically inactive or absent between subgroups have episodically become active.
3. Typically active factors are creating differential effects not experienced previously.

The fact that the range charts are in control indicates that the first possibility is unlikely. It is unlikely, but not impossible, that one or more of the process causes responsible for cycle-to-cycle variations changed the average but did not change the range for that subgroup. It is more likely that one or more of the process causes not captured within subgroups is responsible. "Between-subgroup" causes are ones having the capability of affecting a cluster of parts simultaneously, significantly decreasing or increasing the subgroup average, $\bar{X}$, and leaving unchanged the deviation or range of the parts. The need to focus on possible causes acting between subgroups is one of the reasons why attempts to clarify and define those process factors acting within and between subgroups must be emphasized.

Conclusion 3

There is more part-to-part variability than implied by the average range for any one or for all positions. Consequently, $\bar{R}/d_2$ does not describe total process variability since the $\bar{X}$ chart indicates additional variation above what is captured by $\bar{R}$. Total variation is larger still since position-to-position deviations are also revealed by the different centerlines on the average charts.

Conclusion 4

Since each R chart is in control, cycle-to-cycle variations are consistent at each of the positions. However the effect of cycle-to-cycle causes may not be the same for all positions. The $\bar{R}$'s for each of the positions are, respectively, 1.73,

1.58, 2.2, and 1.9. These different values by position for average ranges may provide evidence that within-position variability is measurably different. Similarly, the centerlines on the $\bar{X}$ charts appear to be quite different, with $\bar{\bar{X}}$'s of 24.45, 26.51, 25.85, and 27.38, respectively for positions 1 through 4. Different averages by position would also provide evidence of distinct subprocess differences. Such differences would result in more process variation than would otherwise exist. To determine whether the differences in the averages of both ranges and subgroup averages indicate real process differences, a statistical comparison of these is given in the following paragraph.

One statistical method for assessing the equality of average ranges involves constructing one range chart for all positions, computing an overall centerline based on all ranges, calculating an upper control limit, and making an assessment of "control." A plot of the ranges from all four positions is shown in Figure 6.16. The average of all position ranges is 1.87, which results in an upper control limit of 4.26. There are no individual ranges outside the limits; the statistical interpretation of this phenomenon is that all positions essentially have the same within-position variability and that the value of 1.87 represents the effects of cycle-to-cycle causes for all positions.

The observed differences among the $\bar{R}$'s can be interpreted statistically as being due to random process fluctuations. Figure 6.16 also contains an overall $\bar{X}$ chart; the upper and lower control limit have been calculated from the common $\bar{R}$ value of 1.87. Since numerous points fall outside of the control limits, the chart reinforces the conclusion that positions run to different averages. It appears that positions 2 and 3 have about the same average, position 1 is running at a smaller average, and position 4 seems to be operating at a higher average.

The observed differences in the subgroup averages for the four positions must be interpreted with caution. The individual position averages, with the possible exception of position 1, are not in control and therefore their values for $\bar{X}$ are not to be trusted. Since the individual positions do not have consistent averages time to time, the overall process average cannot be reliably determined. As discussed under conclusion 2, attention must be directed toward those causes suspected of creating the inconsistent behaviors observed on the $\bar{X}$ charts of Figure 6.15.

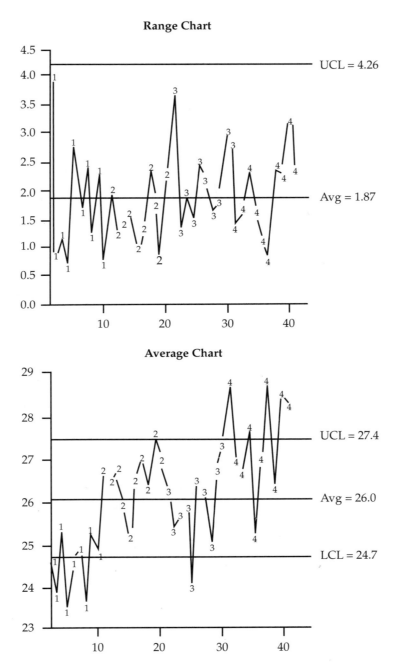

FIGURE 6.16 *Range and X-Bar for the Four Positions*

6.7.7   Alternate Subgroup Structures

There are several reasons for examining the other two subgroup structures identified earlier. Among the three alternate subgrouping plans, a preferred structure has been presented, rationalized, and interpreted. Examination of the other two structures has two distinct advantages:

1. A discussion of the other two structures provides a means for gaining insight into the way in which the determination of within- and between-subgroup causes will influence what is learned from $\bar{X}$ and R charts.
2. Many alternative sampling plans and subgroup structures can be designed. In practice, it is not always clear which one or ones might be more useful for process control or investigation. It is typically the case that sample plans or the subgroup structure, or both, require modification as work proceeds. The different process views from these other plans may support different process insights. Examination of several subgroup structures may be useful because of the alternative views of process behavior that are provided.

Subgroup Structure 2

Using the same sampling plan and the same data, subgroups are formed with position-to-position causes within subgroups. The data of Table 6.1 were recorded in agreement with this structure. Table 6.3 presents these data along with the ranges and averages for each of the 40 subgroups. When parts from different positions are used to form subgroups, cause for within-subgroup variation can be subtle. Deviations among the four parts for each subgroup are thought to be due to measurement and position-to-position causes. Data would only be subgrouped in this fashion if there were valid reasons for believing that position variation is random or that any fixed differences among the positions are not important and discovery or recognition of such differences is not critical to answering specific process questions.

Deviations among the four parts in a subgroup are also influenced by those causes creating cycle-to-cycle differences. However, it will not be possible to describe the magnitude of the cycle-to cycle effect since positional differences are included within the subgroup. But it must be realized that the cycle-to-cycle causes are active in affecting each and every subgroup.

For this structure, only one range chart and one subgroup average chart are needed to provide a display of the process. The range and average charts for this subgroup structure are shown in Figure 6.17. Subgroup ranges and averages, respectively, have been plotted in production sequence. The potential utility of these charts is closely related to knowledge about where and how various process causes may reveal their effects.

TABLE 6.3 *Measurements of Parts with Position–to–Position Variations within Subgroups*

Day	Time	Cycle	Position				Across Position	
			1	2	3	4	Range	Average
1	1	1	22.09	27.57	25.00	29.13	7.04	25.9475
1	1	2	26.00	25.50	26.63	29.50	4.00	26.9075
1	1	3	25.70	27.35	28.46	28.24	2.76	27.4375
1	1	4	24.81	26.17	24.66	28.49	3.83	26.0325
1	2	1	24.46	27.02	25.32	26.13	2.56	25.7325
1	2	2	24.24	25.77	24.45	27.78	3.54	25.5600
1	2	3	24.09	26.72	25.48	26.98	2.89	25.8175
1	2	4	24.87	26.23	25.60	26.62	1.75	25.8300
2	1	1	23.95	27.41	24.75	26.48	3.46	25.6475
2	1	2	24.10	26.35	26.72	27.63	3.53	26.2000
2	1	3	22.77	26.06	25.48	27.15	4.38	25.3650
2	1	4	23.81	27.27	26.09	25.15	3.46	25.5800
2	2	1	25.55	27.04	26.58	28.01	2.46	26.7950
2	2	2	25.28	25.80	25.18	26.86	1.68	25.7800
2	2	3	25.40	25.45	25.30	27.67	2.37	25.9550
2	2	4	25.73	25.47	26.23	28.43	2.96	26.4650
3	1	1	25.08	24.83	25.07	24.81	0.27	24.9475
3	1	2	23.64	24.99	23.54	24.53	1.45	24.1750
3	1	3	22.17	25.73	24.54	25.56	3.56	24.5000
3	1	4	22.24	24.76	22.48	25.63	3.39	23.7775
3	2	1	25.01	25.75	26.57	27.34	2.33	26.1675
3	2	2	24.14	26.46	27.66	26.64	3.52	26.2250
3	2	3	25.35	26.48	26.14	27.16	1.81	26.2825
3	2	4	23.66	27.04	25.47	26.74	3.38	25.7275
4	1	1	26.40	26.49	26.49	30.03	3.63	27.3525
4	1	2	24.01	27.05	25.96	28.34	4.33	26.3400
4	1	3	24.39	26.12	26.98	29.54	5.15	26.7575
4	1	4	24.85	28.59	25.47	27.55	3.74	26.6150
4	2	1	23.20	25.20	24.27	25.82	2.62	24.6225
4	2	2	24.13	27.01	25.76	24.88	2.88	25.4450
4	2	3	23.10	26.36	25.62	27.15	4.05	25.5575
4	2	4	23.39	26.08	23.94	26.80	3.41	25.0525

TABLE 6.3   *Measurements of Parts with Position–to–Position Variations Within Subgroups (continued)*

| Day | Time | Cycle | Position | | | | Across Position | |
			1	2	3	4	Range	Average
5	1	1	25.74	27.81	27.16	26.91	2.07	26.9050
5	1	2	24.64	27.30	24.47	28.86	4.39	26.3175
5	1	3	24.53	27.36	27.60	30.19	5.66	27.4200
5	1	4	26.80	27.99	27.53	27.86	1.19	27.5450
5	2	1	24.71	26.31	25.94	27.35	2.64	26.0775
5	2	2	24.37	28.43	27.45	27.00	4.06	26.8125
5	2	3	24.71	26.11	28.79	28.74	4.08	27.0875
5	2	4	24.95	27.08	27.32	29.33	4.38	27.1700

R Chart

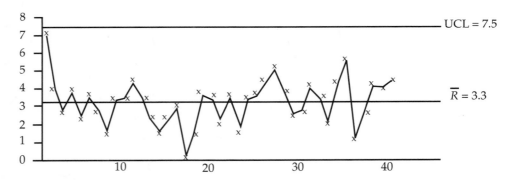

X Chart

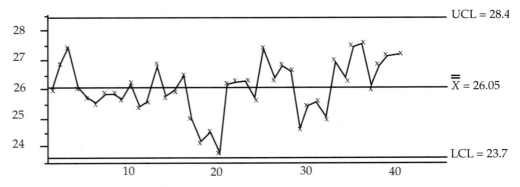

FIGURE 6.17   *Control Charts for Range and Average by Cycle; One Part per Cycle for Each Position*

Among the four identified categories of causes, measurement and position-to-position causes occur within subgroups. Cycle-to-cycle causes and time-to-time causes, *if irregular in their effect*, are structured to occur between subgroups. Discernment of substantial effects from either cycle-to-cycle or time-to-time causes depends on out-of-control signals on the subgroup average chart. The baseline variation against which the possibility of detecting the effects of inconsistent cycle-to-cycle and time-to-time causes is the average effect of position-to-position causes and, to some extent, subtle and often overlooked, cycle-to-cycle variations. The larger the within-subgroup variation is, the more powerful or aberrant must be the effects of other causes before their effects will be noted on the $\overline{X}$ chart.

Since the range chart does not contain any signals suggesting a lack of statistical control, there is some assurance that within-subgroup variations act consistently. On the basis of this chart, process interpretation would claim that position-to-position effects appear to act consistently. The average range, $\overline{R} = 3.3$, reflects the average deviation among four parts, one from each position at each cycle. Because of the subgroup structure, the multiple-source nature of this process has been confined within subgroups. Consequently, distinct differences in averages or internal variability among sources would not only contribute to the magnitude of the subgroup range, but the presence of these differences would also go undetected. Nor will the subgroup average, $\overline{X}$, reveal differences between positions, since the $\overline{X}$'s have been formed by averaging across those differences, if they exist. Distinct inequalities among position averages, if sufficiently large in magnitude, may be revealed on the range chart by a vast majority of points near the centerline.

Two observations about the process can be made by an examination of the $\overline{X}$ chart. There are no points outside of the control limits, indicating that no differential effects due to either cycle-to-cycle or time-to-time causes were identified. However, there is a pattern in the plotted points that does suggest some special cause at work: the last eight data points plot above the centerline on the chart. This suggests that one or more of the cycle-to-cycle or time-to-time causes has exerted a persistent upward effect on the average, that it affected all four positions, and that it may have been active during the last two sample periods. The foregoing remarks are all fairly standard and perfunctory. For gaining understanding in sampling, subgrouping, and their application, it is more worthwhile to compare the charts from this structure with the previously discussed one.

The overall numerical average, 26.05, is the same for the immediate structure and the previous structure based on cycle-to-cycle variations over a fixed time interval. This result is not surprising since the centerline for both $\bar{X}$ charts is the average of the same 160 values. However, comparison of the two range charts reveals opportunity for more distinctive contrasts. The average range for the current structure, $\bar{R} = 3.3$, exceeds the average range from within all positions of 1.87. The differences in the average ranges is due to the different causes acting within subgroups, as stated repeatedly. The obvious reason for the difference is that position-to-position causes exert larger differences among positions than cycle-to-cycle causes create within position. Contrasts in the subgroup average charts for the two subgroup structures begin to demonstrate some of the consequences. The larger within-subgroup variation in subgroup structure 2 results in wider limits on the current $\bar{X}$ chart, 23.7 to 28.4, compared to the limits for subgroup structure one of 24.7 to 27.4, intervals of 4.7 versus 2.7. A consequence is that the chart is less sensitive to between-subgroup variations. The current $\bar{X}$ chart does not call attention to time-to-time causes; the previous $\bar{X}$ charts did. But the questions may deliberately differ. Previously, the focused questions may have been "Are effects of cycle-to-cycle causes as represented within the given time period consistent? Are the discernible episodic events due to time-to-time causes larger than cycle-to-cycle effects contained within subgroups?" The answers were found by subgroup structure 1 to be "yes and yes." The differences in the magnitude of the within- and between-subgroup effects are apparently of process significance. Under different circumstances, pertinent, process questions may be "Are position-to-position effects consistent in the short term, that is, within a cycle? Is there evidence that cycle-to-cycle or time-to-time causes have episodic effects exceeding variations predominantly occurring position to position?" Wider limits may result, as in this instance, in deciding that between-subgroup causes do not create any observable deviations above and beyond that caused by the consistent deviations experienced among positions. It is a demonstration of the adage that effects buried or contained within subgroups will, in general, not be revealed by signals arising between the subgroups. Using the latter structure, an opportunity for attaining reduced process variation, identification, and correction of differences in position averages—perhaps relatively painlessly and economically—would be lost. The process issue is that the relevant questions must be determined first and then subgroup structure is formed to respond to these questions. In terms of statistics, it is seen that the two questions require different subgroup structures. The implied baseline variations against which to measure the effect of episodic events is different for the respective questions.

As a general comment, position-to-position differences may or may not exist in a process; and if they exist they may or may not exceed the variations created by cycle-to-cycle type causes, or for that matter, measurement. The relative magnitude of these effects is process dependent. Such characteristics are among the reasons why it is often necessary to conduct sampling and subgroup trials in order to find a useful data arrangement.

Subgroup Structure 3

The third subgroup structure includes all 16 observations from a sampling period in one subgroup. For convenience, these subgroups are reported in Table 6.4. The cause categories that create variation within the subgroups are measurement, cycle to cycle, and position to position. Time-to-time causes have the possibility of acting between subgroups for this structure; explanation as to unusual movement among subgroup averages would focus on this group of process causes.

The range and average charts for this structure are printed in Figure 6.18. Since the range chart is in control, the process interpretation is that measurement, cycle-to-cycle, and position-to-position effects are consistently realized for the 10 time periods. The $\bar{X}$ chart permits assessment of stability of the averages over time. The point below the lower control limit on the $\bar{X}$ chart suggests that effects due to time-to-time causes have been experienced and that the process average is not predictable, a judgment consistent with that obtained with the first subgroup structure discussed. A further comparison of the present charts with the ones previously developed reinforces some previous points about subgrouping and allows a brief discussion of subgroup size.

It is not appropriate to compare the current average range of 4.8 to the average ranges of 1.87 and 3.3 from the first two structures. The average range has increased for two reasons. First, for an increased subgroup size, the average range is expected to increase for controlled processes. An appropriate comparison can be made by adjusting the average range for subgroup size. The operation is legitimate for controlled processes, which is *not* the case in this situation. But the point is of sufficient interest to make the demonstration worthwhile although of no benefit in this particular process. Adjustment of the average range for subgroup size is done by dividing by d_2, which, for a controlled process, would provide an estimate of a process standard deviation. Table 6.5 gives comparisons for the values of $\bar{R}/d_2$ for the three subgrouping structures. An appreciation of the inflationary effect of including different causal structures within subgroups is provided by this table.

Subgroup structure 3 includes more possible causes acting within subgroups than do either of the previous two structures, resulting in more within-subgroup variation. Aside from increasing the average range, there obviously

TABLE 6.4 *Subgroups by Time Period*

Day	Time	Cycle	Position 1	Position 2	Position 3	Position 4	Time Period Range	Time Period Average
1	1	1	22.09	27.57	25.00	29.13		
1	1	2	26.00	25.50	26.63	29.50		
1	1	3	25.70	27.35	28.46	28.24		
1	1	4	24.81	26.17	24.66	28.49	7.41	26.581
1	2	1	24.46	27.02	25.32	26.13		
1	2	2	24.24	25.77	24.45	27.78		
1	2	3	24.09	26.72	25.48	26.98		
1	2	4	24.87	26.23	25.60	26.62	3.69	25.735
2	1	1	23.95	27.41	24.75	26.48		
2	1	2	24.10	26.35	26.72	27.63		
2	1	3	22.77	26.06	25.48	27.15		
2	1	4	23.81	27.27	26.09	25.15	4.86	25.698
2	2	1	25.55	27.04	26.58	28.01		
2	2	2	25.28	25.80	25.18	26.86		
2	2	3	25.40	25.45	25.30	27.67		
2	2	4	25.73	25.47	26.23	28.43	3.25	26.249
3	1	1	25.08	24.83	25.07	24.81		
3	1	2	23.64	24.99	23.54	24.53		
3	1	3	22.17	25.73	24.54	25.56		
3	1	4	22.24	24.76	22.48	25.63	3.56	24.350
3	2	1	25.01	25.75	26.57	27.34		
3	2	2	24.14	26.46	27.66	26.64		
3	2	3	25.35	26.48	26.14	27.16		
3	2	4	23.66	27.04	25.47	26.74	4.00	26.101
4	1	1	26.40	26.49	26.49	30.03		
4	1	2	24.01	27.05	25.96	28.34		
4	1	3	24.39	26.12	26.98	29.54		
4	1	4	24.85	28.59	25.47	27.55	6.02	26.766
4	2	1	23.20	25.20	24.27	25.82		
4	2	2	24.13	27.01	25.76	24.88		
4	2	3	23.10	26.36	25.62	27.15		
4	2	4	23.39	26.08	23.94	26.80	4.05	25.169

is increased complexity of causes contained within subgroups. Consequently, the relative contribution of cycle-to-cycle and position-to-position causes to total effects observed within subgroups is unknown. If the range chart indicated the presence of abnormal events, the complexity of within-subgroup causes would make the work of determining causes more challenging.

In comparing the charting results, the width of the control limits on the average chart should also be noted. Limits are 25.04 to 27.06, an interval of 2.02, which is smaller than for the other structures. The tight limits are due to the

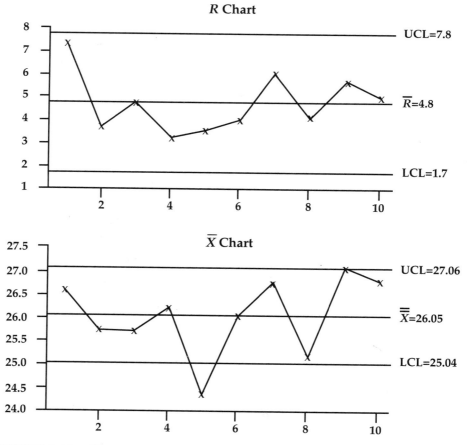

FIGURE 6.18 *Charts for Range and Average; Cycle-to-Cycle and Position-to-Position Causes are Within Subgroups (n = 16)*

TABLE 6.4 *Subgroups by Time Period (continued)*

Day	Time	Cycle	Position 1	2	3	4	Time Period Range	Average
5	1	1	25.74	27.81	27.16	26.91		
5	1	2	24.64	27.30	24.47	28.86		
5	1	3	24.53	27.36	27.60	30.19		
5	1	4	26.80	27.99	27.53	27.86	5.72	27.047
5	2	1	24.71	26.31	25.94	27.35		
5	2	2	24.37	28.43	27.45	27.00		
5	2	3	24.71	26.11	28.79	28.74		
5	2	4	24.95	27.08	27.32	29.33	4.96	26.787

TABLE 6.5 *Subgroup Structure, Causal Effects Within Subgroups, and Average Ranges Adjusted for Subgroup Size*

Subgroup Structure	By Position	By Cycle	By Sample Time Period
Causes within subgroups	Cycle to cycle	Position to position	Cycle to cycle and position to position
$\dfrac{\overline{R}}{d_2}$	$\dfrac{1.87}{2.059} = 0.91$	$\dfrac{3.3}{2.059} = 1.60$	$\dfrac{4.8}{3.532} = 2.33$

effect of increased subgroup sizes on decreasing the random variation that can be expected in values of $\overline{X}$. The net result is that although within-subgroup variation was increased by the choice of subgroup structure, the large subgroup size provided sufficient power to overcome that increase and resulted in detecting significant time-to-time variations. However, the large subgroup size also "hid" position-to-position variation within subgroups because this contributor to process variation was not detected on the control charts. The issue of subgroup size should not be considered separately from deciding on subgroup structure. Since the first subgroup structure provided the best understanding of process variation, increasing the subgroup size might be useful to consider in connection with the use of this structure. The subgroup size of structure 1 could be increased in at least two ways. The user could expand the base time interval and increase subgroup size by increasing the number of sampled cycles. That would typically have the effect of increasing the within-subgroup variation because the list of possible

Principle 5

Never knowingly create subgroups so as to include fixed differences within a subgroup.

Principle 6

Sources of variation affecting within-subgroup variation should be consistent with sources that affect differences in the subgroup averages.

6.9 APPLICATION PROBLEMS

6.9.1 The situation described in this problem is the filling of 10-ounce juice boxes. Ten different kinds of juice may be filled by a 12-headed filling machine. Fill weights from this machine have been erratic with many underweight boxes needing to be removed from the packaging line. A list of reasons for variation in the fill weights is being developed, with the current list as follows:

- within-head variation,
- between-head variation,
- the average fill from the machine may drift over time,
- juice is spilled as the boxes move along the conveyor belt prior to being sealed, the conveyor belt does not run at a constant speed; it may be slowed down or sped up,
- operators adjust the equipment if they believe the fill is too much or too little,
- some juice types are more "pulpy" than others, and
- the same juice type may be more "pulpy" at some times than others.

Three different sampling plans and subgroup structures are described. For each of the three, answer the following three questions:

a. What sources of variation impact subgroup ranges?
b. What sources of variation are reflected between subgroups?
c. What learning objectives would be supported by this subgrouping plan?

Subgroup Plan 1

Twice each shift, for 10 consecutive shifts, measure the net weight of the 12 boxes filled by a cycle.

Subgroup Plan 2

Randomly choose five containers from the 12 filled during a cycle. The five net weights will form a subgroup. Collect one subgroup on each shift of 20 consecutive shifts.

Subgroup Plan 3

Each of the 12 heads is labeled. For two consecutive cycles of the filling machine, determine the net weights of the two boxes filled by head 1. Do this for all 12 heads. Four or so hours later (on the same shift) obtain 12 more subgroups by the same method. There will now be 24 subgroups available to study process variation.

6.9.2 Scuff plates are the plastic plates at the base of the door frame in automobiles. They are manufactured in an injection molding process. Two clamps (one on each of two machines) are used to make these plates. Each clamp has two cavities, one for the left-hand plate and one for the right-hand plate. An injection gun is hooked up to each clamp. The clamps close and the molding material is injected into the cavities. When the clamp opens, two parts are released. The clamps and injection of molding material are operated by automatic controls. Once the parts are removed from the cavities, an operator must clean the cavities, if necessary, to ensure the quality of the next run. Each clamp is run by one operator throughout an entire shift. Some rotation of operators occurs between shifts.

A critical characteristic in the molding process is the length of the parts. Each part has five pinholes to allow for automobile assembly; part length is evaluated by measuring the distance from the end of the plate to the fifth pinhole. The following sampling and subgrouping plan is proposed: For each machine, measure three successively manufactured left-hand parts and right-hand parts every two hours throughout eight successive shifts. Four separate range and X-bar charts will be constructed. There will be range and X-bar charts for right-handed parts from machine 1 and a separate pair of charts for left-handed parts from machine 1. The same two sets of charts will also be constructed for machine 2.

a. How would differences in the average dimension of right-hand and left-hand parts be noted on the four sets of control charts?
b. How would different within-machine variations affect the four sets of charts?
c. Will it be possible to note any differences between shift operation? What kinds of differences might be detected?
d. If molding materials affect part dimensions, how will these effects influence the behavior of the range and X-bar charts?

7

Control Charts for Variables Data: Moving Range and Individuals Charts

In the previous chapters on control charts for variables data, X-bar and R charts were used to gain process information from variables data that are arranged in rational subgroups of two or more observations. In the example on soap container volumes from Section 5.5.2, twice each shift for 16 shifts, weight measurements were taken on the net contents of two consecutive containers from the same head for each of four heads. Subgroups of size two were formed from these data. Information on sources of process variation was then obtained by knowing what sources of variation were likely to affect within-subgroup variation and what sources would be active between subgroups. These same ideas of strategic sampling and subgrouping formation were further explored in Chapter 6. The present chapter is devoted to considering process situations where it is either not practical, possible, or instructive to form rational subgroups of two or more measurements on a particular process characteristic. In these situations only one measurement is available to represent a particular process or product characteristic and it is not logical to form subgroups of more than one observation. However, in these situations the same kinds of process learning are required. We still need to know about the stability and magnitude of process variation and process average and to learn about sources of variation that influence these characteristics. This chapter discusses the use of moving range, individuals, and moving average charts as methods for gaining this required process knowledge.

7.1 PROCESS DATA WITH LOGICAL SUBGROUPS OF SIZE ONE

In many processes, the measurement of a single quality characteristic, such as a process input or product characteristic, logically dictates a subgroup size of one. These are processes where one measurement represents a condition for a given time period, such as when one measurement represents a characteristic of an entire batch. In such a situation, there is no rationale for forming subgroups. Here are three other common situations where only one observation is available to represent a given set of conditions:

1. The variability occurring within a set of conditions at one time period is not realistically useful for defining variation over time.
2. Obtaining measurements for the characteristic to be studied is expensive.
3. Measurements require long periods of time to obtain.

An example of the first situation in which the variability occurring in one set of circumstances is not useful in defining variation over time is based on a process for baking cakes. In the process for baking a cake, there is a need to understand why one cake that is mixed and baked at one time differs from another cake baked by the same person at a later time. If the characteristic of concern is texture, then measuring the texture of several pieces of the same cake is of little help in judging the stability and magnitude of the differences between the cakes over time. Variations within the same cake do not constitute cake-to-cake variations. Within-cake variations may be mostly due to measurement error rather than to the process for making cakes. If unknown, measurement error needs to be understood and its magnitude calculated, but that variation is not useful in assessing the effects of short- and long-term variations in cake texture. Therefore, a reasonable approach for studying the variability in textures between cakes is to obtain one measure of cake texture for each cake baked.

Other examples of situations in which only one number describes a given set of conditions are

*   the study of moisture content in a process that produces large, homogenous batches of material, and
*   the study of process parameters such as temperature, pressure, and amperage.

Examples of situations where measurements are expensive or require long periods of time to obtain are

*   average weekly overtime hours per full-time employee,
*   daily utilization rates of equipment or other resources,
*   accounting data on shipments and orders,
*   monthly number of items produced per direct labor hour,
*   monthly deviations of actual sales from forecast, and
*   chemical levels in treated water that are available in two- to four-hour increments.

In each of the preceding examples, only one observation is available to represent a given set of circumstances. However, in all of the situations represented by the examples, there is a managerial need to examine the predictability, the magnitude of variation, and the average levels of critical process parameters. Also, in order to improve these processes, it is important to understand the causes of variation and to evaluate the effect of management or engineering changes. Therefore, control charting techniques are needed in these situations that have only one measurement of a particular characteristic at any one condition.

7.2   EVALUATING PROCESS BEHAVIOR WITH MOVING RANGE AND INDIVIDUALS CHARTS

When subgroups consisted of two or more observations, short-term variability was evaluated by computing the subgroup ranges. When only one measurement is available at any one condition, the range cannot be computed as the difference between the largest and smallest observations within a subgroup. To be able to obtain information on the stability and magnitude of the short-term variation, a moving range is used. A moving range is defined to be the absolute value of the deviation between two consecutive observations. The average deviation (i.e., the average moving range) is used to calculate limits for a chart of the individual observation. Thus, analysis of the moving ranges allows for

1. a study of the short-term variation and
2. provides a baseline against which long-term process movements can be evaluated.

The calculations involved in the construction of the moving range and individuals charts are demonstrated using the data in the following example.

7.2.1   Construction of Moving Range and Individuals Charts

Data taken from a batch process that produces a sterilized, concentrated baby formula are initially collected to provide a baseline of current process performance. Although there are several important product characteristics that could be studied, the characteristic we have chosen to study in this example is the Brookfield viscosity in centipoise. The specifications for this product characteristic are 900 ± 100. The viscosity for 20 consecutive batches of formula is measured. The sampling scheme for these measurements requires that one specimen (sample) be taken from a randomly selected position in a completed batch. Thus, each reported number represents a finished batch. The resulting measurements are provided in Table 7.1.

TABLE 7.1 *Viscosity Readings*

Batch	Viscosity	Moving Range
1	805	—
2	976	171
3	901	75
4	929	28
5	927	2
6	942	15
7	904	38
8	804	100
9	874	70
10	944	70
11	850	94
12	941	91
13	992	51
14	795	197
15	952	157
16	832	120
17	809	23
18	878	69
19	936	58
20	888	48
Totals	17,879	1477

The individual viscosity measurements, one per batch, constitute the numbers plotted on the "individuals" chart. A measure of baseline variation is required to analyze the stability of sources contributing to variation in viscosity measurements. Moving ranges of the individual measurements provide this baseline variation. Moving ranges are first calculated, the moving range chart is constructed, and the stability of the short-term variation reflected by these ranges is determined.

As indicated earlier, the moving range is the absolute difference between two consecutive, individual values. Therefore, the first moving range (MR) recorded in Table 7.1 is the difference between the first and second viscosity measurements. This first moving range is calculated as:

$$MR = 976 - 805 = 171$$

In calculating the moving ranges, the larger of the two consecutive observations is written first to obtain the absolute deviation between the two numbers. The second moving range is calculated in a similar manner, using the observations from the second and third batches:

MR = 976 − 901 = 75

With the exception of the first and last number in the data set, each observation is used twice in calculating moving ranges. For example, the observation on the tenth batch, 944, is compared with 874, the preceding batch result, to obtain MR=70, and again with the immediately following batch result, resulting in MR=944−850=94.

Once the moving ranges have been calculated, these values are plotted onto the moving range chart. There is one less moving range than there are individual values in the data set. The average value for the moving range is computed as is any other average—by summing the moving ranges and dividing their sum by the number of moving ranges. For these data on viscosity measurements, the average moving range is:

$$\overline{MR} = \frac{1477}{19} = 77.7368$$

Since the calculation of the individual moving ranges involves two observations, the control limits for the chart for moving ranges are computed based on a subgroup size of $n=2$. From the table of factors for use with X-bar and R charts (Table D.2 in Appendix D) there is no value for D_3 when $n=2$. Thus, there is no lower control limit for the chart of moving ranges. The upper control limit is computed in the same way as one would compute an UCL for a range chart for subgroups of size two. For this example, the upper control limit is:

$$UCL_{MR} = D_4\overline{MR} = 3.267(77.7368) = 253.97$$

The completed moving range chart is provided in Figure 7.1. All values on the moving range chart are within the control limits, providing evidence that short-term variation is stable across the base time period of the study. The runs tests for abnormal patterns in the data should not be applied to the moving range chart in checking for stability of short-term variations. Moving ranges are calculated from consecutive observations and therefore share an observation in common with another moving range. This characteristic creates a lack of independence in the values of the moving ranges, a mathematical prerequisite for appropriate use of the usual runs test.

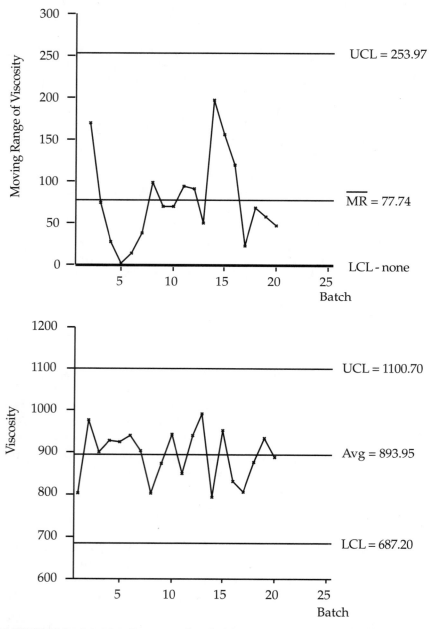

FIGURE 7.1 *Moving Range and Individuals Charts of Viscosity Readings*

The average moving range, $\overline{MR}$, captures the magnitude of the short-term variation. When the moving range chart is in control, $\overline{MR}$ is used to provide a basis for judging the stability of the individual readings of the batch process. If the short-term variation captures all the variations affecting the individual batch viscosity readings, then the formula for computing the standard deviation of individual readings is:

$$\frac{\overline{MR}}{d_2}, \quad \text{where } d_2 = 1.128 \text{ for n} = 2$$

The limits on the individuals chart are based on this short-term component of variation. If additional sources of variation occur over the long-term, as defined by the time span covered by the collected data, this additional variation should show up as an out-of-control signal on the chart of individual values. When the MR chart is in control, the control limits for the individuals chart are computed using the following formulas:

$$UCL_x = \overline{X} + 3\left(\frac{\overline{MR}}{d_2}\right)$$

$$LCL_x = \overline{X} - 3\left(\frac{\overline{MR}}{d_2}\right)$$

Where the d_2 value of 1.128 is used. This is the d_2 value for subgroups of size two and its use is consistent with the use of D_4 for subgroups of size two. For the individual readings on viscosity, the centerline and upper and lower control limits are:

$$\overline{X} = \frac{17879}{20} = 893.95$$

$$UCL_x = 893.95 + 3\left(\frac{77.7368}{1.128}\right) = 1100.70$$

$$LCL_x = 893.95 - 3\left(\frac{77.7368}{1.128}\right) = 687.20$$

The completed chart of individual viscosity readings also appears in Figure 7.1. All of the points lie within the control limits. A runs test is recommended for individual charts. In this instance, an application of the runs test does not indicate any nonrandom behavior in the points. Based on these 20 batches, the process would be judged to be in control.

7.2.2   **Process Characterizations**

Since the process appears to be in control with respect to both short-term and long-term variation, the ability of the process to meet requirements can be evaluated. The process standard deviation can be estimated by:

$$\hat{\sigma} = \frac{\overline{MR}}{d_2} = \frac{77.7368}{1.128} = 68.92$$

Process specifications on this performance characteristic, Brookfield viscosity, were stated as being 900 ± 100, with a lower specification limit of 800 and an upper specification of 1000. The current process average of 893.75 is on the low side of 900, the targeted process value, but may be judged as being satisfactory, given the magnitude of the reported standard deviation of 68.92. However, given the variability in this process, it is obvious that batches will be produced with viscosity measurements outside the required interval of 800 to 1000. The control limits, 687.2 and 1100.7, represent values that are at the current process boundaries. With an average of about 894 and a standard deviation of 69, this process will occasionally produce batches having a viscosity measurement as low as 690 or so and batches having viscosity values as high as 1100. The deviation from the anticipated low to high values is:

$$1100.7 - 687.2 = 413.5$$

This deviation greatly exceeds the engineering tolerance of 200, indicating that the process is not capable of consistently producing according to requirements. For this process, the natural tolerance, NT, is calculated to be:

$$NT = 6\hat{\sigma} = 6(68.92) = 413.52$$

The value for the natural tolerance corresponds to the difference between the upper and lower control limits on the chart for individual values.

7.2.3   **Interpreting the Information Provided in MR and Individuals Charts**

As with all control charting techniques, successful use of the moving range and individuals charts requires that a sampling plan be developed that is based on knowledge of causes that potentially affect the process or product characteristic being studied. For sampling, measuring, and control charting efforts to be productive, they must be planned so that the resulting information can be interpreted in terms of suspected causes of variation and provides information on future direction of improvement efforts.

For the process producing baby formula, the sampling design included the selection of one sample from each completed batch and the determination of one measurement on viscosity for each sample. Thus, each individual observation represented the batch from which the measurement was taken. To interpret the charts with regards to potential sources of variation, it is necessary to understand what sources are captured in the charted values.

Since each moving range was based on two individual readings of viscosity, measurement error will contribute to the variation seen in the moving ranges. Each of the two individual readings used to compute a moving range represented a different batch of formula. These moving ranges will capture the effects of within-batch variability as well as differences between the batches. In summary, the effects of three important sources of variation are captured in the moving range chart for viscosity readings on baby formula. These three sources of variation are

- measurement,
- within-batch, and
- batch-to-batch variability.

To determine the magnitude and reasons for measurement variation, it is necessary to perform studies on the measurement process used to provide viscosity readings. The data collection and methods for studying the variation due to the measurement process are discussed in Chapter 9. The other sources of variation mentioned represent two general classes of causes that are based within the manufacturing process:

1. *Causes that make one batch differ from another.* A common reason for batch-to-batch differences is changes in raw materials. If amounts or properties of raw materials used in the process vary from batch to batch, this could result in batches differing from each other. Other process changes, such as personnel, equipment, or methods changes, also tend to contribute to differences between batches.

2. *Causes that tend to result in nonhomogeneous batches.* In making any one particular batch, there are typically sources of variation that create within-batch variations in addition to and independent of those sources that create batch-to-batch differences. Examples of possible causes of within-batch variability would be temperature gradients within a batch or incomplete mixing in the batch. The within-batch variability will be of a certain magnitude for any one batch and the magnitude of the within-batch variation may be different from one batch to another. The random sample selected from the batch means that the sources driving the variations within batches will be reflected in the measurements made on this process.

To effectively manage process variation, those working on improving the process are responsible for

- knowing about the occurrence of a special cause and when it happens, and
- identifying and addressing the reasons for its occurrence.

Carrying out these responsibilities will require an understanding of how the sources of variation (measurement, within batch, and between batch) will impact the moving ranges and individual values. The moving ranges are affected by both within- and between-batch sources of variation as well as measurement variation. The average value of the moving range indicates the average deviation from one batch measurement to another for this process. The upper control limit for the moving range chart provides information on the magnitude by which measurements from consecutive batches could differ from each other. Points on or above the upper control limit for the moving range chart reveal abnormally large shifts or changes in variation from one batch measurement to the immediately following batch. In view of the sources of variation affecting the moving ranges, such a large shift or change could be a result of the following:

1. The measurement process has either become more variable or its level has changed.
2. The homogeneity of batches has changed so that within-batch variation has increased.
3. The level of viscosity from one batch to the next has changed.

If the moving range chart is in control, control limits are placed on the individuals chart. These limits provide information to judge whether the process is subject to variations exceeding those represented by the average moving range which make measurements on consecutive batches differ. Variations in individual values have already been described by considering what would create differences between consecutive individual observations. Consequently, out-of-control points on the individuals chart may be accounted for by the same three reasons listed earlier. The individuals chart may also provide evidence of long-term changes in process variation; gradual increases or decreases, which would go unnoticed on a moving range chart, may be detected on the individuals chart by one or more points outside the control limits or by failing the runs test. Reasons that could account for long-term changes would be

1. a gradual change in the level of the measurement process or
2. a gradual increase or decrease across batches in viscosity levels.

In designing sampling and subgrouping plans for subgroups of two or more, an effective strategy for process study was to deliberately choose some sources of variation to only be active between subgroups; with this strategy, the X-bar chart would provide evidence of the effects of these sources. By analogy, the moving range and individuals charts separate sources of variation that may be felt to be short term, reflected on the moving range chart, with those whose effects tend to be felt over a longer time frame and so evidenced on the X-bar chart.

Given the information about measurement, within-batch variation, and between-batch variation that can be gained from the moving range and indiviuals charts, the effective use of these charts to gain process knowledge can be considered. As with any of the charting techniques, process knowledge and experience with respect to work practices, material properties, and equipment characteristics will dictate what can be learned about the process. For example, it may be known that measurement variation is quite small and that within-batch variation in viscosity is also likely to be negligible. In this case, the moving range and individuals charts would provide information on sources creating variation between batches. On the other hand, if both within- and between-batch sources of variation are likely to have measurable effects, the distinct nature of these two sources cannot be understood by a sampling plan that only takes one sample from each batch. Separating these sources of variation will require other sampling and subgrouping schemes; this topic is further explored in Chapter 8.

7.3   A DISCUSSION OF ISSUES ASSOCIATED WITH THE USE OF MR AND *X* CHARTS

The basic situation in which moving range (MR) and individuals (*X*) charts are used to analyze process data is one in which there is no rational process basis for establishing a subgrouping scheme of two or more observations. Without a basis for a subgrouping plan, there is no basis for establishing a baseline variation against which to evaluate the stability of process variation over time. A compromise is provided by using two consecutive observations to capture variation that might be reasonably used to baseline the process. The absolute deviation between two consecutive observations is the moving range, and the average value of the moving ranges for a controlled process is used to provide an estimate of process variation.

Two of the key issues regarding the use of MR and X charts center around other methods often suggested for capturing baseline variation. As discussed in Sections 7.3.1 and 7.3.2, these alternative methods are not recommended.

The two methods discussed as alternative ways to capture baseline variation are

1. Using the standard deviation of all observations to establish baseline variation or
2. Arbitrarily grouping the data in distinct subgroups of size two, three, or more and then using R and X-bar charts.

Additional issues discussed in this section are the assumption of normally distributed observations and the moving average chart as an alternative to the individuals chart.

7.3.1 $\overline{\text{MR}}$ Versus the Standard Deviation to Establish Baseline Variation

In the previous example, measurements were taken on batch viscosity. Because each batch was made and finished under a given set of process conditions, only one observation is used to represent each batch. The moving range is used to provide information on process variation. If one or more process causes substantially change from one batch to the next, the moving range will reflect the effect of such changes. The moving range is also influenced by effects taking place during the completion of one batch. An operational understanding of process factors is essential for defining and understanding the sources that contribute to the value of the average moving range.

Methods other than the moving range are frequently suggested for baselining process variation. For example, to study variation in viscosity for a batch process, it may be suggested that one observation be taken from each of 25 batches, with the batches being obtained over a lengthy production period. The average and standard deviation of all 25 numbers can be calculated and control limits for the individuals chart determined from these results. Calculating control limits from an overall standard deviation computed from data collected over a considerable time period is not good practice and should not be used. If the process is subject to numerous and erratic causes of variation, the effects of these causes will be captured in the calculated standard deviation. Limits on an individuals chart that are based on the overall standard deviation will not provide a useful basis for judging process stability since the width of these limits reflects not just common cause variation but also those erratic causes creating additional variation. The inadequacy of this practice is illustrated using the data set of Table 7.2. These data represent viscosity measures for batches of baby formula, collected over a different time period than the data in Section 7.2. For discussion purposes, a limited number of observations is used.

TABLE 7.2 *Viscosity Readings*

Batch	Viscosity	Moving Range
1	930	—
2	926	4
3	940	14
4	900	40
5	890	10
6	880	10
7	850	30
Totals	6,316	108
Averages	902.3	18

The recommended procedure for analyzing the data of Table 7.2 is the moving range and individuals charts. For this data set, the centerline and upper control limit of the moving range chart are computed as:

$$\overline{MR} = \frac{108}{6} = 18$$

$$UCL = 3.267(18) = 58.8$$

Since none of the moving ranges falls above the upper control limit, the centerline and control limits for the individuals chart can be calculated. The average viscosity for these measurements is 902.3. The upper and lower control limits for the X chart are computed as:

$$UCL_X = 902.3 + 3\left(\frac{18}{1.128}\right) = 902.3 + 47.88 = 950.18$$

$$LCL_X = 902.3 - 3\left(\frac{18}{1.128}\right) = 902.3 - 47.88 = 854.42$$

The seventh batch has a viscosity reading of 850, which is below the lower control limit. Thus, there is evidence that the process is inconsistent with respect to process average.

Using the data in Table 7.2, the standard deviation for all seven viscosity measurements is calculated and used to judge process stability. When using the overall standard deviation as a means for analyzing statistical control, the standard deviation of all observations is used to compute the control limits for the individuals chart. For the data in Table 7.2, the standard deviation, s, of the seven viscosity measurements is 31.988. Limits for the X chart are computed using the following formulas:

$$UCL_x = \overline{X} + 3(s)$$
$$LCL_x = \overline{X} - 3(s)$$

The following table provides the values for the upper and lower control limits for both methods:

Limits for X Chart Based on $\overline{MR}$	Limits for X Chart Based on s
$UCL_x = 902.3 + 3\left(\dfrac{18}{1.128}\right) = 950.18$	$UCL_x = 902.3 + 3(31.988) = 998.26$
$LCL_x = 902.3 - 3\left(\dfrac{18}{1.128}\right) = 854.42$	$LCL_x = 902.3 - 3(31.988) = 806.34$

When the limits on the X chart are constructed using s, the process is judged to be in control. However, as previously noted, the process was judged to be out of control when it was analyzed using limits constructed with the average moving range. By using the standard deviation of all of the observations in the data set, the effects of all sources of variation acting within and between batches as well as across all batches, have been included in the "baseline" variation. If special causes are acting on the process, the control limits are much wider when constructed using s than they are when the average moving range is used. These wider limits reflect the effects of causes operating throughout and across the process. Consequently, it becomes more difficult to detect the presence of special causes.

A similar rationale has been used when constructing the two sets of limits in the preceding table. Both $\overline{MR}/d_2$ and s are called "standard deviations." For both cases, three of these "standard deviations" were added and subtracted to $\overline{X}$ in order to determine control limits for the individuals chart. The basis for stating that it is incorrect to use the second set of limits, those based on s, to put limits on the X chart is due to the different process variation, which is reflected in these two different "standard deviations." For the data in Table 7.2, $\overline{MR}/d_2 = 15.96$ and $s = 31.988$. These values differ because the causes reflected in these values differ. The standard deviation of 15.96 reflects the effect of measurement variation, within-batch variation, and variation due to differences in consecutive batches. The s value of 31.98 reflects a much longer and more complex list of possible sources of variation. In addition to those sources affecting $\overline{MR}/d_2$, s is affected by sources impacting all batches over a longer time period and by the presence of any special causes in this time period. The presence of these latter causes in the present cases resulted in the standard deviation for the complete data set larger than the $\overline{\text{MR}}/d_2$ value. By using the average moving range to put limits on the individuals chart, the resulting chart is more likely to reveal the presence of additional long-term variation or special causes acting over the time period in which the data are collected. On the other hand, if s is used to put limits on the individuals chart, these additional sources of variation are included in the baseline variation. Thus, the detection of these additional sources as special causes is impaired.

7.3.2 Formation of Subgroups with the Individual Measurements

Another method commonly used for control charting individual measurements is to group the data in distinct subgroups with an arbitrary size of two or more. The analysis for process stability would then proceed by using R and X-bar charts. Again, this approach is not recommended. If the measurements require a long period of time to obtain, then an obvious disadvantage of this approach is the long wait required before plotting the data. More importantly, however, is the issue that the effects of causes of variation may be hidden or lost with this approach.

1. Similar to the effects of using s to place the limits on the individuals charts, the grouping of individuals data into illogical *subgroups of size three or more* results in a larger estimate of baseline variation. When additional and intermittent sources of variation are captured in the subgroups, the likely result is a large average subgroup range. Any expansion of time used to establish baseline variation means that potential sources of variation that may act intermittently are captured in that baseline variation. Therefore, when the average subgroup range is used to place limits on the average chart, the limits may be insensitive to the presence of any special causes. The effects of causes may be hidden, but would be revealed through the use of the moving range and individuals charts.
2. If *subgroups of size two* are formed, less information about process variation will be had. The truth of this statement can be seen by considering the effect of breaking the data of Table 7.1 up into 10 subgroups of size two. Instead of having 19 moving ranges, only 10 ranges would be used to provide an estimate of baseline variation. Nine comparisons of viscosity from one batch to another would have been "lost." Nine opportunities to identify special causes acting on the moving range chart would be lost; and the amount of information used to estimate baseline variation in order to place limits on the X chart would be reduced.

When there is no rational basis for subgrouping the data, it is best to recognize that limitation in studying variation, make a reasonable compromise, and proceed. A reasonable compromise is to use the range of two consecutive observations in establishing a baseline variation.

7.3.3   Assumption of Normality and the Use of the Moving Average Chart

Most of the control charting methods, including p and np charts, c and u charts, and X-bar charts, are based on the fact that, for a process that is operating in control, the p, np, c, u, and X-bar values have an approximately normal distribution. The explanation of this fact (called the central limit theorem) is contained in Appendix A. However, an approximate normal distribution may not be a reasonable model for describing the distribution of individual values collected on a controlled process. For example, times until the occurrence of an event (such as a machine breakdown, replacement of a cutting tool, or completion of a filtration cycle) are often better described by a skewed distribution, such as the one shown in Figure 7.2. Also, numerical measures for truncated processes, such as moisture measures where values below or above a threshold are impossible, are not typically described by a normal distribution. In any case, where the data are not approximately normally distributed, the use of 3σ limits on an individuals chart is not appropriate.

In addition to the difficulty created when 3σ limits are not appropriate for a measured characteristic, another issue to consider is that, unless a process is

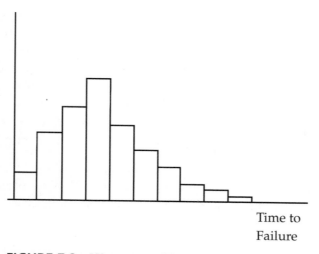

Time to
Failure

FIGURE 7.2 *Histogram of Times to Failure of an Electric Component*

in control, no reliable data can be gathered from the process to determine the distribution of process measurements. Because the process is not known to be in control, it is an error in reasoning to believe that a process distribution exists or that reliable information as to histogram shape can be determined from the ongoing process. Consequently, the decision that individual readings cannot be evaluated using 3σ limits is not completely based on statistics. It relies on knowledge of the physics, chemistry, or engineering of the process characteristic in question. If, based on process knowledge, it is decided that process measurements cannot be expected to be approximately normally distributed when the process is in control, one alternative to the individuals chart is the use of a moving average chart.

The construction of a moving average chart is illustrated using the viscosity data contained in Table 7.1. Individual viscosity measurements are reproduced in Table 7.3, and the two-item moving averages for this data set are also reported. The moving average, $\overline{MX}$, is simply the average value of two consecutive observations. Thus, the first moving average in Table 7.3 is obtained as the average of the first two observations:

$$\overline{MX} = \frac{805 + 976}{2} = 890.5$$

The second moving average is the average of the second and third observations, and so on. As with the calculation of moving ranges, each observation, except for the first and last, is used twice in the computation of the moving average values. Therefore, with 20 observations, there are 19 moving averages.

The stability of the viscosity readings produced by the process can be analyzed using MR and $\overline{MX}$ charts. (This process was judged to be predictable when MR and X charts were used to evaluate control.) The MR chart is contained in Figure 7.1, and is repeated in Figure 7.3. The moving average chart is constructed using the same equations as for an X-bar chart. The centerline and control limits are found by:

$$\overline{\overline{MX}} = \frac{\sum_{i=1}^{19} \overline{MX}_i}{19} = \frac{17032.5}{19} = 896.45$$

$$UCL_{\overline{MX}} = \overline{\overline{MX}} + A_2 \overline{MR} = 896.45 + 1.880(77.7368) = 1042.60$$

$$LCL_{\overline{MX}} = \overline{\overline{MX}} - A_2 \overline{MR} = 896.45 - 1.880(77.7368) = 750.30$$

Since each moving average is the average of two numbers, the control chart constants for $n = 2$ are used.

The moving average chart in Figure 7.3 has all points within the upper and lower control limits. Just as with the moving range chart, the usual runs test is not appropriate because the moving averages are not independent of one another. In summary, when the moving range and moving average charts were used, the batch viscosity process was again judged to be stable. The moving range chart, as indicated earlier, did not provide any new or different perspective on variability. However, the moving average chart may, when the values have a skewed or truncated distribution, provide a different judgment of process stability than does the individuals chart.

TABLE 7.3   *Viscosity Readings and a Two-Item Moving Average*

Batch	Viscosity	Moving Range	Moving Average
1	805	—	—
2	976	171	890.5
3	901	75	938.5
4	929	28	915.0
5	927	2	928.0
6	942	15	934.5
7	904	38	923.0
8	804	100	854.0
9	874	70	839.0
10	944	70	909.0
11	850	94	897.0
12	941	91	895.5
13	992	51	966.5
14	795	197	893.5
15	952	157	873.5
16	832	120	892.0
17	809	23	820.5
18	878	69	843.5
19	936	58	907.0
20	888	48	912.0
Totals	17,879	1477	17,032.5

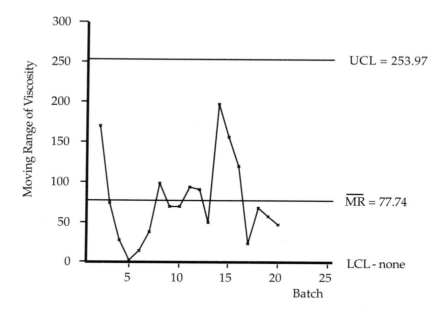

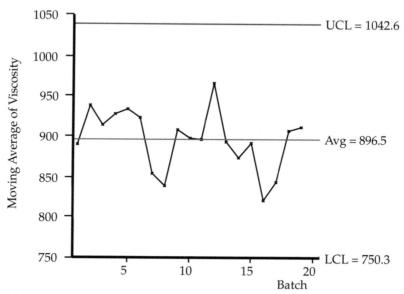

FIGURE 7.3 *Moving Range and Moving Average Charts of Viscosity Data*

The moving range and moving average charts have been introduced as an alternative to the moving range and individuals charts for processes for which, when in control, measurements are not likely to be normally distributed. Other comparisons between the use, interpretation, and sensitivity of the two chart pairs can be made.

Advantages of the Individuals Chart

In addition to being easier to use and interpret, the individuals chart also has the advantage of signaling that there may be trends, cycles, or other patterns in process data. Because movement in the points on the individuals chart corresponds to movements in the measurements taken on the process, patterns observed in the individual measurements can be descriptive of process behavior. The patterns or runs test can be applied to the individuals chart. In addition, other abnormal patterns in the individual values may lead to useful speculations about sources of process variation. If an "up-and-down" or "see-saw" type pattern were observed on the individuals chart, then speculations about what causes might be responsible for producing large then small measurements across time could occur. Future data collection could confirm or not the speculated reasons. When using the moving average chart, such behavior might not be noticeable since the individual measurements are averaged across time periods in the formation of the chart. A further advantage of the individuals chart is that process level and variation are observed on the charts, and the results can be visually compared against requirements.

Advantage of the Moving Average Chart

The major advantage of the moving average chart is that it is more likely to detect small process shifts. As discussed in Appendix A.4, sample averages vary less than individual measurements. For a controlled process, if individuals have a standard deviation of σ, the standard deviation of sample averages for samples of size n will be $\sigma\sqrt{n}$. This fact means that the moving averages will "smooth out" some of the variation observed in individual measurements, making process changes that affect the level of process measurements easier to detect.

7.4   APPLICATION OF MR AND X CHARTS IN PROCESS STUDY

In the process for producing chocolate syrup, one of the production stages includes a cooling operation. One of the critical process parameters currently being studied at the plant is the consistency of throughput rates. In the past, the filling operation has been subject to numerous upsets and stoppages. A work group was assigned to focus on reducing the stoppages of the filling equipment. To understand the relationships between the major stages of the syrup production process, the group's first step was to create a flowchart of the process. Their flowchart of the chocolate syrup process is provided in Figure 7.4. After constructing the process flowchart, the group decided to collect data on the reasons for stoppages of the filling line. A *c* chart on the occurrences of filling line stops was found to be in control. Therefore, the group constructed a Pareto diagram of the reasons for the filling line stoppages. This Pareto diagram is provided in Figure 7.5.

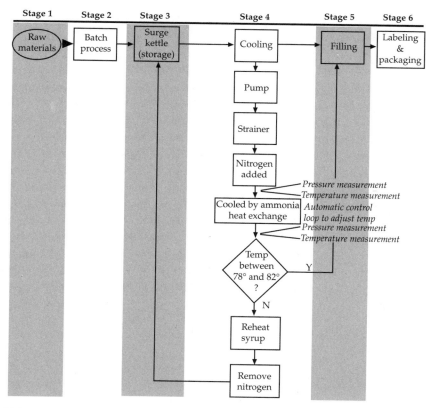

FIGURE 7.4 *Flowchart of the Syrup Production Process*

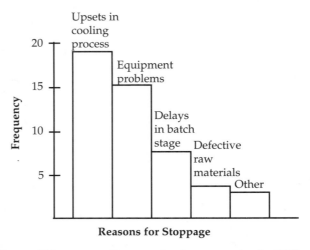

FIGURE 7.5   *Pareto Diagram on Reasons for Stoppages in the Filling Line*

From this work, the group identified that the major cause of the stoppages was upsets in the cooling process. A variety of causes exists for upsets in the cooling lines, and, in fact, the reason for any one upset is most likely due to some interaction between these causes. The filling of the syrup into plastic containers requires that the syrup be cooled to a specified range of temperatures, 78°F to 82°F. This temperature range is necessary to maintain required viscosity values for the syrup, which are essential for correct functioning of the filling equipment. Two parallel lines are in place for cooling the syrup coming from production to the specified levels required by the filling line. These two lines are needed because the syrup is bottled in two container sizes. When the smaller size container is being filled, one cooling line is adequate for cooling the volume of syrup needed. When the larger size container is being filled, both lines must be in operation. To obtain more detailed information on the operation of the cooling lines and the key process parameters of the cooling process, the work group developed a diagram that captures the major phases of the cooling operation. This diagram is provided in Figure 7.6.

Syrup flows into the surge kettle at an elevated temperature from "making" kettles in the production line. The two cooling lines (marked A and B in Figure 7.6) exist not just to lower the temperature of the syrup, but also to ensure that the syrup that goes into the filling machine has a temperature within the range of 78°F and 82°F. The cooling must lower the temperature to this specified range within a predictable amount of time in order to achieve consistent throughput at economic production rates. When the syrup enters a cooling line, a positive displacement pump moves a specified amount up to a heat exchanger, indicated by NH3. Just prior to entering the heat exchanger, in-line sensors measure the temperature and pressure of the syrup at one-minute intervals. In addition, a sample is taken every 15 minutes so that syrup density can be measured. If the density is judged to be too high or too low, then the amount of nitrogen being put into the syrup is adjusted. In the heat exchanger, the syrup is surrounded by coils of liquid ammonia, which cool the syrup to the correct temperature range. Pressure and temperature are again measured after the syrup leaves the heat exchanger. If either of these values falls outside of specifications, the syrup is diverted to the recycle kettle, where it is reheated and the nitrogen removed. The reworked syrup is then returned to the surge kettle.

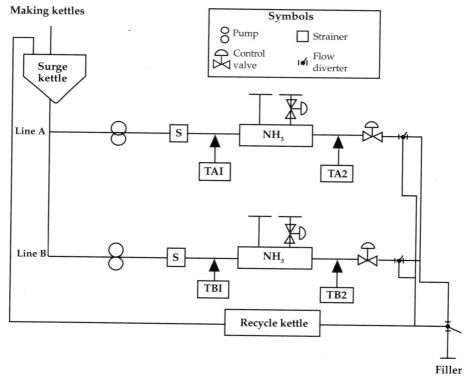

FIGURE 7.6 *Diagram of the Cooling Operation*

After the group had discussed the operation of the cooling line, they realized that there are numerous process parameters to be managed in order for the cooling line to deliver syrup in a consistent, predictable fashion to the filling line. They decided that the first process parameter to study would be the temperature of the syrup as it enters and leaves each of the cooling lines. This parameter was chosen since, when the temperature does not fall within the required range, the syrup is diverted to the recycle line. When this happens, the filling equipment is stopped. Therefore, temperatures out of range affect the ability of the production department to complete its schedules, thereby decreasing throughput capacity in any given time period.

In the illustration in Figure 7.6, the two places on each of the two lines where the temperature of the syrup is measured have been noted. TA1 denotes the temperature of the syrup on line A before cooling, and TA2 denotes the temperature of the syrup after it exits the heat exchanger from line A. TB1 and TB2 denote the same for line B. Temperature measurements are made automatically every minute and stored in the computer, which manages the heat exchanger control loops. To obtain an increased understanding of the current operation of the heat exchangers, measurements from these four recorded temperatures (TA1, TA2, TB1, and TB2) were obtained. Once the data were obtained for analysis, the group questioned how to appropriately examine this type of data. They realized that there are several different ways, depending on process behavior and the ability of personnel and equipment to address certain issues. In fact, they tried several approaches to analyzing the data before finding one that provided them with the information they needed to make process improvements.

The first approach they tried was to compute an average value for entering and exiting temperatures each hour. Because there is one average each hour, the appropriate statistical technique for evaluating the information gleaned from the process would be the moving range and individuals chart. The moving range chart would represent hour-to-hour fluctuations since values within each hour would have little influence on the values of the moving ranges. This approach would be useful for examining the effects of variation sources that tend to enter the process and stay for extended spells.

Next, the group tried the use of an average value for each shift, rather than one for each hour. The moving range chart in this instance would primarily reflect shift-to-shift differences. However, averaging across all the readings within an hour or a shift could smooth the data so much that no signals would remain to indicate variability. The calculation of an average value, say, each hour would tend to "hide" the variations within each hour. Thus, based on a lot of preliminary work and data analysis for this process, the group agreed that they should begin studying the minute-by-minute observations.

Data were obtained on temperatures for incoming and outgoing temperatures for each minute within an hour. The data for the most recent hour of operation during which both cooling lines were operating were charted using moving range and individuals charts. Figure 7.7 contains moving range and individual charts for the incoming temperature, TA1, on line A, and Figure 7.8 contains these same charts for line B. Figures 7.9 and 7.10 show the control charts of the outgoing syrup temperatures for lines A and B, respectively. A discussion of these charts led the group to several conclusions about the process being studied:

Observation 1

> Based on the information on the MR charts of incoming temperatures (TA1 and TB1), the syrup that enters both heat exchangers has stable short-term variation (minute to minute). However, the individual temperatures show quite a bit of drift as evidenced on the X charts of Figures 7.7 and 7.8. The similarity in the charts in these two figures is to be expected, since materials enter both lines from the same surge kettle at approximately the same times.

Observation 2

> Both heat exchangers appear to remove the drift in the heat of the incoming syrup. The X chart of Figure 7.10, showing temperatures after the syrup exits the line B heat exchanger, is in control. However, on line A, it appears that the rise in incoming temperature has been overcompensated for, because there is a run of length 18 below the centerline on the X chart in Figure 7.9.

Observation 3

> The temperature measurements after the heat exchanger on line A show more short-term variation than those on line B, as noted by comparing the $\overline{MR}$'s of the moving range charts in Figures 7.9 and 7.10. The difference in the magnitude of these variations (0.23 versus 0.075) was judged to be of practical process importance.

Observation 4

> The data and data patterns reported for this hour of production were typical of process results during hours of production when the process was not experiencing upsets. It will also be necessary to study and collect information about the process during times of upsets and stoppages.

Observations 2 and 3 led to the planning of an investigation into the differences between the variation observed on the two lines. Both a large short-term variation and drifts in the temperature level on line A would contribute to driving the temperature of the syrup outside the specified range. The group believed that there were two possible causes for this difference in variation between the two cooling lines. One suspected cause was that the measurement device used to measure temperatures on line A is less precise than the one used on line B. (An elaboration of the effect of measurement variation on process variation is discussed in Chapter 9.) Thus, a next step in the process improvement efforts was to be an investigation into the accuracy and precision of the measurement process. The second cause that was believed to need further investigation was differences in the operation of the feedback control loops, which serve to adjust automatically the temperature of the ammonia used to cool the syrup.

The analysis of the set of data used to construct the charts in Figures 7.7, 7.8, 7.9, and 7.10 does not answer all of the questions about the variability in the temperature measurements. The initial question that was raised concerned recycling of syrup caused by temperatures outside of the specified range. Although the work performed allowed for the ability to maintain less variation in temperature measurements, there was no investigation of the process during times in which the temperature went outside the specified range, as noted in observation 4. Observation 2 led to further speculation about the ability of the control loop to handle incoming temperature variation. Is it known that the variation observed in the incoming temperatures is an adequate representation of the amount that is typically observed in the process? If larger incoming variations were experienced, would the heat exchanger on line B still be adequate to handle these variations? The fact that these incoming temperatures are unstable, as evidenced by the individuals charts, means that the actual level of variation cannot be confidently predicted. Further data collection on the process, possibly on the temperatures in the surge kettle and information on what affects these temperatures, is required.

The output or result variable that was studied in this example is the syrup temperature at the filling operation. Statistical charts were used to examine the ability of a physical mechanism, the engineering control loop, to maintain consistency and stability in this end result. The application reinforces recommendations in this book about the study of the behavior of inputs over time, their effects on an end result, and that simple statistical charts can aid in this study. Also, without process knowledge and experience, the charts provide little help. Process interpretation of the results requires the use of knowledge and judgment not found in statistics, but gained through experience and the careful study of a process.

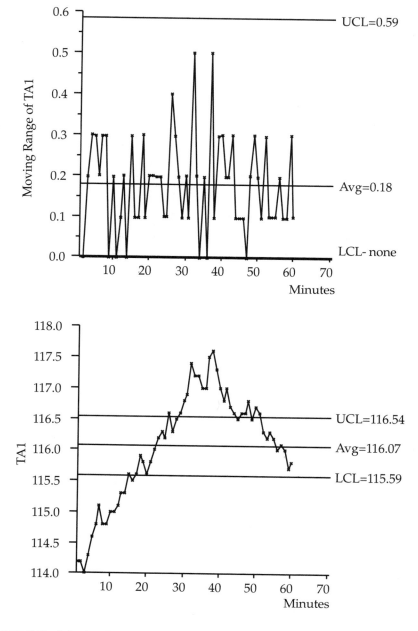

FIGURE 7.7 *Moving Range and Individuals Charts of Incoming Temperature on Line A (TA1)*

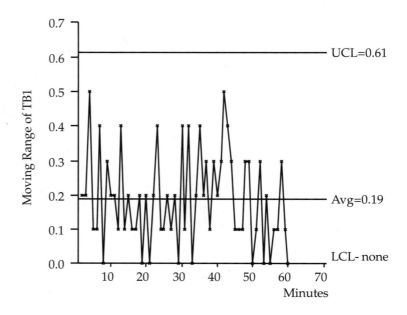

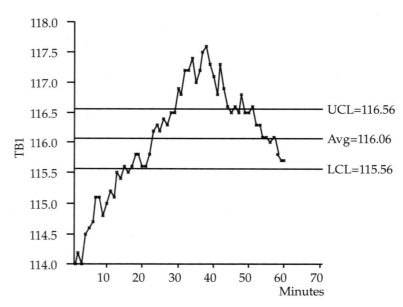

FIGURE 7.8   *Moving Range and Individuals Charts of Incoming Temperature on Line B (TB1)*

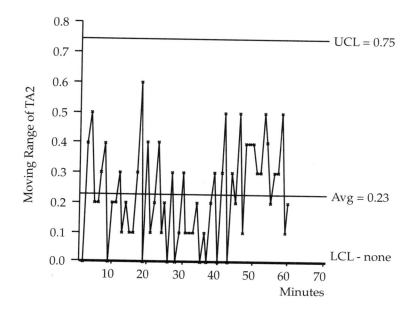

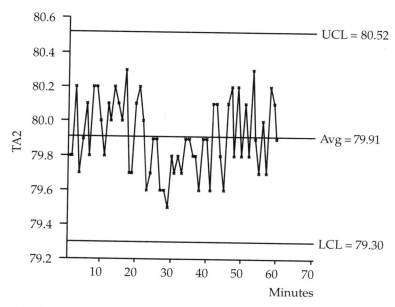

FIGURE 7.9   *Moving Range and Individuals Charts of Exit Temperature on Line A (TA2)*

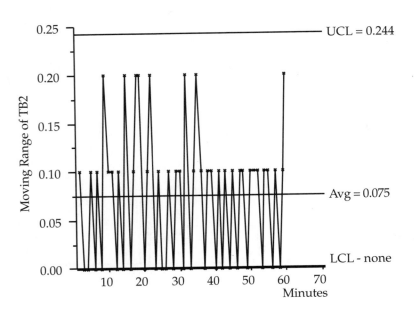

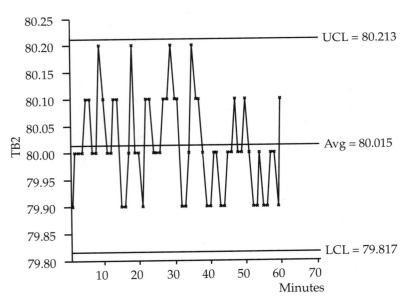

FIGURE 7.10 *Moving Range and Individuals Charts of Exit Temperature on Line B (TB2)*

7.5   IDEAS FOR THE EFFECTIVE USE OF MR AND *X* CHARTS

As previously discussed in this chapter, it is not always possible to create useful subgrouping plans of two or more observations. When there is no process basis for forming rational subgroups, the moving range and individuals charts for process evaluation are often the only reasonable alternative to range and X-bar charts. These situations occur routinely in practice. There will perhaps be only one reading per batch, one measurement value per hour, one result each day, week, or month. The scarcity of data and the difficulty of forming a sampling plan does not eliminate the need to make process evaluations. In such situations, the moving range and individuals charts are helpful in assessing the magnitude and stability of variability and in evaluating the predictability and level of the average of a process.

As with all of the techniques in this book, no attempt should be made to make a particular process evaluation fit a certain statistical technique. The nature of the measurement, sampling scheme, and data collection methods should always be examined, questioned, and understood before attempting to apply a particular statistical procedure. The focus of any work using statistical methods for process improvement should be on the following items:

- the development of an understanding of the process issue;
- the consideration and development of appropriate characterization of the process issue, including the use of flowcharts to pinpoint points in the process that impact the process issue and the development of cause and effect diagrams to describe potential sources of variation;
- the development of metrics, measurements, and sampling plans that will provide insight into the issue and verify process knowledge; and
- the establishment of a factual basis from which appropriate decisions can be made and actions can be taken to improve the process.

With these objectives in mind, various statistical procedures can be considered to determine an appropriate one for the issue under study. The following ideas provide some additional thoughts about the effective use of moving range and *X* charts and about situations in which these charts are useful.

> *Idea 1: The process variations captured by the moving ranges should provide for a process-based description of short-term process variation.*

When moving range and individuals charts are used to study processes where only one measurement is available to represent a particular situation or set of circumstances, the moving ranges serve as a means of describing short-term variation. In other words, the absolute differences between consecutive measurements provide information on sampling time to sampling time variation. The usefulness of this information depends on the sampling times used, as illustrated in the following examples.

Case 1

Temperature measurements are recorded each minute. The moving ranges of these measurements will capture minute-to-minute variation in temperature measurements. Common cause sources contributing to baseline variation could be understood in terms of changes in the instrument, material being measured, etc., which might occur within a minute.

Case 2

Temperature measurements are recorded at sporadic points in time. Some days several measurements are made each hour, while on other days only two or three measurements are recorded. In this instance, common cause variation will be difficult to describe, since the intervals covered by the moving ranges are of varying widths. Some intervals would be affected by changes in raw materials; others would not. Some intervals would capture large differences in environmental conditions; others would not. Process personnel may still find the study of a moving range chart useful to see if inconsistency in the variation captured by the moving ranges is present. If there is evidence of special cause variation in the process, the times at which the measurements were made must be known in order to identify the sources of that variation.

Case 3

From a group of consecutively produced batches, one measurement is taken from each batch. Baseline variation in this example will capture batch-to-batch (as well as within-batch) variation. Some of the batches were produced last week; some were produced yesterday; and some were produced today. The moving range for the difference between the last batch in a day and the first batch on the following day will also potentially be subject to setup variation that would not be observed between batches produced on the same day. A study of the consistency of the moving range chart will provide guidance as to whether a day-to-day effect results in additional batch-to-batch variation.

Idea 2: Data reported across multiple, broad conditions may need to be disaggregated in order to obtain useful information concerning sources of variation acting on a process.

Many of the typical measurements on productivity, costs, or efficiencies are reported on a daily, weekly, or monthly basis. Because there is only one number to represent a given condition per time period, the moving range and individuals chart is a reasonable statistical method to use for analysis of these data. In considering the interpretation of such charts, it is necessary to consider the structure of the data. For example, production line efficiencies are data of this nature. Line efficiencies are often maintained on a 24-hour basis and are computed by:

$$\text{line efficiency} = \frac{\text{actual production}}{\text{target production}} \times 100\%$$

Here, there is only one number to represent a 24-hour production period. The selection of a 24-hour production period over which to report efficiencies may rest only on tradition or usual practice. It is possible that the three shifts that make up the 24 hours act quite differently in terms of efficiencies. Useful clues for increased process understanding may be occurring within the 24-hour time frame over which the data are aggregated and reported. It may be more useful to report the data each shift, rather than across all shifts. It is useful to keep the idea of disaggregation in mind when examining data that have been reported across multiple conditions. If the efficiency data were reported on a shift basis, additional information might be gained by breaking that data down by hours, or by looking at similar data by workstation, product type, or other characteristic. The general recommendation is to know the time and conditions across which data are aggregated. When data are collected across many conditions, the aggregated data will result in moving range values that provide little insight into the causes of short-term process variation.

Idea 3: Moving range and individuals charts are useful in studying machine setup procedures.

After setups, a common practice is to take the first few parts that run off of a machine and measure the critical part characteristic to ensure a correct setup. "Machine runoff studies" typically display and analyze data in one of two ways. One method frequently used is to compute an average and standard deviation of all parts and draw a histogram of the measurements. Any evidence of instability is completely lost by this method of analysis. Another method commonly used is to group the data into arbitrary subgroups of four or five part measurements. Subgroup ranges and averages are computed, plotted, and examined. For the reasons discussed in Section 7.3.2, however, this arbitrary grouping of data is not recommended.

Since there is no rational basis for forming the subgroups, it is recommended that the moving range and individuals chart be used for evaluating numerical results collected under these conditions. The first parts produced after process setups can be measured and the data values used, over numerous setup attempts, to analyze the current ability to set up the process to the correct target. The moving ranges of the data permit an evaluation of the variability of repeated attempts to make a target value and the individuals chart permits an examination of closeness to target.

Idea 4: The moving range and individuals chart can also be applied, at times, to data on fraction nonconforming.

Moving range and individuals charts are useful for studying the fraction nonconforming when the subgroup size, n, is very large. For example, inspection data might include results on daily fraction nonconforming, which are based on production rates in the thousands. Traditional p charts would not be helpful in assessing process behavior. With very large values for n, the distance between the upper and lower control limits will be very small, and the p chart will display a large number of points outside of the control limits. Thus, from a practical viewpoint, the chart is not useful in evaluating predictability of process results. The charts are reporting that each day is different than another day, but to say that all days are significantly different does not allow for a focus on the critical factors that create distinct differences in process results. In such situations, it is more reasonable to apply the moving range and X chart to the data to provide information on which days are sufficiently different. Then, it becomes possible to isolate the possible causes for the differences. When used to analyze this type of data, the moving range reflects day-to-day differences. The control limits on the individuals chart, based on the average moving range, can help identify those days with distinct differences, favorable or unfavorable.

7.6   SUMMARY

The moving range and individuals charts provide a method for evaluating process data when no basis exists for forming a rational subgroup, since one measurement represents a set of conditions for a given time period. The moving range chart provides information on the stability and magnitude of short-term variation. If the short-term variation is determined to be stable, then the individuals chart can be used to evaluate the long-term process behavior. Since the limits on the individuals chart are based on the average moving range, the level of short-term variation represented in the moving ranges provides a baseline against which the long-term process movements can be evaluated. If sources of variation occur over the long term, in addition to the sources of short-term variation, then this additional variation will show up as an out-of-control signal on the chart of individual values. As with all statistical charting methods, successful use of the moving range and individuals charts requires the existence of knowledge of causes that potentially affect the characteristic being measured.

7.7   PRACTICE PROBLEMS

7.7.1   A divisional directive to improve cycle time is received by the manufacturer of a complex medical device. To determine a baseline for the process, the time required to complete the most recent 36 jobs has been recorded and is provided in Table 7.4.

a. Some of the moving ranges in Table 7.4 have been deleted. Complete the column of moving ranges.
b. Complete the moving range chart in Figure 7.11 and, if appropriate, the individuals chart. Is the time it takes to complete a job stable over time?
c. If appropriate, provide an estimate of the standard deviation of completion times.

7.7.2   The logistics manager for an organization is interested in improving the organization's ability to provide the correct materials when needed to the manufacturing floor. With the help of automated manufacturing records, the manager is able to "pull" the percentage of incomplete or short kits that were sent to the floor for the last 25 weeks. These historical data on kit shortages are reported in Table 7.5.

a. Does this organization show consistent provision of raw material to employees?
b. Which other functional areas of responsibility should be involved in the improvement of this process?

TABLE 7.4   *Job Completion Times*

Job No.	Completion Time	MR	Job No.	Completion Time	MR
1	950	—	19	902	—
2	901	49	20	883	19
3	884	17	21	892	9
4	908		22	903	11
5	925	17	23	916	13
6	902	23	24	920	4
7	895	7	25	919	1
8	908	13	26	904	15
9	908	0	27	921	17
10	921		28	916	5
11	915	6	29	908	8
12	909	6	30	899	9
13	950	41	31	895	4
14	877	73	32	912	
15	950	73	33	950	
16	908	42	34	900	
17	907	1	35	898	
18	950	43	36	909	

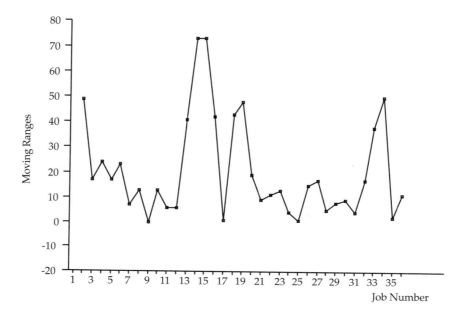

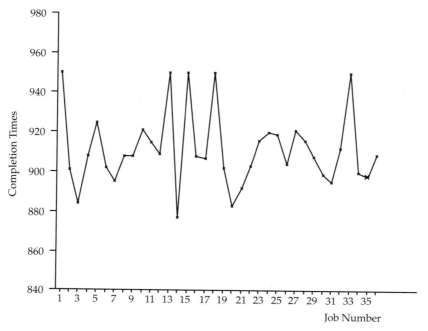

FIGURE 7.11 *Plots of Moving Ranges and Individual Values for Job Completion Times*

TABLE 7.5 *Percentage of Incomplete Kits*

Week	% Kits Short	Moving Range
1	0.78	—
2	0.75	0.03
3	0.67	0.08
4	0.94	0.27
5	0.80	0.14
6	0.33	0.47
7	0.87	0.53
8	0.33	0.53
9	0.53	0.19
10	0.68	0.16
11	0.55	0.14
12	0.56	0.01
13	0.83	0.28
14	0.64	0.19
15	0.80	0.16
16	0.50	0.30
17	0.80	0.30
18	0.64	0.16
19	0.45	0.19
20	0.76	0.31
21	0.81	0.05
22	0.72	0.09
23	0.62	0.10
24	0.60	0.03
25	0.22	0.37

7.8 APPLICATION PROBLEM

The data set in Table 7.6 contains the weekly deviations between sales forecasts and actual sales of cases of a powdered soap product. (These data were referred to in Chapter 5, Section 5.6.2.)

a. Construct a moving range and individuals chart for these data. Comment on the ability of the organization to forecast sales.
b. What does the average moving range measure in this application? What does X-bar measure? What is the interpretation of a positive value of X-bar?
c. Suppose our forecasting capability were very good. How would you expect values for deviations to behave? Suppose forecasting ability were improved still further. What would you expect the deviations to look like? What would happen to the values for the average moving range and for the average deviation?
d. What is the percentage of weeks in which sales forecasts would be less than actual sales? Assuming that production were able to supply a forecast of the number of cases required, how much inventory would you recommend be kept on hand to ensure that product demand can be met?

TABLE 7.6 *Weekly Deviations Between Sales Forecasts and Actual Sales*

Week	Deviation	Moving Range	Week	Deviation	Moving Range
1	64	—	14	51	18
2	81	17	15	41	10
3	20	61	16	82	41
4	35	15	17	12	70
5	-10	45	18	28	16
6	58	68	19	48	20
7	37	21	20	34	14
8	45	8	21	77	43
9	27	18	22	42	35
10	53	26	23	37	5
11	71	18	24	78	41
12	-28	99	25	27	51
13	69	97		1079	857

8 Subgrouping and Components of Variation

The study of processes for the purpose of improving their performance is usually a two-stage effort. The first stage is a control stage. As discussed in previous chapters, the first responsibility in process study is to determine how to operate the process so that it consistently produces outputs that meet customer requirements. This statement implies that the causal factors and the nature of their effects on process performance are understood. It also implies that the process can be managed in a stable fashion in the presence of these influential factors. Data collection and analysis must support an understanding of how raw material properties, machine settings, operator adjustments, run rates, and other causes affect that which is produced by a process.

After a demonstrated ability to run the process in a stable, consistent fashion, the next responsibility is to identify changes to the process that improve the effectiveness and efficiency of process performance. Evaluating the effectiveness of a process means that what is provided by the process may need to be evaluated and changed. At other times, it may be appropriate to study how efficiently the process provides the required outputs. The directions set by management for improving the process are based on an understanding of what customers value, how the process delivers that value, what characteristics of process performance might be of more value, and how those characteristics are created. This stage of process improvement includes selecting better target values for influential variables and reducing the variation in those influential variables in order to reduce the variation in process outputs. Data collection and statistical analysis will only help these efforts if they provide direction for identifying and understanding those influential factors that affect process performance. Therefore, the value of statistical analysis is dependent on several factors:

- the process knowledge of those using the methods,
- the rationale behind and the purpose of the analysis,
- the alignment of the study with the objectives concerning improvement of the systems that support quality, delivery, and cost of product or service,
- the timeliness and relevancy of the data collected, and
- the sampling or subgrouping strategy by which data on the process are collected.

Given these factors, it is quite obvious that statistical analysis can only provide direction and useful information if the intent is understood prior to collecting process data and if this intent guides the way in which the data are collected. Thus, the purpose and the subgrouping strategy are major determinants of the quality of information provided by a statistical analysis. The purpose of this chapter is to discuss subgrouping methods that help us understand the contribution of the influential factors to the overall variation in process results.

8.1   UNDERSTANDING THE COMPONENTS CONTRIBUTING TO TOTAL VARIATION

The degree of success in working to reduce process variation depends on the ability to understand and manage the factors contributing to the total variation observed in the process. This understanding requires confirmed knowledge of the nature and magnitude of the sources, or components, contributing to total variation. The subgrouping and analysis techniques in this chapter focus on situations where hierarchical levels of variation must be studied in order to learn the manner in which causes contribute to the total variation in a process. Many manufacturing processes are usefully studied by using hierarchical process levels as a framework for studying the effect of various process causes. The following sections discuss two examples where developing an understanding of the relative contributions to total variation guided the efforts to reduce variation in the process outputs.

8.1.1   Identifying the Components of Variation in the Thickness of Gelatin Capsules

In the production of gelatin capsules used by pharmaceutical firms, a critical characteristic is the variation in the wall thickness of the capsules. This variability in wall thickness is a practical issue, especially from the customer's point of view. Large variability in the wall thickness requires that the process produce capsules with a high average thickness in order to minimize occurrences of breaks or tears in the gelatin capsules. Also, the large variability in wall thickness affects the efficiency of the customer's filling equipment. At one level of consideration, the equipment that forms the gelatin capsules might be targeted as the reason for the observed variability. However, the forming of the capsules is at the bottom of a cascading series of factors which may each have a measurable effect on the average and variability of thickness. The flowchart provided in Figure 8.1 helps us to understand this hierarchical view of factors causing the large variability in wall thickness. A hierarchical process view supported by appropriate data collection can be a powerful illuminator of the possible causes of variability.

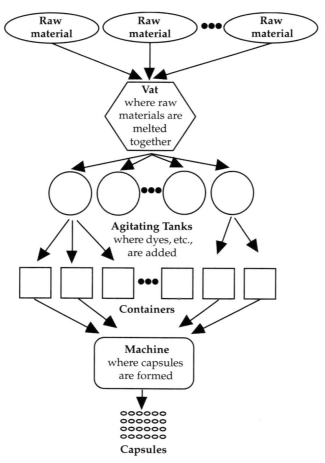

FIGURE 8.1 *Macro Process Flow of Capsule Production Process*

As illustrated in the flow diagram of Figure 8.1, raw materials used in production of these capsules are *melted* together in a large vat. From each vat, several *tanks* of materials are obtained. In the tanks, further processing, such as the addition of dyes, occurs. Although the tanks are continually agitated, materials in different *locations* within a tank may receive different amounts of processing. The processed materials in the tanks are then emptied into containers, which are eventually used by the machine that produces capsules. Of course, the *operation* of the machine will have cause factors that will induce variation in capsule thickness. Finished capsules from this operation are then *measured* for thickness.

Another representation of how the variation present in this process can be studied by considering the hierarchical nature of the sources of variation affecting wall thickness is provided in Figure 8.2. To understand the variation imparted by the machine that forms the capsules from melted material, multiple capsules would need to be selected from material from the same location within a tank. The variation occurring within a tank could be studied by collecting capsules produced by material from different locations within a tank. Multiple tanks would need to be included in the study if variation imparted by the processing that occurs in the tanks is to be understood. Finally, capsules from different melts would need to be studied if the variations from one melt to the next are to be studied. All process causes at these different activity levels have the potential for affecting the total, observed variability in wall thickness.

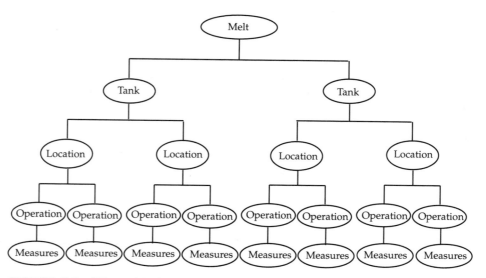

FIGURE 8.2 *Hierarchical Levels of Variation Affecting Capsule Wall Thickness*

An important step toward understanding the potential causes of variation affecting process results is the construction of a cause and effect diagram. This diagram provides a way to organize the many causal factors that potentially affect the variation in a process. When the potential causes are hierarchical in nature, it is important to tie the potential factors to the "level" of the hierarchical process description. Figure 8.3 provides a cause and effect diagram listing the many sources that could affect variation in wall thickness. To develop an understanding of the effect of these sources on thickness, it is necessary to collect information on the process to reveal their effects. For example, one of the causes of variation in capsule wall thickness listed on the cause and effect diagram is the gelatin, a raw material. Therefore, to study the effect of variation in gelatin on thickness, the variations across different lots of raw materials must be considered. In terms of the process hierarchy, these variations will show up as variations from melt to melt. Another factor that could affect thickness of the capsule is the dye used to color the capsules. The effect of dye will show up in the tank-to-tank variations. Of course, tank-to-tank variations will also include variations imparted by other processing that occurs within tanks. Since at each level, many factors could be contributing to the variation at that level, it is useful to consider the causes of variation with respect to the process "level." This breakdown by level of the overall list of potential causes is also illustrated in Figure 8.3. The construction of these different cause and effect diagrams helps to understand how sources of variation at each "level" may affect the capsule production process. These diagrams are also helpful for providing direction on appropriate subgrouping strategies for evaluating the consistency and magnitude of the different levels of variation.

As seen in the cause and effect diagrams of Figure 8.3, many different sources of variation contribute to the melt-to-melt and tank-to-tank variations. Thus, a single data collection strategy will probably be insufficient to identify the contribution of any one particular cause of variation. At the tank level, some of the variation in wall thickness might be due to the methods used for processing the material in a tank. In addition, some of the variation observed may be due to what could be referred to as long-term machine operation. This long-term component of variation would be due to differences in machine behavior from the time one tank is processed to the next. The objective of the subgrouping plan is to provide data that point to the particular level where most of the variation seems to be created. Although the data plans are confounded as to causes, specific work at the identified level, often complemented by statistical design of experiments (introduced in Chapter 10), can unravel the causal thread and provide information on influential variables.

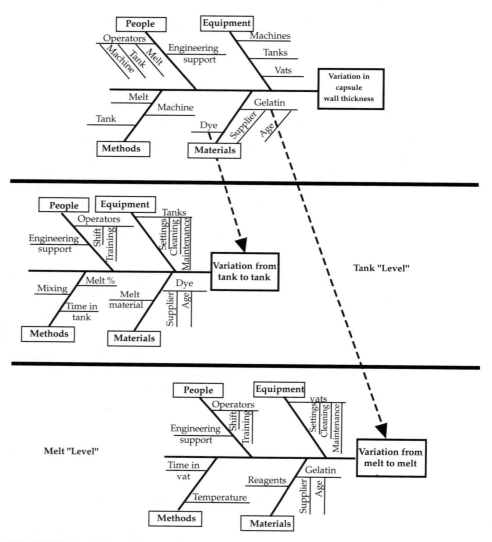

FIGURE 8.3 *A Breakdown of the Causes of Variation in Capsule Wall Thickness Using Cause and Effect Diagrams*

Work to reduce variation in capsule thickness will require that those sources contributing to variation be identified and their effect known. The intent to collect data to study this process should be clearly understood in terms of these possible sources. Which of these sources contributes a large part of the observed variation? How does this variation present itself? Is it consistent over time and conditions or larger at some times than others? The answers to these questions will provide focus for where and how to begin work to reduce process variation.

8.1.2   Hierarchical Levels of Variation in the Production of a Porous Membrane

As illustrated in Figure 8.4, porous membranes are used in filtration devices to filter various solutions. In the production of these membranes, a characteristic of critical interest is the variation in pore size of the membrane. Because the pore size affects the application of the filtration devices, customers expect uniform pore sizes that are at the target levels. These membranes are produced in a batch process. The raw materials are put into a kettle, which contains a solution at an elevated temperature. As the temperature is reduced, polymer drops out of the solution and forms thin membrane sheets, which are called lamina.

Variability in one or more critical raw material properties used to create the membrane may be responsible for much of the variation in pore size. To direct work for reducing variation, the nature of that contribution to the total variation in pore size needs to be studied. In this instance, the material used to produce membranes is received in large containers. Is the variation in pore size largely affected by the differences from one container to another? Is there something about the material within each and every container that results in variability in the pore size? Is the variability in pore size caused by the manufacturing process itself? An understanding of the nature of the contributions of within-container and between-container sources of variation would help answer these questions. This situation can be described by a two-level hierarchy, as illustrated in Figure 8.5. To understand the contribution of within-container sources of variation, multiple measurements on laminas made from the same container would be required. To understand the variation in pore size created by container differences, measurements from laminas made from different containers would be needed.

The following section discusses a method for collecting and analyzing data to understand the contributions different hierarchical levels make to total variation. Although the example used to illustrate the study of components of variation comes from a batch process, the value of the method is not confined to batch processes. Rather, the sequential nature of process operation often means that the contribution of sources of variation might usefully be determined by considering a hierarchy of causes acting on the process and collecting subgroups of data with respect to that hierarchy in order to reveal the relative contributions of causes to total process variation.

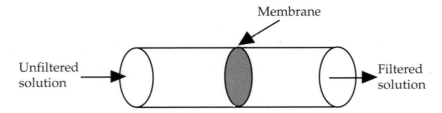

FIGURE 8.4   *Porous Membrane Used in Filtration Devices*

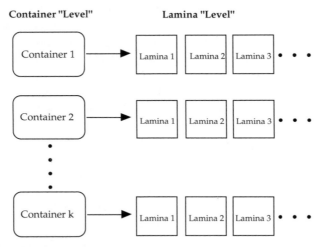

FIGURE 8.5   *Description of the Two Hierarchical Levels in Membrane Production*

8.2   STUDY OF THE COMPONENTS OF VARIATION IN A BATCH PROCESS

A work team has been organized to study the variation of a batch process. They have constructed a process flowchart and identified the quality characteristics of interest for the product. One of the characteristics thought to affect final product quality is viscosity of the batch at an intermediate stage of processing. The specified range for the viscosity measurement is from 83 to 85 with a target of 84. At the time work on the process began, little was known about how well the viscosity was being maintained. The process manager, with supporting individuals from engineering and operations, decided to determine the current process status by taking one sample from each batch in order to evaluate the current level of variation and the current average of viscosity measurements. Data of this type are typically taken for evaluating a batch for an acceptance or rejection decision, for possible corrective actions to the process, or to provide documentation should questions arise later. The procedures adopted by this group were chosen as a result of the purpose for using the data, their display, and the reporting format. The initial purpose was to use the data to determine current process performance for viscosity. In particular, they were concerned with the observed amount of variability in viscosity over time, the level of viscosity, and the consistency of these process characterizations. Another purpose for the data collection was to study the effect of current control parameters and methods on the output result, viscosity. A third purpose was to acquire the information needed to achieve a reduction in viscosity variability and, perhaps, more closely achieve the desired average.

Data for the initial 20 batches studied are given in Table 8.1. Measurements on viscosity result in variables data. Because one measurement is taken per batch and because there is no process reason for grouping batches to create homogeneous subgroups, moving range and individuals charts are used for statistical evaluation of process stability and to describe process characteristics. Figure 8.6 contains the moving range and individuals charts constructed from these data. Because both charts indicate stable process results, viscosity measurements across batches are judged to be consistent.

TABLE 8.1　*Viscosity Measurements from a Batch Process*

Batch	X	MR
1	84.6	—
2	84.6	—
3	84.1	.5
4	83.7	.4
5	84.0	.3
6	84.1	.1
7	84.1	—
8	83.1	1.0
9	83.0	.1
10	84.3	1.3
11	83.9	.4
12	83.7	.2
13	83.9	.2
14	84.4	.5
15	82.6	1.8
16	84.2	1.6
17	83.7	.5
18	84.8	1.1
19	83.8	1.0
20	84.3	.5

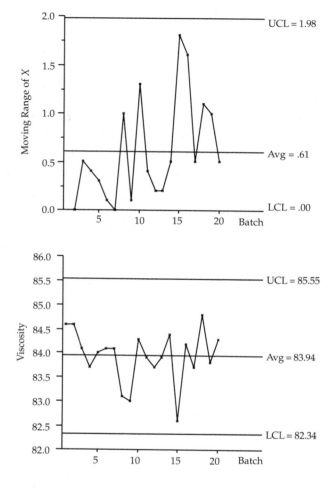

FIGURE 8.6 *Moving Range and Individuals Charts on Viscosity Measurements*

The charts of the viscosity measurements report that those causes currently known about are being consistently managed to provide consistent results in viscosity. An estimate of the standard deviation of viscosity is:

$$\hat{\sigma}_t = \frac{\overline{MR}}{d_2} = \frac{.605}{1.128} = .536$$

The stability of the process and the magnitude of this standard deviation permit an assessment of capability relative to requirements. However, it is crucial that managers, engineers, and operators know how to interpret this number and how to identify the reasons for the magnitude of the estimated standard deviation. Each moving range value is found from one specimen having one laboratory analysis from two consecutive batches. In that regard, note that the estimated standard deviation has the subscript t for "total." Because moving range values are being created by causes revealing themselves in measurement variation, in within-batch variability, and in those causes tending to make one batch distinct from another, the total variation is affected by both within- and between-batch sources of variation. The subscript is used as a reminder of this fact. As a consequence of the subgrouping procedure used to collect the data in Table 8.1, the estimated standard deviation represents the effect of all of these sources. The natural tolerance, NT, of the process is found to be:

$$NT = 6(.536) = 3.216$$

The specified range for viscosity is 83 to 85; that is, the engineering tolerance, ET, is only 2. Thus, the process variation is much wider than the stated requirements for viscosity. Because of this large variation, the team working to reduce the variation in viscosity must identify the sources contributing to this large variation. Before the results of that work are reported and discussed, a plausible statistical view of how respective hierarchical levels contribute to total variation is presented in the following paragraphs.

8.2.1   Sources Affecting Total Variation in a Batch Process

To direct improvement efforts aimed at reducing the variation in viscosity readings, the work team needed more specific information about the sources affecting total variation. Figure 8.7 is a graphical illustration of two possible scenarios. In the scenario on the left, there is little difference in the viscosity measurements for batches 1, 2, and 3. Each of the three batches has about the same spread and the same average. The total variation in the combined output from the three batches is primarily because of the large variation *within* each of the three batches. In the scenario depicted on the right, there is much less within-batch variation for each of the three batches depicted, but there is a larger difference between the three batch averages. The variation in the combined output in scenario 2 is approximately equal to that for scenario 1. However, in scenario 2, the explanation for the large total variation in the combined output is because of the large differences *between* batches.

Understanding how variation sources, within and between batches, are contributing to the total variation is important in directing where work efforts should be concentrated in order to reduce total variation. The intent to "decrease variation" is a large and general statement. Without clues as to where, when, and by how much things change, it may be very difficult to identify causes creating most of the variability. Different causal structures may

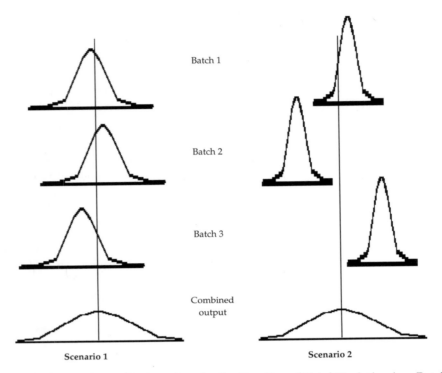

Batch 1

Batch 2

Batch 3

Combined
output

Scenario 1   Scenario 2

FIGURE 8.7   *Two Explanations for the Creation of Total Variation in a Batch Process*

be responsible for within-batch variation as compared to between-batch variation. In scenario 2, if large resources were focused on making batches more homogeneous, little practical reward would come from the work, even with 50% or more reduction in that component of variation. Similarly, in scenario 1, little benefit would result from working to make all batches to the same average. For example, mixing practices might be responsible for variations observed within a batch, but have little effect on observed differences in the level of viscosity from one batch to the next. Large variations in the amount of a viscous ingredient added in the initial stage of production of each of the batches might result in large differences in viscosity from one batch to the next without significantly impacting within-batch variation. Therefore, if scenario 1 in Figure 8.7 summarizes the impact of within- and between-batch variation on total variation, attention would be directed toward those sources contributing to the large within-batch variability. Similarly, if scenario 2 described the relative contribution of within- and between-batch sources to total variation, work would be directed towards making batches more alike on the average.

The task of determining the effect of within- and between-batch sources of variation requires the development of a data collection, or subgrouping, strategy to capture the effects of these two sources of variation. Investigation of the magnitude and stability of within-batch variation requires multiple readings of viscosity from each of several successive batches. Also, in order to assess deviations and stability from one batch to another, readings from many batches are needed.

8.2.2   Assessing the Stability and Magnitude of Within-Batch Sources of Variation

Multiple samples from each of 20 successive batches must be collected in order to characterize the magnitude and stability of within- and between-batch variations. Determining how the samples are to be obtained from each selected batch is a prerequisite to the actual collection of samples. At least two samples from each batch are needed to study the magnitude and stability of within-batch variability. How should they be selected? There are two possible selection criteria. One is to select the samples randomly, and the second criteria is to take the samples from specific locations. The method for choosing the samples must be based on what is currently known or suspected about the nature of within-batch variation. For example, if it is suspected that settling of the material may cause viscosity to vary top to bottom in the vat from which the samples are to be drawn, then positional differences should be reflected in the selection of samples. In the present situation, no such knowledge about the occurrence of variation is at hand. Consequently, the work team decided to collect five samples from randomly selected locations in each of the batches.

The meaning of the word *random* is important. *Random* does not mean that the samples are collected at the convenience of the person drawing the sample, nor does it mean that the samples are selected without a plan. Instead, *random* means that a mechanism is used so that each location of material in the batch has an equal chance of being chosen. With no knowledge of how variation within a batch is presenting itself, a random sampling scheme is used to ensure that the variation occurring in the batch is represented by the subgroup selected from within a batch. The data analysis for this subgrouping scheme is based on the assumption that random samples have been obtained from the low level of the hierarchy. The situation where fixed positions are to be studied requires a different method of analysis. (Discussion of this type of analysis is delayed until Section 8.4.)

In this study, five samples are taken from each selected batch and a measurement of viscosity is made on each sample. These five individual readings, which are not reported, determine a subgroup at the low hierarchical level. The averages and ranges of each subgroup are recorded in Table 8.2. Before considering an analysis of these data, it is useful to consider the sources of variation that are captured by the ranges. Since each of the five measurements in a subgroup comes from the same batch, the magnitudes of the ranges reflect those sources of variation active *within* a batch of material. Large measurement variation will also act to increase the magnitudes of the ranges, as discussed in Chapter 9. Of course, the amount of variation within a batch may not be consistent from one batch to the next. A range chart will be required to decide whether the within-batch component of total variation is stable across the 20 batches included in the study. The range chart for these data appears in Figure 8.8.

TABLE 8.2 *Subgroup Averages and Ranges of Viscosity Readings*

Batch	Average	Range
1	84.04	0.4
2	83.98	0.4
3	83.63	0.7
4	84.57	0.8
5	83.53	0.8
6	84.19	1.2
7	84.30	0.7
8	83.68	1.3
9	84.00	0.4
10	84.48	0.9
11	84.26	0.8
12	83.36	0.5
13	83.97	0.8
14	84.67	1.1
15	83.87	0.8
16	84.13	0.5
17	83.73	0.9
18	83.76	1.0
19	83.94	0.5
20	83.41	1.0

Because the range chart in Figure 8.8 is in control, there is no evidence of inconsistent within-batch variation. Thus, an estimate of the standard deviation for within-batch variation can be found. This component of variability is found by:

$$\hat{\sigma} \text{ within} = \hat{\sigma}_w = \frac{\overline{R}}{d_2} = \frac{.775}{2.326} = .333$$

In order to assist in the proper interpretation of this standard deviation, the subscript w is used to represent within-batch standard deviation. This estimate does not represent the total variation in the batch process; it merely provides an estimate of the within-batch component of the total variation in the process.

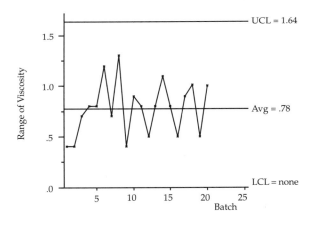

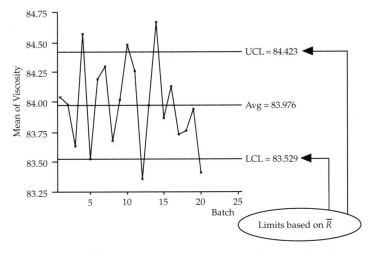

FIGURE 8.8 *Range and X-Bar Charts of Viscosity Readings*

8.2.3   Assessing the Stability and Magnitude of Between-Batch Sources of Variation

The data collection strategy used to generate the data in Table 8.2 also provides information on the effect of between-batch sources of variation when the behavior of the batch averages is analyzed. Because each subgroup average is an estimate of a batch average, the variation in these batch averages will represent variation in the process from all causes. This variation, of course, is due to sources acting both within batches as well as "between" batches. While the ranges examined earlier provided an understanding of the behavior of the within-batch sources of variation, analysis of the batch averages will provide an understanding of the between-batch component of variation.

Control limits for the X-bar chart in Figure 8.8 have been determined by using $\bar{R}$, the centerline on the range chart. We calculate the limits in this way to provide insight into the information that is available through an examination of the X-bar chart. As stated earlier, only within-batch sources of variation are contributing to the magnitude of $\bar{R}$. Consequently, sources of variation in addition to those acting within batches will be revealed by an out-of-control condition on the X-bar chart. Because variation in the X-bars is affected by additional sources of variation (those producing between-batch variation) not captured by $\bar{R}$, the points outside of the control limits indicate the presence of process causes tending to make one batch different from another batch. This conclusion is quite different from saying that the batch process is not producing batches with consistent levels of variation in viscosity. The subgrouping scheme has separated the within and between components. The range chart in Figure 8.8 shows the within-batch variation to be of a consistent magnitude, and the X-bar chart shows that between-batch variations are present. If all points on the X-bar chart had been within the control limits, one would conclude that the between-batch component of variation is negligible in comparison to the within-batch component and all significant process variation is being created by sources that cause nonhomogeneous batches.

At the beginning of this study of a batch process, individual measurements were taken on viscosity, and moving range and individuals charts were used to analyze the data in Table 8.1. This analysis used one analytical result per specimen and one specimen per batch. The variation in these individual readings was subject to both within- and between-batch sources of variation. The moving ranges of the readings captured the short-term variation in these sources (from one batch to the next). The individuals chart of Figure 8.6 was examined for long-term process variation. An out-of control condition on either chart would have indicated some assignable cause of variation occurring at unanticipated times. The second set of data taken on this process, summarized in Table 8.2, was selected to reveal effects of within and between sources of variation on viscosity. Range and X-bar charts were used to analyze these data. The out-of-control conditions on the X-bar chart in Figure 8.8 indicated the presence of between-batch variation in addition to the within-batch variation captured by the ranges.

At this point, an unanswered question is whether or not this between-batch component of variation is consistent across the time that the 20 batches were produced. Is the between-batch component of variation stable over time or is there one or more special causes acting on the batch levels? To answer this question, the control limits on the X-bar chart must reflect common cause sources of variation affecting both the within- and between-batch components of variation. To provide this baseline variation, moving ranges of the batch averages are used. The magnitude of the moving ranges of the batch averages reflects both the within-batch and the short-term between-batch sources of variation. Table 8.3 contains the batch averages as reported in Table 8.2 as well as the moving ranges of these batch averages.

Letting k represent the number of subgroup averages, and using constants for samples of size $n = 2$, the calculations for the moving range chart of the batch averages in Table 8.3 are:

$$\overline{MR}_{\overline{X}} = \frac{\sum_{i=1}^{k-1} \overline{MR}_{\overline{X}}}{k-1} = \frac{9.21}{19} = .485$$

$$UCL_{MR} = D_4\,\overline{MR} = 3.267(.485) = 1.58$$

$$LCL_{MR} - \text{none}$$

Figure 8.9 contains the completed moving range chart. Since the moving range chart is in control, it can be concluded that short-term between-batch variation appears to be stable. The estimated standard deviation for the batch averages is found as:

$$\frac{\overline{MR}}{d_2} = \frac{.485}{1.128} = .430$$

TABLE 8.3   *Batch Averages and Moving Ranges of the Batch Averages*

Batch	Average	Moving Range
1	84.04	—
2	83.98	.06
3	83.63	.35
4	84.57	.94
5	83.53	1.04
6	84.19	.66
7	84.30	.11
8	83.68	.62
9	84.00	.32
10	84.48	.48
11	84.26	.22
12	83.36	.90
13	83.97	.61
14	84.67	.70
15	83.87	.80
16	84.13	.26
17	83.73	.40
18	83.76	.03
19	83.94	.18
20	83.41	.53

Using the average moving range and this estimate of the standard deviation of the batch averages, control limits for the batch averages are calculated. Because each moving range is calculated by computing the difference between two numbers, the control chart constants for $n=2$ are used. The centerline for the X-bar chart is computed as the average of the X-bars:

$$\overline{\overline{X}} = \frac{1679.50}{20} = 83.98$$

The control limits for the X-bar chart are found by adding and subtracting three standard deviations of the batch averages from $\overline{\overline{X}}$:

$$3(0.430) = 1.29$$

$$\text{UCL}_{\overline{X}} = \overline{\overline{X}} + 3\left(\frac{\overline{\text{MR}}}{d_2}\right) = 83.98 + 1.29 = 85.27$$

$$\text{LCL}_{\overline{X}} = \overline{\overline{X}} - 3\left(\frac{\overline{\text{MR}}}{d_2}\right) = 83.98 - 1.29 = 82.69$$

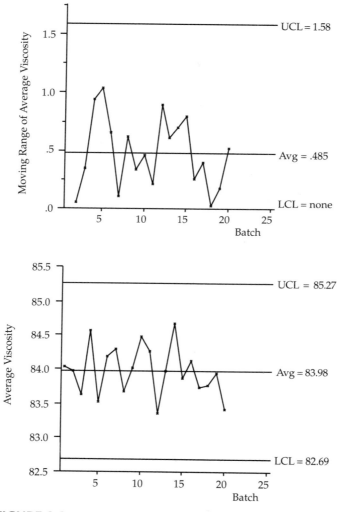

FIGURE 8.9 *Moving Range and X-Bar Charts of Viscosity Readings*

The completed *X*-bar chart, shown in Figure 8.9, is in control. Thus, it can be concluded that the between-batch component of variation is stable over time.

Comparing the *X*-bar charts of Figures 8.8 and 8.9, we note that the *X*-bar chart of Figure 8.8 resulted in numerous points falling outside of the control limits, whereas the *X*-bar chart in Figure 8.9 is in control. It is important to understand that having points out of control on the subgroup average chart in a process investigation (as in Figure 8.8) is not an adequate reason for adopting the techniques presented to baseline the between-batch variation. Points outside of control limits always indicate active causes that are not captured within subgroups. If there are process causes acting systematically at the level of subgroup averages, then an evaluation of the magnitude and consistency of these causes is appropriate. However, if there is no process rationale for process causes acting at the subgroup level, then out-of-control points on the X-bar chart simply signal special causes. No additional purpose is served by using moving ranges of these averages to "widen" the limits on the X-bar chart. Effective use of these statistical methods require that

- the process studied is one where the hierarchical levels present a realistic view of process behavior, and
- the subgroups are taken from the low level of the process hierarchy.

Without this basic understanding the techniques discussed in this chapter can easily be erroneously applied.

Having determined that the between-batch variation is stable, it is necessary to have an estimate of that component of variation in order to assess the relative contribution of between-batch sources of variation to total variation. This estimate, $\hat{\sigma}_b$, is based on the estimated standard deviation for the batch averages (computed earlier). The formula for estimating the standard deviation of the between-batch component of variation is:

$$\hat{\sigma}_b = \sqrt{\left(\frac{\overline{\text{MR}}}{d_2}\right)^2 - \frac{\hat{\sigma}_w^2}{n}} \, ,$$

where n is the number of observations used to calculate each batch average ($n = 5$ in this application).

The within-batch standard deviation, $\hat{\sigma}_w$, was previously estimated to be 0.333. So, the between-batch standard deviation is calculated as:

$$\hat{\sigma}_b = \sqrt{(.430)^2 - \frac{(.333)^2}{5}} = .4034$$

This between-batch standard deviation represents process causes acting above and beyond those causes acting within batches.

8.2.4   ## The Contribution of Within- and Between-Batch Sources of Variation to Total Variation

An estimate of the total variation of the batch process can be made using the estimates of within- and between-batch variation. This estimate is based on the fact that the total variance of the batch process is found by summing the variances of the within- and between-batch components. Thus, the standard deviation for the batch process is found by:

$$\hat{\sigma}_t = \sqrt{\hat{\sigma}_b^2 + \hat{\sigma}_w^2} = \sqrt{(.4034)^2 + (.333)^2} = .523$$

The standard deviation, $\hat{\sigma}_t$, was found by first estimating the two components contributing to the total process variation and then using these two components to arrive at an estimate of total process variation. This estimate of process variation captures the effect of all process variables, which are, of course, the same ones whose effects were estimated by the average moving range from the data presented in Table 8.1 and charted in Figure 8.6. Those data were collected by a simpler sampling plan, which resulted in one observation per batch. Although estimating the same quantity, the advantage of the more complex sampling plan is that the respective effect of process sources acting within and between batches can be quantified. Knowing the respective magnitudes enables an objective judgment as to what process variables need to be addressed in order to reduce total variation. This knowledge is not available with the results of the simpler subgrouping plan.

Another way of displaying the relationship between total batch process variation and the within- and between-batch components of the total variation is to write the relationship in terms of the total variance being the sum of the between- and within-batch variances. In other words, the relationship can be described as:

$$\hat{\sigma}_t^2 = \hat{\sigma}_b^2 + \hat{\sigma}_w^2$$

The values of these respective component variances are:

$$\hat{\sigma}_b^2 = (.4034)^2 = .1627$$

$$\hat{\sigma}_w^2 = (.333)^2 = .1109$$

$$\hat{\sigma}_t^2 = .1627 + .1109 = .2736$$

The description of total variation in terms of the variance is needed in order to quantify the relative contribution of the components to the total variation. In this example, the within contribution accounts for 40.5% of the total variation. This percentage is found by:

$$\frac{\hat{\sigma}_w^2}{\hat{\sigma}_t^2} \times 100\% = \frac{.1109}{.2736} \times 100\% = 40.5\%$$

Since about 40% of the total variation can be attributed to within-batch variation, 60% of the total variation can be accounted for by the differences between batches. This statement implies that reducing the variation in viscosity readings between batches promises a greater reduction in the variation of the batch process than working on within-batch sources of variation. The work team addressing the variation in viscosity readings is now directed toward examining the causal structure affecting batch-to-batch variation.

The analysis of the components of variance for the viscosity readings from a batch process required several stages of plots and calculations. To aid the reader in recalling the flow of the work in analyzing the components of variation for a two-level hierarchy, a flowchart of the steps of the analysis is presented in Figure 8.10. An example providing a step-by-step discussion of this flowchart is provided in the following section.

8.2.5   Two-Level Components of Variance Study for a Crystallization Process

In the production of a dry, bulk material in a chemical processing plant, one of the key characteristics measured at an intermediate stage of the process is the particle size of the material. The uniformity of particle size is thought to be important in later processing steps because of its possible effect on processing time and, consequently, on output quality. The current target of the process is set at 500, and specifications require a minimum particle size value of 300. At the initial stage of study, three primary questions existed concerning the process:

- Is the process being managed in order to meet these requirements?
- Are the specifications adequate to decribe what is needed by the process?
- Do large variations in particle size between batches have an adverse effect on processing time or do variations within the batches affect this time?

To understand the relationship between particle size and processing time, information on variation in particle size within each batch of bulk material as well as on differences between batches is required. Since there was little knowledge of variation within a batch to direct the selection of samples, six samples were randomly selected from each batch of material after the

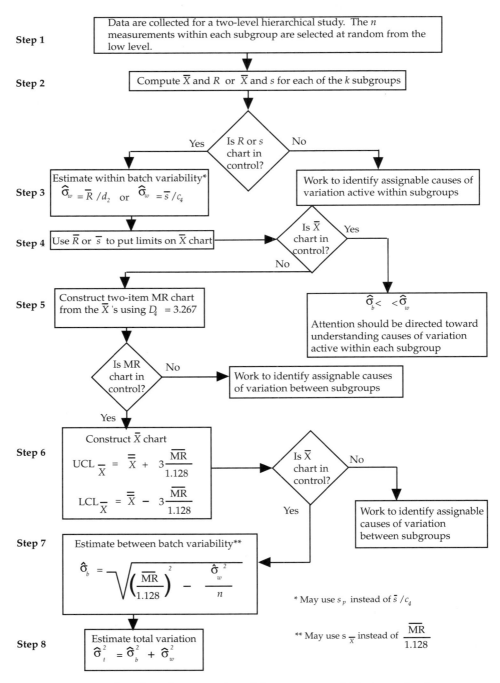

FIGURE 8.10 *Flowchart for a Two-Level Components-of-Variance Study*

completion of the crystallization phase of the process. These six measurements from 25 consecutively produced batches are provided in Table 8.4. The means and standard deviations for these 25 subgroups are also included in the table.

Step 1 of the flowchart (Figure 8.10) for a two-level component of variance study contains a description of the type of data required for analyzing the between and within components of variation contributing to total variation. For the particle size study, the low level investigated is the within-batch variation in particle size. The letter n in the flowchart refers to the number of readings collected from each batch. For the data set of Table 8.4, n is equal to 6.

TABLE 8.4 *Data on Particle Size*

Batch	X_1	X_2	X_3	X_4	X_5	X_6	$\bar{X}$	s
1	565	532	689	468	582	596	572.0	73.3
2	674	539	547	632	645	466	583.8	79.3
3	819	695	627	679	837	666	720.5	86.4
4	362	499	469	342	244	296	368.7	98.6
5	306	475	426	296	273	468	374.0	92.4
6	620	656	469	669	611	457	580.3	93.5
7	323	294	570	433	427	283	388.3	110.2
8	646	606	511	436	579	578	559.3	74.8
9	547	690	556	508	471	600	562.0	76.5
10	734	745	687	709	817	682	729.0	49.8
11	437	447	523	470	425	512	469.0	40.5
12	451	628	548	417	685	671	566.7	113.8
13	586	597	521	624	645	497	578.3	58.0
14	429	408	678	418	455	526	485.7	103.3
15	784	855	710	872	635	622	746.3	107.9
16	425	510	304	334	535	510	436.3	98.7
17	489	602	471	632	453	694	556.8	99.3
18	500	426	287	268	546	315	390.3	117.3
19	563	418	635	567	583	608	562.3	75.7
20	722	761	757	561	680	569	675.0	90.1
21	336	448	420	315	565	415	416.5	89.3
22	501	578	621	496	647	567	568.3	61.4
23	605	862	895	705	769	740	762.7	105.4
24	845	771	865	791	683	762	786.2	65.0
25	432	439	488	572	558	695	530.7	99.4
							13,969.0	2159.9

The letter k refers to the number of entities from the high level that are included in the study. In the present example, $k = 25$ batches is selected. So that the control charts used for analysis provide a reasonable statistical description of process variation, at least 20 batches should be selected. However, more important than the issue of how many batches to select is the issue of how the k batches are selected. The way in which the batches to be studied are selected affects the type of information that is obtained about between-batch variation. The k batches should be chosen so that the sources of variation thought to affect between-batch variation have a chance to be active. Many process investigations on components of variance are ineffective and even misleading because the time frame for the data collection is so short that it does not permit major variables to become active. For example, if it is believed that a change in raw materials affects between-batch variation, then batches chosen that do not include batches from different raw material sources will not display the effect of this potential source of variation on between-batch variation.

Once the data are available for analysis, *step 2* on the flowchart is to construct either a range chart from the ranges or an *s* chart from the standard deviations of the k subgroups. The range or *s* chart is used to evaluate the stability and magnitude of the within-subgroup component of variation. If this chart is out of control, then work needs to be directed at discovering why within subgroup variation is inconsistent. A cause and effect diagram, which captures sources of variation affecting within-subgroup variation, can be explored to provide direction for investigating which sources may be acting intermittently to create unstable variation within the batches. An *s* chart has been used (Figure 8.11) in the particle size study to explore the behavior of within-batch variation.

Since the *s* chart is in control, it is appropriate to estimate the within-batch component of variation, as indicated in *step 3*. The estimated standard deviation for within-batch sources is computed as:

$$\hat{\sigma}_w = \frac{\bar{s}}{c_4} = \frac{86.396}{.9515} = 90.80$$

Another method for determining an estimate of the within-subgroup standard deviation is also referred to in step 3. This estimate, s_p, is the method used by most software programs that provide data summaries of the type of hierarchical data under discussion. (These data summaries are also referred to as nested studies.) The pooled variance, s_p^2 is determined by averaging the variances of all subgroups. For the data on particle size, the pooled variance is found to be:

$$s_p^2 = \frac{\sum_{i=1}^{k} s_i^2}{k} = \frac{196{,}766.60}{25} = 7{,}870.66$$

The pooled standard deviation, s_p, is the square root of this quantity. In the present example, s_p is 88.72. This estimate is close to the one found using $\bar{s}$, as it should be since the pooled standard deviation is simply another method for estimating the within-subgroup standard deviation. Although the two estimates will differ by small amounts, the pooled standard deviation is often a preferred estimate due to its statistical properties.

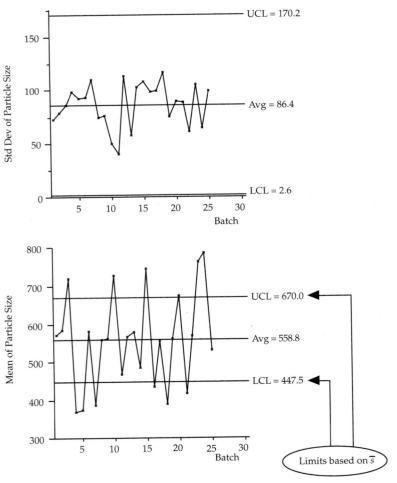

FIGURE 8.11 *Standard Deviation and X-Bar Charts of Particle Sizes*

Of more importance to the group working on the particle size process is the magnitude of the within-batch standard deviation, which was estimated to be 90.80. If each and every batch is centered at an average of 500 (i.e., there is no between-batch variation), the within-batch standard deviation is so large that there will be particles whose sizes fall below 300, the lower specified limit. The distribution in Figure 8.12 provides a pictorial illustration of the effect of the large within-batch standard deviation on the ability of the process to deliver the stated requirements. If it is determined that particle size should remain above 300 with an average of about 500, then additional work is necessary to reduce the variation in size observed within each and every batch.

If the *s* (or range) chart in step 2 had been out of statistical control, then the conclusion would be that there is no single process for creating within-batch results. With that judgment, it would be misleading to provide an estimated standard deviation for the within-batch component of process variability. Lack of consistency suggests that at times there is one process for manufacturing results within batches, while at other unpredictable times, there is another process with different practices in managing materials, equipment, or methods that leads to different within-batch variability.

Since the total process variance is the sum of the component variances (i.e., $\hat{\sigma}_t^2 = \hat{\sigma}_b^2 + \hat{\sigma}_w^2$), if one of the components contributing to total variation is not a credible number, then the total also lacks credibility. Stability of within-subgroup variations must be achieved before reduction in one or more of the components can reliably begin. From another perspective, it is inappropriate to calculate limits for the batch averages if within-subgroup variation is inconsistent. There are reasons for not constructing the X-bar chart

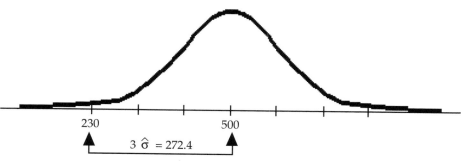

FIGURE 8.12 *Possible Distribution of Particle Sizes for a Batch with an Average Particle Size of 500*

when the range chart is out of control. First, statistical theory does not provide a way for evaluating the stability of batch averages when within-subgroup variation is unstable. The second reason is more pragmatic and process based. Because the within-subgroup variation is inconsistent, the personnel working on the process have already been provided a direction for work. The current responsibility is to discover and correct the reasons for the unstable, within-subgroup variation.

The X-bar chart in Figure 8.11 is the chart specified by *step 4* of the flowchart. Since the limits on this chart are based on $\bar{s}$, these limits represent the variation that will be observed in the X-bars if no between-batch variation is present. An X-bar chart that is in control indicates that between-batch variation is almost negligible when compared to within-batch variation. However, the X-bar chart in Figure 8.11 is out of control. This X-bar chart verifies the hypothesis that process causes are operating above those which generate within-subgroup variation. These causes contribute significantly to process variation. As with any other common cause system, these sources may be affected by sporadic, abnormal behaviors. Thus, it is important to evaluate the stability of these sources of variation.

Step 5 of the flowchart is to construct a moving range chart from the subgroup or batch averages. The value of D_4 for samples of size 2 is used to calculate the upper control limit, since there are two averages used to calculate each moving range. The moving range chart is constructed in order to evaluate the stability of the short-term, between-subgroup variation. Figure 8.13 contains the completed MR chart. Since this chart is in control, it is concluded that the short-term between-batch variation in particle size is consistent across the time studied. If the moving range chart had been out of control, then work would be directed toward identifying why, at some times, one or more batches are very different from the others with respect to average particle size. Also, if the moving range chart had been out of control, then it would be inappropriate to construct the X-bar chart.

Since the moving range chart is in control, it is appropriate to construct the X-bar chart in *step 6* of the flow diagram. This X-bar chart provides information on the stability of long-term, between-subgroup variation. For example, if there are trends or cycles in the batch averages, this X-bar chart will show evidence of this special cause. The X-bar chart for the batch averages is provided in Figure 8.14. Since there are no out-of-control signals, then it is appropriate to estimate the between-batch variability. If the X-bar chart had shown evidence of special causes, then the next step would be to identify the reasons for the out of control condition.

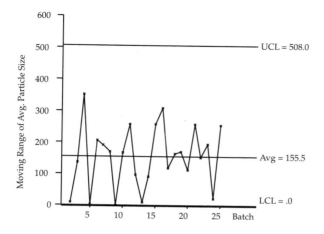

FIGURE 8.13   *Moving Range Chart of the Batch Averages*

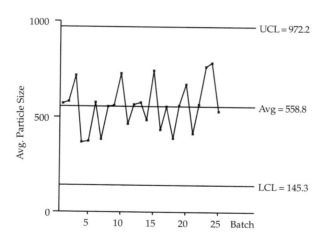

FIGURE 8.14   *X-Bar Chart of Average Batch Particle Sizes*

Because the X-bar chart in Figure 8.14 is in control, *steps 7* and *8* of the flow-chart can be completed. These steps provide formulas for calculating the between and total components of variation. Knowledge of the magnitude of between-subgroup variation and its percentage contribution to total variation is helpful in prioritizing work for improving the process. An estimate of the between-subgroup variation using $\overline{MR}$ to capture the variation in the X-bars is provided by:

$$\hat{\sigma}_b^2 = \left(\frac{\overline{MR}}{d_2}\right)^2 - \frac{\hat{\sigma}_w^2}{n} = \left(\frac{155.521}{1.128}\right)^2 - \frac{7870.66}{6} = 17,697.25$$

$$\hat{\sigma}_b = \sqrt{17,697.25} = 133.03$$

A second method for estimating the between-subgroup variation using the calculated variance of the X-bars, $s_{\bar{x}}^2$, is also provided. This method provides an estimate close in magnitude to the one obtained by the method that uses $\overline{MR}$. This second method is presented since it is the method used by most statistical software packages. For this example, the estimate of between-subgroup variation obtained by using $s_{\bar{x}}^2$ is:

$$\hat{\sigma}_b^2 = s_{\bar{x}}^2 - \frac{\hat{\sigma}_w^2}{n} = 15,670.21 - \frac{7870.66}{6} = 14,358.43$$

$$\hat{\sigma}_b = \sqrt{14,358.43} = 119.83$$

For the data on particle size, the between-batch component of variation is even larger than the within-batch component. Previous analysis of the within-batch component resulted in the conclusion that, even with no between-batch variation in particle size, the within-batch variation was so large that the process could not be targeted at 500 and still produce particle sizes above 300. Now it is observed that an even larger problem is created by the between-batch variation. The group addressing the issue of the effect of particle size on later cycle time found that the process was creating much more variation in particle size than had been anticipated. The next phase of work was to investigate what effect each of the components, within and between variation, had on later cycle time, and to use that knowledge as a basis for deciding where the next steps for reducing the variation in particle size should begin: on the within-batch component or the between-batch component of the total variation.

8.2.6   **Summary Comments on Two-Level Hierarchical Studies**

Two-level hierarchical studies are extremely useful in efforts to reduce the total variation in process results. These techniques for studying components of variation allow work to be directed toward the levels in the process that contribute the largest proportion of variation to the total observed variation. As previously stated, hierarchical studies on components of variance are useful in many processes in addition to batch processes. For any successful application of these techniques, several essential criteria are necessary:

1. A subgrouping plan must be in place with two or more observations forming a subgroup.
2. A range (or standard deviation) and average must be computed for each subgroup.
3. Process reasons exist to explain variation. Some of these sources act long term and tend to affect all materials or parts. These sources tend to drive the subgroup average significantly up and down, but do not tend to make materials different that are created "close" together. Other process sources tend to act only in the short term, dominating the creation of within-subgroup variation. These sources also affect all parts and materials.

Subgroup averages reflect the effects of long-term sources of variation. When subgroup averages also capture the effects of short-term variation, then these averages consistently display more variation than the magnitude predicted by the average value for subgroup range. This phenomenon was seen in the preceding example when samples from a batch were selected at random. When this occurs, there are process reasons for why some sources of variation reveal themselves in variation displayed over a different time interval rather than within subgroups.

From these examples and discussions, it is obvious that the classification of process variations with the words *common cause* depends extensively on the subgrouping plan adopted for process study. The collection of subgroups to study a process determines the division of common cause sources into hierarchical levels. For such a subgrouping plan, some common cause sources will only be active within subgroups, while other common cause sources will only act to create between-subgroup variation. The use of the components of variation technique for studying process variation relies on process knowledge in order to speculate about the sources of variation that may impact the process, as well as to design subgrouping plans that will provide useful information on the effect that these sources have on process output.

8.3   EXAMPLE OF A MULTIPLE-LEVEL SUBGROUPING PLAN AND ANALYSIS

The two-level components of variance model can easily be extended to multiple-level hierarchical situations. An illustration of a three-level study is provided by the study of the manufacture of a porous membrane. As illustrated in Figure 8.4, porous membranes are used in ultrafiltration devices for straining, filtering, and removing impurities from certain products. Pore size and uniformity of pore size are critical parameters of the membrane. In this manufacturing process, polymer is put into a quasi-solution at an elevated temperature. As the temperature is reduced, the polymer drops out of solution and forms the membrane, called a lamina, at this manufacturing step. Pore size is fixed at this stage in the manufacturing process. Because of increasing customer requirements for more stringent uses, the manufacturer needs to produce a more uniform hole size in the pores of the laminas formed from this process.

The first stage of working on this process to achieve more uniform pore sizes focused on obtaining answers to the following questions:

- Are the important parameter settings being consistently achieved?
- Are the materials received actually on target and within specifications?
- Are the recommended methods consistently practiced?

Answers to these questions were obtained and verified by tracking process variables by several of the control charting procedures discussed in previous chapters. This stage of evaluating current process performance was necessary to be able to achieve uniform pore size. However, even with consistency in methods and parameter settings, information on where the majority of the variation in pore sizes is created is needed. These are the next questions that need to be addressed:

- Do those important process variables which are known to be consistently practiced have a consistent effect on the average and variability of pore size?
- What is the relative effect of a process variable and where is it revealed, within lamina or between lamina?
- Although current methods and settings are consistently achieved, these practices are insufficient for achieving present or future required characteristics of pore size. What parameters need additional technology, work, and attention in order to achieve more uniform pore size?

Different process variables will have different relative importances at different process levels. For example, a parameter setting may not affect the uniformity in pore size within a lamina, but, as the parameter changes or interacts with other variables, it may be a predominant cause of the variability in pore size over longer periods of time.

The data collected at the current stage of investigation were meant to provide information about the effect of current methods, practices, and parameter settings on uniformity of pore size. For example, if most of the variability in pore size is occurring within the formed lamina, then it is likely that process equipment or methods are creating the observed variability since there are no significant material variations in the amount of material required to make one lamina. It is also probable that no significant equipment variations or changes in work practices take place during the time required to make one lamina. Thus, if most of the variation is occurring within a lamina, then it is reasonable to suspect the likely sources of variation as being current equipment settings, ability to hold a given equipment setting for even a short period of time, material flow rates, inappropriate control over the temperature drop or similar process factors. On the other hand, if most of the variability is being created by differences between laminas, then it is reasonable to consider alternate sources of long-term variation that could be contributing to variability in pore size. These sources of long-term variations may include material properties, inconsistent setup or maintenance practices, or equipment changes. If a determination can be made as to where the variation is being created, either within or between lamina, then the work can be successfully focused to reduce the variation in pore size. Thus, data need to be collected and analyzed, the sources of variation isolated, and the effects of the sources either confirmed or refuted.

8.3.1 Subgrouping Plan for a Three-Level Components of Variation Study

Polymer is received in railcars from which many membrane laminas are made. The primary sources of variation to be studied have been identified:

1. the within-lamina variation, representing variation in the process,
2. the between-lamina, but within-railcar variation, representing repeatability of the process, and
3. the variation between laminas produced from different cars, representing material variations.

Large material variations *within* a car can also result in increased variability in pore size in the within- and between-lamina components of variation. The manufacturer believes that the within-railcar material variation is insignificant, but that the magnitude of material variation between cars has an effect on pore size. The conclusions reached in the following data analysis are based on these speculations, and will therefore depend on the soundness of this manufacturer's process knowledge and on the soundness of preceding data collection and analysis.

To study the sources of variation contributing to variation in pore size, three laminas are randomly selected from all laminas made with material from the same railcar. Each selected lamina is divided into a grid, and five squares are randomly chosen from this grid. Figure 8.15 is a pictorial illustration of this sampling plan. The average pore size and the range for the five squares are obtained. (Past work has demonstrated the reliability and constancy of the measurement process.) The data from 10 consecutive railcars and necessary calculations are summarized in Table 8.5.

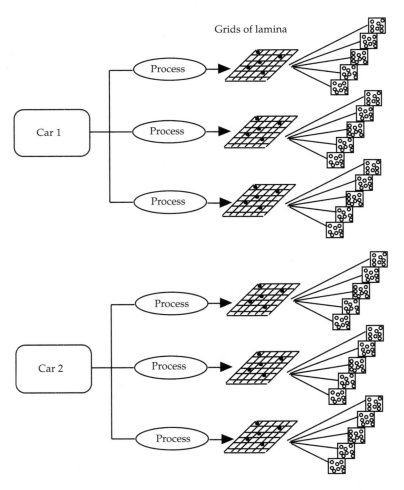

FIGURE 8.15 *Illustration of the Subgrouping Plan for the Membrane Process*

TABLE 8.5 *Average Pore Sizes for Five Grids*

Car (I)	Lamina (II)	Within Lamina		Between Lamina		$MR_{\bar{Y}C}$ (VII)
		$\bar{Y}_L$ (III)	R_{WL} (IV)	$\bar{Y}_C$ (V)	R_{BL} (VI)	
1	1	0.6022	0.169			
1	2	0.5750	0.160	0.58140	0.0352	
1	3	0.5670	0.174			
2	1	0.5604	0.105			
2	2	0.5628	0.138	0.54047	0.0646	0.04093
2	3	0.4982	0.090			
3	1	0.5918	0.116			
3	2	0.5894	0.118	0.57780	0.0396	0.03733
3	3	0.5522	0.132			
4	1	0.5918	0.121			
4	2	0.4892	0.164	0.53067	0.1026	0.04713
4	3	0.5110	0.123			
5	1	0.5616	0.198			
5	2	0.5082	0.101	0.55447	0.0854	0.02380
5	3	0.5936	0.092			
6	1	0.5146	0.150			
6	2	0.4652	0.142	0.49367	0.0494	0.06080
6	3	0.5012	0.220			
7	1	0.5486	0.116			
7	2	0.6052	0.045	0.57833	0.0566	0.08466
7	3	0.5812	0.183			
8	1	0.5130	0.092			
8	2	0.5276	0.121	0.53300	0.0454	0.04533
8	3	0.5584	0.079			
9	1	0.6132	0.071			
9	2	0.5696	0.132	0.60653	0.0672	0.07353
9	3	0.6368	0.096			
10	1	0.5590	0.083			
10	2	0.5694	0.128	0.58507	0.0678	0.02146
10	3	0.6268	0.077			

8.3.2   ## Analysis of Three-Level Components of Variation Studies

The columns of numbers in Table 8.5 have been labeled with roman numerals to aid in the analysis of the information in these data about the membrane process. A description of the data in each of these columns is provided below:

I The numbers in this column identify the car of material.

II The numbers listed in this column correspond to the three separate laminas (coded as 1, 2, and 3) all made from the material within the respective cars.

III This column contains the 30 averages for the 30 laminas collected in the study. The differences between all 30 of these numbers can be explained by the sources of variation, which contribute to variations in laminas from the same car as well as more long-term car-to-car differences.

IV Each number in this column is the range, R_{WL}, in pore sizes taken from five randomly selected squares within the same lamina. There are 3 laminas from each of 10 cars for a total of 30 laminas and, therefore, 30 ranges. (The range chart displayed in Figure 8.16 is a plot of these 30 within lamina ranges.)

V Column V contains results on average pore size for each car. These numbers may be calculated in one of two ways. There are three laminas from each car with measurements on pore size for five areas for each lamina having been taken; the average pore size for the car can be calculated from these 15 individual observations. The averages in column V could also be found from the numbers reported in column III, each being the average pore size over five grids. The average of these three numbers for car one is:

$$\frac{.6022 + .5750 + .5670}{3} = .58140$$

Note that the numbers in column V summarize average pore size for laminas made from material from the same car.

VI The ranges in column VI are the ranges of the three lamina averages ($\bar{Y}_L$) for laminas made from material from the same car. (These ranges are plotted in the control chart shown in Figure 8.17.)

VII The numbers in this column are moving ranges of the averages in column V. Consequently, these moving ranges capture short-term car-to-car variations. A moving range chart of these values will provide a means of evaluating the stability of the short-term, between-rail car variation.

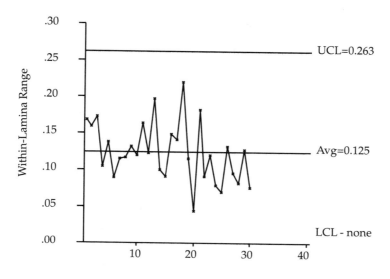

FIGURE 8.16 *Range Chart for Within-Lamina Variation*

Given this description of how the numbers in each of the columns are obtained and what they represent, the analysis of this set of data for studying the components of the total variance in pore size progresses as follows:

First, to study the stability of the within-lamina component of variation, a range chart is constructed of the 30 within-lamina ranges provided in column IV. This range chart is provided in Figure 8.16. The upper control limit uses D_4 for $n = 5$; there is no lower control limit. An examination of this range chart provides information on the stability of the within-lamina component of variation.

Because this range chart of the within-lamina ranges is in control, we conclude that within-lamina variation is stable. Therefore, the next step is to estimate the within-lamina variation from the average range. Again, the value of d_2 for samples of size $n = 5$ is used to estimate this component of variation. The within-lamina variation is estimated by:

$$\hat{\sigma}^2_{WL} = \left(\frac{\overline{R}}{d_2}\right)^2 = \left(\frac{.12453}{2.326}\right)^2 = .0086634$$

With an estimate of the within-lamina component of variation, the next step is to examine the ranges of the averages of the three laminas selected from each of the 10 railcars. Through this examination, the short-term, lamina-to-lamina variation within a car can be separated from the between-car variations. These 10 ranges of the averages, contained in column VI, are plotted in the control chart of Figure 8.17. The control chart constant, D_4, is used for samples of size $n=3$, since three averages are used in computing each range. The stability and average size of these ranges are important for insight and judgment. Stability contains information about the consistency across all cars of the effect of process and material effects for laminas made from the same railcar's material.

Since this range chart for between-lamina variation is in control, then the between-lamina variation can be estimated. The magnitude of the average range, $\bar{R}_{BL}$, reflects variations within-laminas and variations between laminas within the same car. How much of that reported variability is due to between-lamina causes as distinct from within-lamina causes can be estimated by using the formula provided below. The first term in the formula, $(\bar{R}/d_2)^2$, captures the variation occurring between lamina averages from the same car. As previously noted, this variation includes not only variation between laminas, but it also includes the effects of within-lamina variation. The second term in the formula, $\hat{\sigma}^2_{WL}/n$, is the amount of variation in the averages that can be explained by the variation occurring within laminas. The within lamina variance is divided by $n = 5$, the number of measurements used to calculate each lamina average. Thus, an estimate of the between-lamina variation is calculated as:

$$\hat{\sigma}^2_{BL} = \left(\frac{\bar{R}_{BL}}{d_2}\right)^2 - \left(\frac{\hat{\sigma}^2_{WL}}{n}\right) = \left(\frac{.06138}{1.693}\right)^2 - \left(\frac{.00286634}{5}\right) = .00074117$$

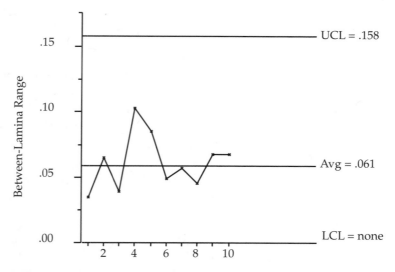

FIGURE 8.17 *Range Chart for Between-Lamina Variation*

With estimates of both the within-lamina and between-lamina components of variation, the next step of the analysis is to study the stability of the short-term, between-car variations. This analysis is done using the moving ranges of the averages, provided in column VII. A moving range chart of these values is provided in Figure 8.18. This moving range chart is in control, so we conclude that the short-term, between-car variation in average pore size is consistent across the 10 cars examined in this study. Hence, it is appropriate to examine the average pore size, car by car, for stability. The average chart in Figure 8.19 is a plot of the values in column V of Table 8.5. This average chart is used to investigate the presence of a long-term component of between-car variation using $\overline{MR}_{Y_c}$ as the baseline variation. The calculations for the control limits for the average chart of Figure 8.19 are:

$$UCL = \overline{\overline{Y}}_c + 3 \left(\frac{\overline{MR}_{\overline{Y}_c}}{d_2} \right) = 0.558 + 3 \left(\frac{.048}{1.128} \right) = .686$$

$$LCL = \overline{\overline{Y}}_c - 3 \left(\frac{\overline{MR}_{\overline{Y}_c}}{d_2} \right) = 0.558 - 3 \left(\frac{.048}{1.128} \right) = .430$$

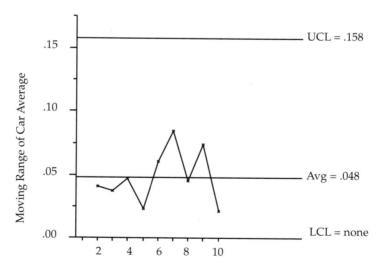

FIGURE 8.18 *Moving Range Chart for Car Averages*

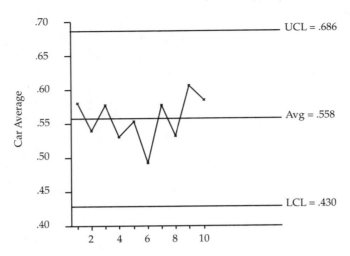

FIGURE 8.19 *Average Chart for Car Averages*

Since the average chart in Figure 8.19 is in control, we conclude that the averages in this chart show no indication of an additional long-term component of between-car variation. The car-to-car variations captured by the moving ranges can now be used to estimate the between-car component of variation. The formula for calculating this component is provided here. The first term in this formula, $s_{\bar{Y}_C}^2$, is the calculated variance of the averages in column V. The term b refers to the number of laminas selected from each car, which is three in this case. The term n represents the number of grids chosen from each lamina, so $n=5$ for this set of data.

$$\hat{\sigma}_C^2 = s_{\bar{Y}_C}^2 - \frac{\hat{\sigma}_{BL}^2}{b} - \frac{\hat{\sigma}_{WL}^2}{b \times n} = .001141188 - \frac{.00074117}{3} - \frac{.00286634}{15} = 0.000703$$

Our analysis of the data on average pore size has shown that each of the three components of variation is stable across the time of the study. If this had not been the case, then process personnel would have been directed to examine the nature of the instability and remove the special causes discovered. For example, if the range chart in Figure 8.16 had been out-of-control, then the reasons for inconsistent variation in pore sizes within a lamina would need to have been identified. Also, if the range chart in Figure 8.16 had been out of control, the analysis of the between-lamina and between-car sources of variation would have been inappropriate. The next step would have been to determine and remove the reasons for the instabilities in the within lamina variation before studying the between-lamina and between-car variations. On the other hand, if the within lamina variation was found to be consistent, but the between lamina variation, car to car, was not stable, then work efforts would be directed toward identifying the source of instability between laminas made from the material from the same car.

Since the three components of variation were, of course, found to be consistent, the total variation in the process can be determined using the following formula:

$$\hat{\sigma}_t^2 = \hat{\sigma}_{WL}^2 + \hat{\sigma}_{BL}^2 + \hat{\sigma}_C^2 = .00286634 + .00074117 + .000703 = .00431051$$

An examination of this total variation in average pore size shows that the major contributor of variation to total process variation is the within-lamina variation. This information determines the future work that is needed to reduce the variation in pore size. There are undoubtedly several process factors that contribute to within-lamina pore size variation. Knowledgeable, experienced individuals can surely name many of these variables, some of which probably involve material properties. By knowing where in the process most of the variation is occurring, it is possible to focus attention on a more restricted set of potential variables. Follow-up work is now necessary to identify sources more specifically, to verify the effects of those sources, and to determine the economic and technical feasibility of eliminating the effects of one or more of these sources.

8.3.3 Alternate Subgrouping Plans

In the membrane process, the subgrouping plan was structured to provide information on the process issues believed to be relevant. This subgrouping plan determined the three levels of variation studied—within-lamina, between-lamina, and car-to-car variation. For different objectives, a different subgrouping plan would be necessary. For example, a study on the membrane process might have had the following objectives:

- To confirm or negate the belief that a major portion of the variation in pore size is due to the measurement process for determining average pore size.
- To obtain information on within-lamina variation in order to understand the effects of equipment and methods on the variation.
- To study variation between cars to understand the effects of different raw materials on the variation in pore size.

Given these objectives, the following subgrouping plan might be adopted:

1. Obtain at least one lamina from each selected car.
2. From each lamina, select several grids.
3. Take at least two repeated measurements on pore size from each grid.

With this subgrouping plan, variation due to differences between cars of material can be studied by examining the variation in the average lamina measurements for each car. Within-lamina variation can be studied by examining the variation between average pore size for each grid within a lamina. The component of variation imparted by the measurement process can be studied by comparing the repeated readings. To summarize, the three levels of variation explored by this subgrouping plan are

- between-car (or material) variation,
- within-lamina (or processing) variation, and
- measurement variation.

At different times, different sources of variation will need to be explored. The choice of the subgrouping plan determines which sources can be examined, and the process variations to be investigated determine the subgrouping plan.

8.4   STUDY OF FIXED EFFECTS IN A HIERARCHICAL MODEL

The methods discussed in the previous sections apply to situations where random sampling was used to collect data about different hierarchical levels of a process. The random sampling scheme was chosen due to the knowledge, or lack of knowledge, of process behavior. In other situations where an understanding of the levels of variation in a process is needed, a more directed sampling scheme may be required. A situation may require taking measurements from specific locations at specific times. When such a sampling scheme is used, a different method of analysis is required. The following study of within- and between-rack variation in an electroplating process illustrates this idea of studying fixed effects in a hierarchical model.

8.4.1   Study of an Electroplating Process

In the manufacture of copper electroplated stainless steel blades, one of the current concerns is the quality of the electroplating process. A critical characteristic in determining this quality is the thickness of the electroplating. Specifications for this thickness are 35KÅ ± 3K. The blades are plated in racks, each of which holds 200 blades. Process investigation began with the development of a flowchart of the electroplating process (Figure 8.20). Data collection initially consisted of selecting three blades from each of three consecutively produced racks and measuring the thickness on each of the three parts. From this study, the following information was obtained:

1. Using the current measurement methods and subgrouping plan, the process is judged to be stable.
2. The process level averages 35.15K Å, which is close to target.
3. The process standard deviation is estimated to be 2.76K Å. Thus, the NT is 6(2.67K) = 16.02K Å. The process is not capable, since the natural tolerance exceeds the engineering tolerance of 2(3K) = 6K Å. These results are consistent with process yields.
4. To meet specifications, process variation must be reduced.

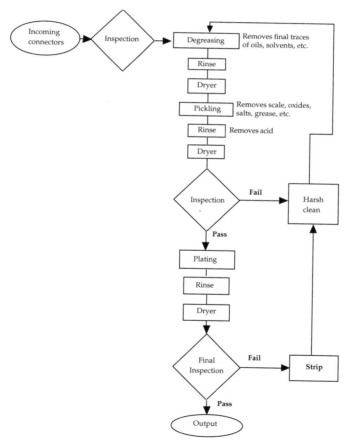

FIGURE 8.20 *Flowchart of the Electroplating Process*

The data suggest that most of the process variability occurs within racks. The level of the process remains consistent over time, indicating little long-term or rack-to-rack variation. However, the subgrouping plan used did not allow for the study of within-rack variability.

Although the process average is essentially on target, the process produces product outside of specifications because of the large process variability. Even though the process average of 35.16K corresponds almost exactly with the target of 35K, the process will deliver many blades outside of the stated specifications due to the large variation existing under the current conditions. To obtain guidance on how to reduce this large process variation, a cause and effect diagram of the reasons projected for variation in thickness was developed (Figure 8.21). To study the effect of key supposed reasons on process variation, the cause and effect diagram must be examined to identify sources of variation that may be influential in creating variations that reveal themselves in different ways.

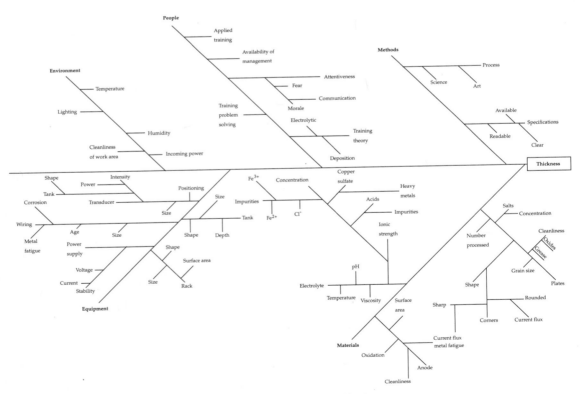

FIGURE 8.21 *Reasons for Variation in Copper Plating*

Three categories of variation could be identified as contributing to total process variation in plating thickness:

- within-part variation
- within-rack variation, and
- between-rack variation.

Consideration of these levels of variation leads to a host of possibilities for why the within-subgroup variation was found to be large in the initial stage of process investigation. For example, the concentrations and impurities of the electrolytes could vary within the tank used in the plating process. These variations would lead to different thicknesses on blades within the same rack. The different positions of the blades in the racks could result in varying degrees of thickness because of the nature of the electrolytic process. The cleanliness of the work area or of the incoming blades could affect the variation within parts, since foreign material on a blade could result in inconsistent variation across that blade. Many additional reasons for within-part and within-rack variation can be suggested. A narrowing of this suggested list would be possible if the stability and magnitude of these levels of variation were known. For example, if within-blade variation is inconsistent between racks, sources of variation that might affect some racks of blades and not others would be investigated. If both within-blade and within-rack variation are consistent but the within-rack variation is the larger contributor to total variation, reasons for consistently large levels of variation within racks will be sought.

With the intention of providing direction for investigating causes of thickness variation, the management group decided to collect data on the variation within racks. Prior to the current investigation, the consistency of thickness within a blade was explored and found to be consistent over time and small in magnitude. Consequently, efforts are focused on characterizing within-rack variation. To understand this component of variation, multiple blades from a number of racks need to be selected. The main issue, now, is how to choose the blades from within the racks, since differences in plating thickness within a rack can occur in different ways. It might be that all locations within a rack have essentially the same average, but have significantly different variability. On the other hand, the respective locations may have distinctly different averages with variability around that average being about the same from one location to another. An understanding of the manner in which within-rack variation occurs requires further investigation.

Because it is thought that thickness differences may be associated with different locations in the rack, a random selection of blades from a rack would not be appropriate. Consequently, blades were selected specifically to explore fixed differences at different locations. Because it is believed that thickness differences may exist between blades at the two ends of a rack and between middle and end blades, five rack positions were chosen for study. These positions are shown in Figure 8.22, which illustrates the sampling scheme used in this study. Three parts were selected from within the five preselected locations within a rack. Twenty racks were sampled over a period of 10 days. Each selected part was measured one time at a randomly chosen location on the part. Thus, from each rack, 15 different thickness readings were recorded. The readings from the first two racks are provided in Table 8.6.

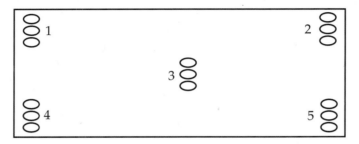

FIGURE 8.22 *Sampling Scheme for Studying Within-Rack Variation in Electroplating Process*

TABLE 8.6 *Thickness Readings of Electroplated Blades*

Rack	Position		Thickness		$\overline{X}$	R
1	1	33.05	35.21	33.88	34.05	2.16
1	2	33.29	31.05	32.41	32.25	2.24
1	3	37.54	36.25	37.09	36.96	1.29
1	4	32.13	34.56	32.99	33.23	2.43
1	5	33.74	34.17	33.21	33.71	.96
2	1	33.66	33.17	34.28	33.70	1.13
2	2	30.63	32.57	31.44	31.55	1.94
2	3	38.12	37.36	37.03	37.50	1.09
2	4	34.39	32.24	32.42	33.02	.18
2	5	34.94	34.45	33.46	34.28	1.48
3	1	•	•	•	•	•
3	2	•	•	•	•	•
•	•	•	•	•	•	•

It would be possible to calculate the range of these numbers as a means of evaluating the within-rack variation. For example, the range of the 15 thickness readings from rack 1 would be 37.54 – 31.05 = 6.49. However, summarizing the information from the study of within-rack variation in this fashion would be a tactical error. In calculating the range of the 15 numbers from the five different positions in rack 1, the information on difference by position is lost. A more insightful means of analyzing these numbers is to examine the variation and level of thickness separately for each of the five positions. In Table 8.6, a range and average have been calculated for each subgroup of three readings which represents one of the five rack positions. Thus, the behavior of thickness at each different position can be examined.

The ranges displayed in Table 8.6 are influenced by those causes contributing to the within-rack component of variation. However, the sources of variation acting to affect the ranges are those that impart variation in a spatially small region. For purposes of discussion, the variation observed within a rack at a particular location will be referred to as within-location variation. The sources creating within-location variation may affect different locations in the rack in a different manner, or they may act differently during the production of some racks and not others. The 100 ranges (20 from each of five positions) are charted in Figure 8.23. The identity of the location in the rack corresponding to each range has been maintained by using the number of the location to plot the range instead of a point. The ranges from the five positions in the first rack have been plotted, then those from the second, etc. The centerline on the range chart is the average of all 100 ranges. The fact that this range chart is in control indicates that there are no observable differences in within-location variation among the five positions and that there is no evidence of inconsistent within-location (WL) variation in the 20 racks studied.

Because the R chart in Figure 8.23 is in control, an estimate of the contribution of the within-location variation can be calculated as:

$$\hat{\sigma}_{WL} = \frac{\overline{R}}{d_2} = \frac{1.609}{1.693} = .95$$

This estimate of within-location standard deviation also captures within-part variation, since the thickness readings were obtained by measuring a random location on each part. Earlier, the process (i.e., total) standard deviation was stated to be 2.67, a value considerably higher than this estimate of $\hat{\sigma}_{WL}$. Thus, unless the process has changed significantly between the time of the two studies, a major portion of process variation beyond the within-part and within-location components has not been accounted for.

An examination of the X-bar chart of Figure 8.23 provides an explanation for the observed differences in magnitude of the two reported standard deviations. The location averages have also been plotted in time order, and the location numbers have been used to maintain location identity. All of the averages for location 3 fall above the upper control limit, indicating that thickness readings tend to be higher, on average, at location 3. Some of the averages for location 2 fall below the lower control limit, so the process appears to deliver parts with a lower average thickness at location 2.

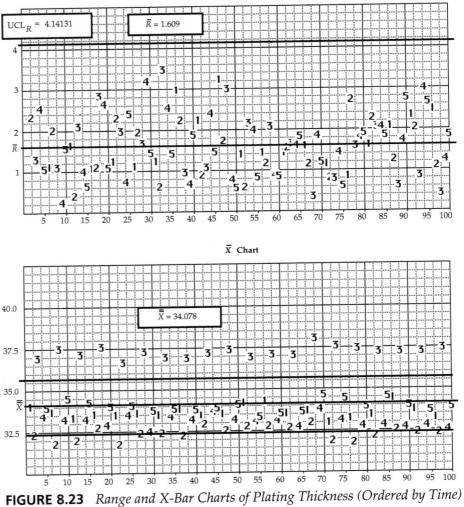

FIGURE 8.23 *Range and X-Bar Charts of Plating Thickness (Ordered by Time)*

The differences for the five locations are more clearly seen by plotting the ranges and X-bars by position while maintaining the time order across the positions. The range and X-bar charts for the same data ordered by time within position are contained in Figure 8.24. In the range chart of Figure 8.24, the range of the three numbers from position 1 of the first rack is plotted, followed by the range for position 1 of the second rack, and so on. Then the 20 ranges, in time order, for the 20 subgroups corresponding to position 2 are plotted. This same ordering is used on the X-bar chart.

The range chart provides no additional information over what was gained by the earlier range chart. However, the differences in the average thickness at the five different locations are more clearly seen on the X-bar chart of

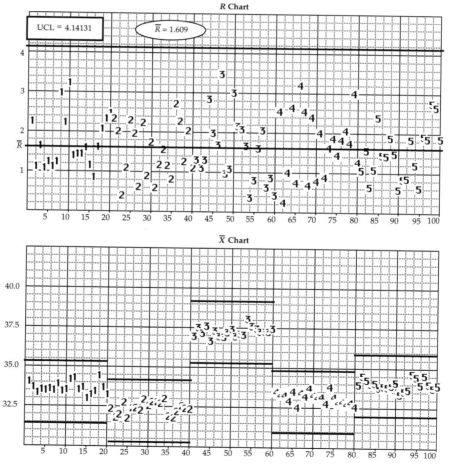

FIGURE 8.24 *Range and X-Bar Chart for Plating Thickness (Ordered by Time Within Position)*

Figure 8.24, where separate control limits have been calculated for each different position. So, for position 1, the overall average of the 20 X-bars was found. Then, $A_2\overline{R} = 1.023(1.609) = 1.936$ was added and subtracted from this average for position one. The same procedure was followed to arrive at limits for the remaining positions. Since the five positions were seen to have the same within-location variability, the same constant, 1.936, was added and subtracted to each of the overall position averages to obtain the control limits.

From Figure 8.24, additional information about the differences in average thickness for the five positions can be obtained. Since none of the average thickness readings falls outside of their respective control limits, the average thickness at each position is determined to be consistent across the 20 racks studied. This information provides additional direction for discovering the causes for the location-to-location differences in averages. Investigation of these causes should focus on those sources of variation that occur consistently, rack after rack, causing higher levels of copper plating at the middle and third positions and lower levels of plating at the end position (position 2).

The group performing the study of the electroplating process thought that the more likely explanation of the differences in average thickness was due to the tank design, in particular, to the placement of the ultrasonic transducers. The location of the transducer was changed and additional runs were made on the rebuilt tank. Data, collected using the same subgrouping strategy defined earlier, are provided at the end of the chapter in the Section 8.6 application problem. It is left as an exercise to evaluate whether the new tank design has lowered the variation in plating thickness by reducing the differences in average thickness across locations within a rack.

8.4.2 Studying Deterministic Effects in a Batch Process

This example from the production of Curemo, a granular antibiotic sold to veterinarians, is based on the need to evaluate a deterministic effect in a batch process. At a final stage of the process, the active ingredient, in powdered form, is mixed with a dry, inert ingredient whose function is to provide bulk to the finished product. After the product is mixed, it is transported to the filling line where it is filled into plastic bottles, subject to the availability of the filling machine. Records of material properties added at this final stage of the batch process are kept. These records include

- measurements on the purity of the final active ingredient,
- the amount, by weight, of active ingredient added, and
- the amount, by weight, of inert ingredient added.

A measurement is also made after the material has been mixed. Moving range and individuals charts of the percentage of inert ingredient in a sample of material have been routinely maintained in the past. These charts indicate that the percentage of active ingredient in the batch process is in control with an average of 74.8 and a standard deviation of .35.

Recent initiatives have called for a broader examination of the percent inert ingredient in bottles purchased by the company's customers. The control charts discussed earlier evaluate the level and variation of the inert ingredient prior to the filling operation. However, there is a suspicion that the filling machine may impart additional variation to the process, which would result in increased variation in the percent inert ingredient in bottles of the product. Consequently, an examination of the percent inert ingredient in filled bottles is to be undertaken.

How the samples of filled bottles were selected was driven by knowledge of how the filling operation is likely to add variation to the mix of ingredients. An entire batch of Curemo is loaded into a large hopper on the filling machine, and the bottles are then filled one at a time. It is suspected that during the course of the filling some settling of ingredients is induced as the material flows out of the hopper. If this is the case, bottles filled later in the operation will tend to have a larger percent of inert ingredient than bottles filled at an earlier stage. Consequently, the decision was made to select three bottles from each batch of material, one just after the filling operation has begun, one at about the halfway point in the filling of a batch, and one in the final stages of filling. The percent inert ingredient found in the three bottles selected from the 20 most recently filled batches appears in Table 8.7.

As with the study of particle size in Section 8.2.5, multiple readings have been selected from within each batch of material. In the particle size study, six samples were randomly selected from each batch and the standard deviation of the six samples in each subgroup was calculated in order to provide information about the within-batch variation. In the present example, where three bottles have been selected from each batch, calculating the standard deviation or the range of the three would not provide the kind of information sought about within-batch variation. Bottles filled at three different times from within each batch have been taken to determine if the percent inert ingredient in bottles varied according to the time at which the bottles were filled. If the filling process does result in different levels of inert ingredient at different times, these differences will help explain the creation of within-batch variation. These differences, which would contribute to increased variability, will be noted by distinctly different averages at the respective sampling times across all batches. In other words, an analysis by time of fill would not only indicate the presence of a within-batch component of variation affecting total variation in filled bottle percentages, but would also provide additional guidance on how this variation was created.

The data in Table 8.7 are arranged according to the time at which the bottles were selected. Bottle 1 denotes the first bottle selected from a batch when the filling process had just begun, and bottle 3 denotes the last one selected when the filling of the batch is close to completion. Each time (or location in the batch) needs to be evaluated with respect to variation and level. To examine the variation at each of the fill times, moving ranges of the measurements on percentage inert ingredient have been calculated. Figure 8.25 contains a plot of the 3 x 19 = 57 moving ranges. The moving range chart is in control. Therefore, two conclusions about the variation observed at the three different times of sampling can be made:

1. It appears that the differences observed from a bottle selected early in a batch to one selected early in the following batch are consistent across the 20 batches studied. A similar statement about the batch-to-batch consistency can be made about the other two times or positions.
2. This short-term, batch-to-batch component of variation is of a similar magnitude for each of the three positions studied.

TABLE 8.7 *Data on Percent Inert Ingredient*

Batch	Bottle 1 X	Bottle 1 MR	Bottle 2 X	Bottle 2 MR	Bottle 3 X	Bottle 3 MR
1	74.1		75.1		76.3	
2	73.2	.9	75.0	.1	75.0	1.3
3	72.7	.5	73.9	1.1	75.4	.4
4	73.0	.3	73.6	.3	74.3	1.1
5	73.5	.5	73.7	.1	74.8	.5
6	73.5	.0	73.9	.2	75.3	.5
7	72.7	.8	74.8	.9	76.5	1.2
8	74.0	1.3	74.1	.7	75.2	1.3
9	73.0	1.0	74.5	.4	76.3	1.1
10	72.4	.6	74.1	.4	74.5	1.8
11	73.1	.7	74.3	.2	75.5	1.0
12	74.5	1.4	75.0	.7	76.3	.8
13	73.3	1.2	74.2	.8	75.4	.9
14	73.6	.3	74.9	.7	75.0	.4
15	72.8	.8	74.6	.3	76.1	1.1
16	73.7	.9	73.9	.7	75.0	1.1
17	73.0	.7	74.2	.3	75.9	.9
18	73.1	.1	75.5	1.3	75.7	.2
19	74.2	1.1	75.0	.5	75.5	.2
20	73.1	1.1	74.1	.9	75.2	.3

An examination of the individuals chart in Figure 8.25 provides an understanding of how fill time creates variation in fill weights beyond that due to the manufacturing process. Bottles filled early have, on average, lower percentages of inert ingredient than bottles filled later.

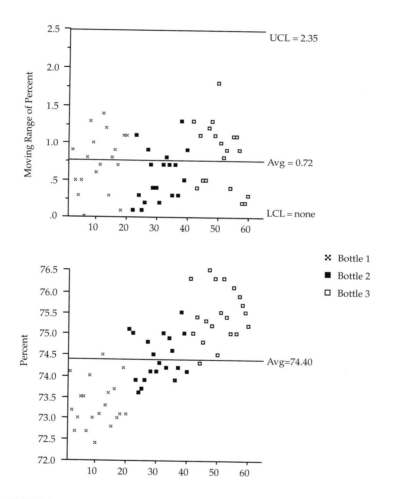

FIGURE 8.25   *Moving Range and Individuals Charts on Percent Inert Ingredient (Analysis by Bottle)*

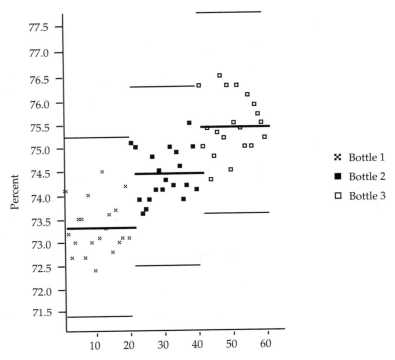

FIGURE 8.26 *Individuals Chart of Percent Inert Ingredient (Limits Calculated by Time of Fill)*

From the individuals chart in Figure 8.25, it was concluded that bottles filled early have, on average, lower percentages of inert ingredient than bottles filled later. Next, in order to study the consistency of the level of inert ingredients across batches, the individuals chart shown in Figure 8.26 was constructed. In this chart, a separate average has been calculated for each different time of fill. Individual control limits for each fill time have been calculated by adding and subtracting

$$3\frac{\overline{MR}}{d_2} = 3\left(\frac{.72}{1.128}\right) = 1.91$$

to each of the three averages. This same value of $\overline{MR}/d_2$ is used for all three sets of control limits, since the short-term batch-to-batch variation was determined to be the same for all three fill times. All of the points on the individuals chart of Figure 8.26 fall within their respective control limits. Thus, the filling process is acting to create different average percentages of fill at the different times and these different averages are being consistently maintained batch to batch.

The data collected on the process for making and dispensing the antibiotic into bottles shows that the percent of inert ingredient increases throughout the filling process. Previously, evaluation of the variation in percentage of inert ingredients had occurred by sampling the batches of Curemo prior to filling. From these data, a standard deviation for the process was found to be .35. When data were collected on the process after filling, it was found that the batch-to-batch variation at a given time of fill was consistent. The batch-to-batch variation at a given fill time is calculated as:

$$\frac{\overline{MR}}{d_2} = \frac{.72}{1.128} = .64$$

This value of .64 does not capture all of the variation that would be observed in the bottles filled, since a significant amount of variation between the average percentages is observed at the different times of fill. The total variation in the percentage inert ingredient after the filling process is considerably larger than .64.

At the very beginning, the standard deviation prior to filling was stated to be .35. After filling, the batch-to-batch variation at a location is .64, so the total variation after filling, which will also include position differences, will exceed .64. Clearly, a significant amount of variation is being imparted to the process by the filling machine. Some of this additional variation is reflected in the different levels observed on the individuals chart. The reasons for the difference between the two standard deviation values, .64 and .35, are more subtle. Even when taken at nearly the same time of fill, bottles from different batches exhibit more batch-to-batch variation in percentage inert ingredient than did the material prior to fill.

The next phase of investigation in the process is to determine how to reduce the variation imparted by the filling phase of the process. Two options for reducing the variation need to be investigated. The first option is the manner in which the active and inert ingredient are mixed. Can this procedure be improved so that the two ingredients "stay mixed" during fill? The second one is to redesign the filling operation. In considering the two options from the customer's viewpoint, the first option was chosen as more appropriate. If the two ingredients did not stay thoroughly mixed during the filling operation, then it would be the case that they would not remain mixed during shipment and storage. Therefore, the next step in the process study is to identify a better method for mixing the ingredients.

8.5   PRACTICE PROBLEMS

8.5.1   A chemical product is produced in large batches. An important characteristic of the product is the level of active chlorine in the material. It is suspected that two major sources of variation in the process exist that affect the level of active chlorine in the product. One possible source of variation is that the mix of ingredients is not the same from batch to batch. Another potential source is that the batches may not be thoroughly mixed. To study the active chlorine levels, three samples are chosen at random from each of 15 successive batches. The levels of active chlorine in the three samples are measured. These values appear in Table 8.8. The range and average of the samples are calculated and plotted on control charts in Figure 8.27.

   a. Which of the two major sources of variation are reflected in the ranges? Which in the averages?
   b. Compute control limits for the range chart. If appropriate, complete the X-bar chart.
   c. Which source of variation might cause the range of variation in the subgroup averages to be larger than that predicted by the control limits on the X-bar chart?
   d. If appropriate, estimate the percent of total variation due to the between-batch component of variation. (The flowchart in Figure 8.10 may be useful here.)

8.5.2   One property of concern in the manufacture of a product is the silicon content, Si. The specification of Si is 1000 ppm ± 225 ppm. Considerable difficulty has been experienced in trying to make the product conform with these specs. Past practice has been to take one portion from each batch and have a single measurement made on the portion by the laboratory. In beginning to address the issue of the variability currently existing in these measurements, the measurement process has been questioned.

To investigate the variability in Si results, five portions have been selected from the 10 most recently produced batches. Each of these portions has been sent for one measurement by the lab. (Each of the batches is thought to be homogeneous in Si content.) The determinations of Si are listed below:

	Batch 1	Batch 2	Batch 3	Batch 4	Batch 5
	1140.1	1058.6	922.6	989.4	1063.2
	1192.8	956.2	901.1	932.1	947.2
	1094.3	1082.4	980.4	1148.2	1031.6
	970.0	1066.2	907.9	1127.0	1163.2
	<u>1045.4</u>	<u>1013.9</u>	<u>896.0</u>	<u>1084.7</u>	<u>1142.5</u>
$\bar{X}$	1088.52	1035.46	921.60	1056.28	1069.54
s	85.8408	51.0655	34.3575	92.4012	87.3815

	Batch 6	Batch 7	Batch 8	Batch 9	Batch 10
	996.7	977.1	1132.8	973.6	1084.7
	943.6	936.6	1157.2	933.1	1143.5
	943.9	1051.2	1114.3	965.5	1067.2
	893.4	941.5	963.2	941.3	983.6
	<u>964.7</u>	<u>976.9</u>	<u>1019.6</u>	<u>997.2</u>	<u>1033.9</u>
$\bar{X}$	948.46	976.66	1077.42	966.14	1062.58
s	37.6392	45.8189	82.4337	25.7364	59.4118

TABLE 8.8 *Active Chlorine Levels in Three Samples from Each of 15 Batches*

Batch	Sample X_1	X_2	X_3	$\bar{X}$	R
1	.0589	.0614	.0578	.05938	.0036
2	.0453	.0483	.0461	.04658	.0031
3	.0540	.0559	.0577	.05585	.0037
4	.0532	.0560	.0538	.05435	.0028
5	.0625	.0570	.0610	.06019	.0055
6	.0409	.0433	.0437	.04261	.0028
7	.0489	.0453	.0525	.04891	.0071
8	.0590	.0579	.0545	.05716	.0045
9	.0504	.0512	.0507	.05076	.0008
10	.0559	.0506	.0562	.05423	.0056
11	.0580	.0572	.0565	.05725	.0016
12	.0529	.0533	.0581	.05475	.0052
13	.0583	.0610	.0580	.05912	.0030
14	.0566	.0508	.0509	.05278	.0058
15	.0605	.0541	.0577	.05745	.0065

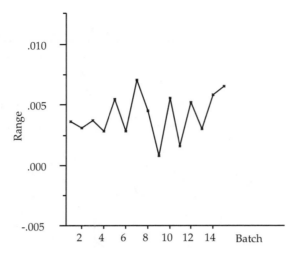

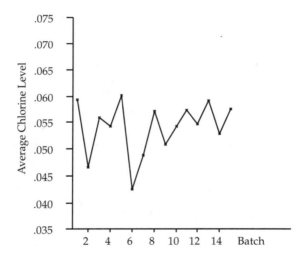

FIGURE 8.27   *Plots of Ranges and Averages of Active Chlorine Levels*

a. What is the source of variation in determinations for any one batch?
b. What effect, if any, does the measurement process have on evaluating batches?
c. What sources of variation affect the batch averages?
d. What is learned about the stability of the process for producing batches from the X-bar chart? What is learned about the measurement process from the X-bar chart?
e. If appropriate, estimate between-batch variation. What percentage of observed variation is due to measurement variation?

8.6   **APPLICATION PROBLEM**

In Section 8.4.1, an electroplating process was described. Data collected on the process showed that there were differences in thickness of plating related to the position of the blade within a rack. The tank in which the plating occurs has been redesigned by changing the placement of the ultrasonic transducers. Runs have been made on the rebuilt tank. Data have been collected using the same subgrouping strategy as before. Range and averages for these data are plotted in Figure 8.28. (As in Section 8.4.1, the numbers 1 through 5 are plotted on the charts to represent position.) Some summary results for the 100 subgroups (20 for each of the five positions) are as follows:

Summary Results by Position After Relocation of Transducers

Position	$\bar{R}$	$\bar{\bar{X}}$
1	1.1860	35.7138
2	1.2275	35.8367
3	1.0940	36.2185
4	1.5615	36.0753
5	1.1450	35.9373

a. What effect has the placement of the transducers had on reducing the within-rack variation?

b. If appropriate, provide an estimate of process variation. How would you assess the current process's capability to meet specifications?

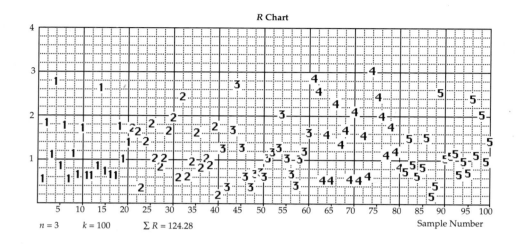

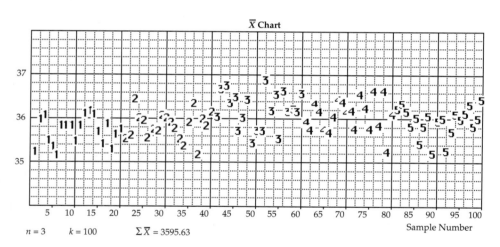

FIGURE 8.28   *Plating Thickness—Ordered by Time Within Position*

9 Measurement Processes

Like a production process, the act of making measurements is a process and must be examined over time and conditions to evaluate how well it delivers its intended output. In the case of a measurement process, the intended output is, of course, a measured value and, as with other processes, variation exists in process output. If a given measurement process is applied repeatedly to the same material, part, or component, the measured result will undoubtedly be different from one time to the next, even though what is being measured is unchanging. Measurement variation arises because measurement is undertaken in different environments, in different circumstances, with different instruments, and by numerous technicians.

Measurement variation creates uncertainty about what is actually occurring in a production process being measured.

1. If a measurement process exhibits unstable variation, then the correct interpretation of measurements from a production process will be compromised. If instability in measurement variation goes undetected and therefore unrealized, the unstable, erratic variation may be attributed to an instability in the production process being measured; inappropriate actions may then be taken on the wrong things for the wrong reasons.
2. If a measurement process has stable, yet large variation, then the ability to understand production process variation will be reduced. If this large measurement variation is not known, the large variation in reported values may be attributed to the production process; again, undue attention may be devoted to unproductive tasks.
3. If a measurement process does not deliver measurements at a consistent level, then the ability to evaluate the average of a production process will not exist and incorrect decisions may be made about a product.

For these reasons, evaluating the statistical reliability of the process used to generate measurements is an unavoidable first step in collecting measurements to describe a production process. An evaluation of a measurement process should address the following concerns:

1. The variation and average delivered by a measurement process should be stable over time and conditions.
2. On average, measured results should agree with a master standard, procedure, or reliable reference value.
3. Differences among operators or laboratories should be evaluated to see if the same materials or parts would receive consistently the same measurement or judgment.
4. The magnitude of measurement variation should be examined to determine if the measurement process can detect important differences among parts or materials.

The methods and examples in this chapter deal with evaluating these four concerns relating to measurement processes. As with the evaluation of a production process, one data set and its analysis cannot address all concerns relating to a measurement process; similarly, there is no one, specific method for evaluating a measurement process. However, a process for evaluating measurement processes can be described. A study of measurement processes includes

1. knowing what is required of the measurement process,
2. describing the way in which the process operates,
3. describing the process flow,
4. considering the sources of variation affecting the process, and
5. obtaining data to determine how well the process delivers what is required.

Adding to the complexity of studying measurement processes and their variation is the fact that the words *measurement variation* are subject to different definitions at different times by different people. At a macro level, for example, measurement variation can mean the variability arising from all instrument–technician combinations as measurements are made on the same material at a production rate across a significant time period, subject to changes in supporting material, equipment, and environmental conditions. At another focus level, measurement variation might simply mean the inherent variation within one instrument or device as used by one person over a short period of time. A variety of needs falls between these two extremes, each requiring its own data collection plan, subgrouping tactics, and analysis methods. This chapter provides the necessary background for developing appropriate measurement studies. Methods for characterizing measurement processes are discussed as well as examples of designing measurement studies to address appropriate process evaluation. However, the chapter does not contain a complete catalog of all possible approaches to measurement studies. Instead, the focus is on developing the knowledge and tools required to design appropriate measurement studies.

9.1 CHARACTERIZING A MEASUREMENT PROCESS

The concepts of measurement variation can be illustrated by the following example dealing with the determination of the weights of bars of soap. The primary attributes of a measurement process are described in this simple example with little emphasis on the operation of the measurement process. However, this elementary measurement process has all of the characteristics of more complex measurement processes. The measurement device used, a scale, is subject to the same questions as other measurement devices. In particular, questions to be addressed about a measurement device follow:

- How closely will measurements from a device repeat themselves? Does it do this consistently?
- Does the device, on average, tend to provide results sufficiently close to the accepted standard? Does it do this consistently?

An effective method for obtaining data to address the above issues is to

- obtain a standard for the average weight of soap that is not changed by time, use, or environment,
- get at least two independent measurements on the standard within a short time span (this requires independent and complete "setups" for each measurement),
- make the repeated measurements over time and circumstance,
- find the range and average for each subgroup, and
- assess the stability and magnitude of the corresponding range and average charts.

As part of the ongoing check of the scale used to weigh bars of soap, a standard weight of five ounces has been obtained. Twice each shift, the standard is weighed on the scale used for weighing the bars of soap. Several purposes are served by collecting these weights over time. One purpose is to check on the magnitude and consistency of the within-scale variation and its effect on measurement variation. A second is the ability to examine the magnitude and consistency of the average determination made by the measurement process. The two readings on the same shift allow assessment of the within-scale variations. Deviations in these two numbers might be attributed to

- the inability of the scale to return the same value for the same material,
- variations due to scale setup procedures,
- positioning of the test unit on the plate, or
- slight changes in the ability of the scale to read to the same average.

A range chart will be used to assess these deviations because the duplicate readings on the standard were obtained by the same individual within a relatively short time period. Consistency of the average weight is checked by the X-bar chart. Because a standard weight was used in the study, the average value delivered by the measurement process can be compared to what should be delivered if the measurement process is found to be in control.

Recorded measurements for the past 40 shifts are reported in Table 9.1. These data have been used to construct the X-bar and R charts in Figure 9.1. The R chart is in control, indicating that variation in the measured weights of the standard within a shift remains consistent across the study duration. Since the X-bar chart is also in control, the average weight read by the scale appears to be consistent across the time of the study. Changes in operators, environment, or methods across time do not result in an unstable measurement process. It is useful to consider what the stability of the measurement process does and does not mean for the ability to use weight measurements from this scale to study the process creating production weights of bars of hand soap.

> **It does mean** *the measurement variation imparted to the reported production values is consistent, although the amount by which the actual weight is increased or decreased is random and is affected in magnitude by within-scale variation.*

> **It does not mean** *the measurement process is sufficiently sensitive to be able to discriminate among bars of soap. Consistency, by itself does not assure discrimination. The discrimination of the measurement process depends on the relationship of the standard deviation of the measurement process to that of the production process. This concept is discussed further in Section 9.3.*

Since the weighing process is stable, a histogram constructed from the 80 measurements of Table 9.1 reliably describes the distribution of variation imparted by this measurement process. The histogram of Figure 9.2 provides a visual representation of the variation that would result when the same piece of material (in this case a five-ounce standard) is weighed repeatedly. An implication is that when a five-ounce bar of production soap is weighed one time, a value ranging from about 4.90 to about 5.10 is possible, with a result close to 5.00 being more likely. Of particular interest is a characterization of the center or average and the spread or range of this distribution. In characterizing measurement processes, the concept of accuracy refers to how close on average the measurement process delivers an accepted value of a known standard. The concept of precision captures the amount of variation delivered by a measurement process.

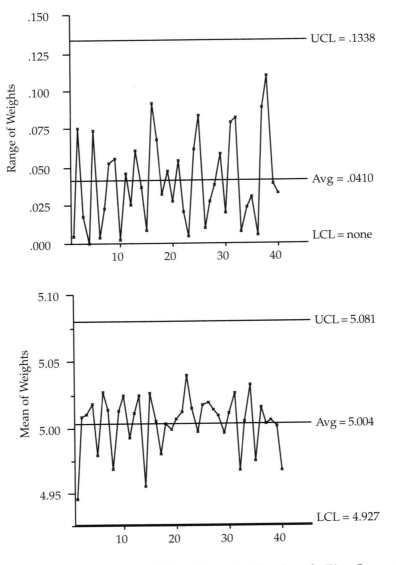

FIGURE 9.1 *Range and X-bar Charts for Weights of a Five-Ounce Standard*

TABLE 9.1 *Measured Weights of Five-Ounce Standard*

Subgroup	X_1	X_2	R	$\bar{X}$	Subgroup	X_1	X_2	R	$\bar{X}$
1	4.948	4.943	.005	4.9455	21	4.985	5.039	.054	5.0120
2	4.971	5.046	.075	5.0085	22	5.029	5.050	.021	5.0395
3	5.019	5.001	.018	5.0100	23	5.017	5.012	.005	5.0145
4	5.018	5.018	0	5.0180	24	5.028	4.966	.062	4.9970
5	4.942	5.016	.074	4.9790	25	5.059	4.975	.084	5.0170
6	5.029	5.025	.004	5.0270	26	5.014	5.024	.010	5.0190
7	5.025	5.002	.023	5.0135	27	5.028	5.000	.028	5.0140
8	4.995	4.942	.053	4.9685	28	5.028	4.990	.038	5.0090
9	5.041	4.985	.056	5.0130	29	5.026	4.967	.059	4.9965
10	5.023	5.026	.003	5.0245	30	5.021	5.001	.020	5.0110
11	5.016	4.970	.046	4.9930	31	5.066	4.987	.079	5.0265
12	4.999	5.024	.025	5.0115	32	5.009	4.927	.082	4.9680
13	4.994	5.055	.061	5.0245	33	5.009	5.001	.008	5.0050
14	4.974	4.937	.037	4.9555	34	5.044	5.020	.024	5.0320
15	5.022	5.031	.009	5.0265	35	4.959	4.990	.031	4.9745
16	5.051	4.959	.092	5.0050	36	5.019	5.013	.006	5.0160
17	5.014	4.946	.068	4.9800	37	5.048	4.959	.089	5.0035
18	5.019	4.987	.032	5.0030	38	5.061	4.951	.110	5.0060
19	4.975	5.022	.047	4.9985	39	5.021	4.982	.039	5.0015
20	5.021	4.993	.028	5.0070	40	4.951	4.984	.033	4.9675

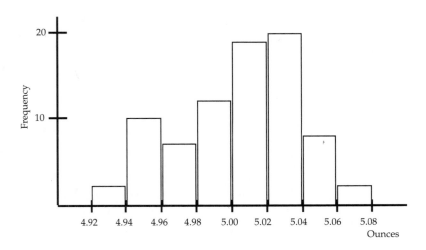

FIGURE 9.2 *Histogram of Weights of a Five-Ounce Standard*

9.1.1   Precision of a Measurement Process

Figure 9.3 contains histograms from four different measurement processes. For each of these processes, a five-ounce standard was weighed repeatedly over a variety of time and conditions and the measurement process was found to deliver stable levels of variation and average. Histograms A and B show that the respective processes generating these measurements have less variability than those which generated C and D. Measurement processes A and B are said to be more *precise* than processes C and D. The precision of a measurement process refers to the level of variation that would be delivered by repeated measurements of the same unit of product or material over a short interval of time.

The statements about the precision of the measurement processes represented in Figure 9.3 must be qualified.

1. Since a single standard of five ounces was used to characterize the measurement process, the precision of the measurement process has only been determined when the weight of the object measured is five ounces. It is entirely possible that, say, measurement process A has significantly larger variation when an object that weighs seven ounces is measured. Thus, if the weights of soap from production vary over a wide range, it should be understood that no knowledge has been provided about the precision of the measurement process over this range. This knowledge can only, in the final analysis, be had by actually evaluating the precision at various levels throughout the range of weights delivered by the soap-making process. In addition, because the measurement precision may change with changing levels of product values, the choice of the weight(s) used for a standard(s) should be selected with care. If only one standard is to be used, a reasonable choice would be to use one that measures close to the target value of the production process.

2. The fact that a known standard, rather than production material, was used to evaluate precision is a cause for questions. In the process of investigating measurement, it may be that greater care than usual was taken in evaluating the standard. In addition, knowledge that the standard weighs five ounces may mean that those weighing the standard are more likely to observe values closer to five ounces than if the weight were unknown, since people often "see" what they expect to see rather than what actually occurs. In both of these cases, the result would be to reduce the magnitude of the reported precision, making it smaller than that realized in actual practice.

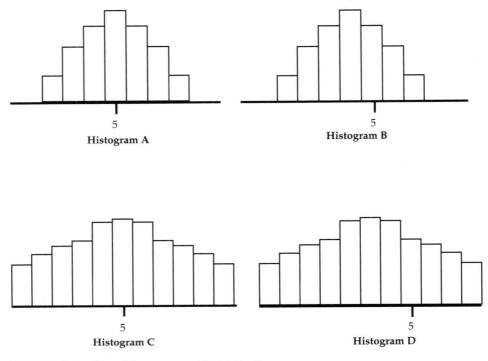

FIGURE 9.3 *Four Measurement Distributions*

The data reported on the measurement of the five-ounce standard in Table 9.1 resulted in stable range and X-bar charts with a value of $\bar{R}$ equal to .04095. Thus, the precision of the process can be described using the estimated measurement process standard deviation. For this example, the precision (sometimes referred to as "measurement error") would be estimated by:

$$\sigma_e = \frac{\bar{R}}{d_2} = \frac{.04095}{1.128} = .0363$$

9.1.2   Accuracy of a Measurement Process

The four histograms of Figure 9.3 also illustrate the concept of accuracy. Histograms A and C both are centered over five ounces, the accepted weight of the standard. Since the average of numerous measurements on the given material is almost the same as the quantity measured, measurement processes A and C are said to be *accurate*. The accuracy of a measurement process refers to the ability of the process to deliver, on average, the recognized value of a standard. The measurement processes that generated histograms B and D are not accurate, and are said to be *biased*. When the measurement process is one as simple as the weighing process discussed here, it will likely be possible to identify the reason for the bias and remove that cause. However, in more complex measurement processes it might be that attempts to remove the bias require drastic changes in the measurement process. In such instances, it is useful to know that a biased measurement process can still be used for evaluating product characteristics. Once the amount of bias is known, the measured values can be adjusted by the amount of the bias.

Just as in the case of measurement process precision, care should be exercised in extrapolating the accuracy of a measurement process when measuring one value (here that value is five ounces) to the accuracy that may or may not be observed when other weights are measured. Again, the accuracy of a measurement process over the range of possible production values can only be determined by actually investigating the measurement process over the indicated range of values.

The centerline from the X-bar chart for the data of Table 9.1 provides an estimate of the average value the measurement process delivers when measuring a five-ounce standard. This value, 5.00355, differs from the actual value of the standard by .00355 units. The magnitude of this difference can be evaluated by considering how large a difference would be anticipated from a measurement process that is accurate. If the measurement process were accurate, the absolute value of this difference should be less than:

$$3\frac{\hat{\sigma}_e}{\sqrt{N}}$$

where $\hat{\sigma}_e$ is the estimated measurement variation and N is the number or observations used to calculate the centerline of the X-bar chart. In the present example:

$$3\frac{\hat{\sigma}_e}{\sqrt{N}} = 3\frac{.0363}{\sqrt{40}} = .0172$$

Since the observed deviation, .00355 is smaller than .0172, it can be concluded that the difference between $\bar{\bar{X}}$ and five ounces can be accounted for by random variation and does not indicate a bias in the measurement process.

9.2   EVALUATION OF A MEASUREMENT PROCESS: REPORTING ON TOTAL LABORATORY VARIATION

The previous section discussed characterizations of a measurement process. The use of these characterizations will require a decision about how best to define measurement variation. For example, if it is necessary to understand how measurement variation affects reported values of, say, batch purity, then an estimate of measurement variation must account for all of those differences in time and circumstances that would affect determinations of the numbers purporting to represent purity. If measurements are made by different people, working at different times, using different equipment, then these sources must be considered as part of the measurement process. If reagents and equipment are used and these too change over time, then the description of measurement variation must recognize these changes.

Measurement processes can be as deceptively simple as making a judgment and recording the result, or they can be extremely complex, involving prescribed multistep protocols for chemical, mechanical, or electronic analysis. Regardless of the sophistication of the equipment and procedures applied to produce a measurement, many measurement processes may be usefully characterized in terms of the following, rather general, sources of variation:

- the *people* who carry out the measurement activity,
- the items measured and the physical *materials* used,
- the *equipment* used to carry out the measurement,
- the *methods* applied in making the measurement, and
- *environmental factors*, such as temperature and relative humidity.

The effect that organizational practices have on a measurement process must also be considered. Personnel selection and training practices, pressure to meet schedules, equipment procurement policies and practices, and a host of other managerial practices create the environment in which measurement takes place and influences the quality of a measurement process.

The following example illustrates the use of previously developed statistical techniques to understand, describe, and evaluate a measurement process. An integral part of this evaluation is the identification of possible sources of variation that affect the measurement process. In the example, measurements are obtained from gas chromatography. However, variation created by the chromatography devices is not the only source creating measurement variation; the measurement technique, or assay, involves a multiple-step procedure for preparing samples prior to their being injected in the chromatographs. The methods, equipment, and materials used at each of these stages can introduce variation into measured values. The development of a flowchart for the measurement process aids in identifying these sources of variation. By defining the steps in the measurement process, points at which variation may be introduced in the process are identified.

9.2.1   Identification of Sources of Variation Affecting a Measurement Process

The measurement study occurred in the analytical laboratory of a chemical manufacturing firm. The laboratory monitors the amount of active ingredient in MGC, an herbicide. An assessment of the stability of the measurement process and an estimate of the magnitude of measurement variation is required in order to determine what impact measurement variation has on interpreting measurements on the production process. The following study was designed to

1. determine if the measurement process operates in a stable fashion,
2. evaluate the magnitude of measurement variation, and
3. determine if the measurement process is accurate.

In this application, *measurement variation* means the variability in readings made on the same test specimen by technicians practicing typical methods and techniques at rated production and volume, employing any and all devices, subject to changing reagents and other materials. The assessment of measurement variation requires that appropriate data be collected to evaluate measurement variation as just defined. Collecting appropriate data will necessitate consideration not only of what sources of variation affect the measurement process, but also how and when these variations might occur. Figure 9.4 contains the flowchart used by the laboratory technicians to describe how determinations of amount of active ingredient are made. It is worth noting that this flowchart is of a measurement procedure already in use. It is surely true that the suitability of the measurement procedure for evaluating the amount of active ingredient was addressed prior to its adoption by the laboratory. The ability of the procedure to detect the correct amount of MGC with small variation was investigated prior to adoption by the company for use by the laboratory. It is not the selection of the measurement technique and the characteristics of that technique which guide selection that are under investigation. At issue in this example is the variation imparted by the people, materials, equipment, and environmental factors when the measurement procedure is used day to day, week to week, and month to month.

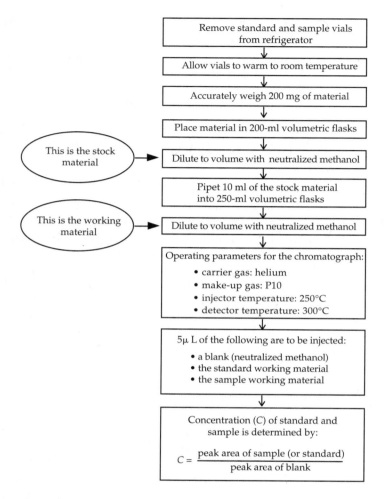

FIGURE 9.4   *Flowchart of MGC Assay*

Some of the sources of variation that could affect the process of using the procedure for measuring the amount of MGC have been identified:

1. *The analytical laboratory contains four different chromatographs, any one of which might be used in the assay for MGC.* Even if the same material were being measured, different instruments might provide different results for several reasons. The different chromatographs might report different levels of MGC. Even if the reported levels were the same, the chromatographs could have differing amounts of variation about an average.

2. *Even if the same device were used for each assay, the device would almost surely result in different measurements from one assay to the next.* Some amount of

within-chromatograph variation will surely be present. The consistency and magnitude of this variation must be investigated.

3. *Any one of eight different laboratory technicians may perform the assay.* Here again, differences within and among technicians in skill levels, physical attributes, and consistency in interpreting and applying methods might contribute to variations in numerical results even when measurements are being made on the same material.

4. *The assay requires a fairly involved sample preparation.* Even if only one technician performed the assay, there would certainly be differences in the sample preparation, which might produce variation in results. Test tubes, beakers, pipettes, and other equipment used will change. Material may not be completely warmed to room temperature. Differences in room temperature may occur. Small differences in the amount of methanol added will occur. These and other differences in the sample preparation methods could result in small or dramatic changes in assay results.

5. *Characteristics of the reagents used to do the assay may change in subtle or dramatic fashion over time.*

6. *At infrequent intervals the equipment itself changes, since the chromatographs have columns that need to be repacked.*

7. *The standard used in the assay is provided by the corporate research facility.* At infrequent intervals, the research facility will provide additional standard material which may differ significantly from the previously used standard material.

There may be other reasons why the measurement process will induce variation; many of these may not have been identified or described in the initial work to define the measurement procedure or may not have been anticipated in specifying and buying equipment, or in determining the content of training programs. The preceding list, incomplete though it might be, provides a basis for identifying potential sources of variation in the measurement process.

Collecting information on the measurement process, as with the study of other processes, begins by choosing a subgrouping plan to capture relevant sources of variation. In the present example, the manager of the analytical laboratory wanted a description of the measurement variation induced by the laboratory. The method adopted to capture this variation was to routinely determine the amount of MGC in a standard known to contain 96 units of MGC. Whenever an assay was run on production samples, an additional determination of the amount of MGC in a sample of the standard material was also made. Since MGC is a stable compound, the actual amount of MGC in the standard sample should not change; variations in the determined amount of the standard should reflect variation in the measurement process.

Another practice sometimes used for estimating measurement variation is to run, say, twenty or more standard samples during a single assay. In other words, the sample preparation of the material for the twenty samples would be done at the same time and determinations of the amount of MGC in the samples would be reported. There would be little value in placing these numbers on control charts, because they are measured over a very short time period; however, the sample standard deviation could be calculated from these numbers. This standard deviation would provide an estimate of within-assay variation, but only, of course, for the one assay involved. Nothing would be known about the stability of the within-assay variation over even a short period of time. The effect of different operators or devices would not be captured by the sample standard deviation. This practice of estimating within-assay variation might be useful for understanding sources of variation in the measurement process, but it clearly cannot be used as a description of measurement variation in a broad sense.

9.2.2   Evaluating Measurement Variation

Table 9.2 contains the measured amount of MGC in the standard for the 30 most recent assays; all of these determinations were made on the known standard. The determinations were, however, made by different technicians, at different times, and on different chromatographs. Consequently, the variations in the numbers in Table 9.2 are due to the measurement process. The Table 9.2 data allow us to answer these questions about the measurement process:

- How much variation exists among determinations made on the same material? Is this variation consistent?
- What is the average delivered by the measurement process? Is it consistent? If consistent, is it in close agreement with the 96 units of MGC contained in the standard?

A quantitative assessment of these questions can be obtained by the use of the moving range and individuals charts. Because the measurements were taken at different times and under different circumstances, there is no rational basis for subgrouping the measurements into homogeneous groups. Thus, a moving range chart is recommended for an assessment of the magnitude and stability of measurement variation and the individuals chart for evaluating magnitude and predictability of the level of the measurement process.

As with other process applications of the moving range, it is important to identify why consecutive observations differ from one another. This information is needed to assess the behavior of the moving range chart. For example, if the moving range chart were out of control, a detailed scrutiny of the list of

TABLE 9.2 *Amount of MGC Measured in Standard Sample*

Date	Assay Value	MR
07/26	95.9	—
07/27	95.7	.2
07/29	96.7	1.0
08/04	95.8	.9
08/05	96.9	1.1
08/07	95.5	1.4
08/10	96.8	1.3
08/10	96.0	.8
08/11	95.6	.4
08/12	96.4	.8
08/14	96.0	.4
08/14	95.2	.8
08/16	96.2	1.0
08/16	96.0	.2
08/16	95.2	.8
08/18	95.5	.3
08/19	96.2	.7
08/20	96.1	.1
08/20	96.7	.6
08/21	96.2	.5
08/22	96.2	.0
08/22	96.5	.3
08/24	95.3	1.2
08/26	96.0	.7
08/27	95.5	.5
08/28	95.5	.0
08/29	96.0	.5
09/02	97.2	1.2
09/04	95.3	1.9
09/05	96.4	1.1

reasons would be made in an attempt to understand the causes for the unstable behavior. Or if the average moving range was judged to be too large in the face of defined needs, then work would be directed toward selecting one or more of the reasons for studied reduction of its effect on measurement variation. In the present instance, because an assay may have been performed by any of several technicians on any of several devices, differences in devices or technicians would contribute to the magnitudes of the moving ranges. Furthermore, changes due to different sample preparation, or setups, would also contribute

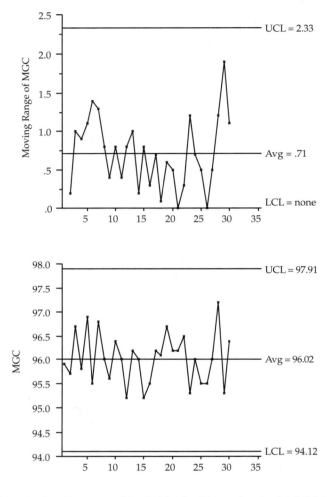

FIGURE 9.5   *Moving Range and Individuals Charts for the MGC Measurement Process*

to the magnitude of the moving ranges. The moving ranges have been plotted on a control chart in Figure 9.5. Since the moving range chart is in control, there is no evidence that assay-to-assay variation is inconsistent across the diverse and changing conditions covered by the collected data. In view of the identified sources of variation contributing to the ranges, it could be concluded that setup variations, operator variations, and device-to-device variations are fairly consistent over the time during which the data were collected.

Because short-term variation, as reported by the moving range chart, appears stable, the stability of the average value across different technicians, different devices, reagent changes, and time can be evaluated. However, before examining the individuals (X) chart, it is again useful to consider what sources of variation would cause the X values to exceed the baseline variation represented by the average moving range. If the reagents used to conduct the assay deteriorated over time, this might result in a trend in the values for the standard. Or, if the column were repacked during the course of the study, it might be that the assay read the standard to a different average value. The completed X chart in Figure 9.5 does not indicate any such trends or shifts in the average reading over time. The measurement process appears to be reading to a stable average of 95.8, a number very close to 96, the actual amount contained in the standard. This is an important finding. It provides the knowledge that across all instruments, technicians, and over time the average reported readings are in statistical agreement. This knowledge in turn says that measurements on production material will not differ due to deviations in ability to measure to the same average.

Since both the moving range chart and the X chart are in control, a standard deviation is estimated from the average of the moving ranges. This estimate would be the value of $\overline{MR}/d_2 = .629$ used to calculate the control limits for the X chart. Because the determinations of MGC in a standard are performed by any analyst on any machine and the values are separated in time by at least a day, the standard deviation describes the measurement variation of the laboratory. It reflects the amount of variation contributed to reported product values that is due to laboratory measurement variation.

9.2.3   Questions and Issues Raised by the Measurement Study

The choice was made to have measurement variation include all of those sources affecting the results delivered by the laboratory. Although this choice has resulted in a description of laboratory measurement variation, it does not provide information on what laboratory characteristics contribute significantly to total measurement variation. In the following section, methods for evaluating the size of measurement variation are discussed. If it is determined that measurement variation is too large, then other data collected on the measurement process could provide additional information about the sources contributing to measurement variation. For example, it may be suspected that the differences between chromatographs contribute significantly to measurement variation. Then future evaluation of the standard (and production samples) might proceed in one of two ways:

1. *Always use the same chromatograph.* If this approach were used, then the estimated "measurement" variation would reflect within-device variation,

as well as variation from materials, reagents, and sample preparation. This practice would not reduce measurement variation if differences in chromatographs did not account for a significant portion of measurement variation. A comparison of this variation with the variation captured by the subgrouping scheme of Table 9.2 would provide information on the magnitude of variation imparted by devices.

2. *Each time a standard sample is prepared, prepare an adequate amount of standard to inject the prepared material into all (or several specific) chromatographs.* A larger effort would be required by this plan since all chromatographs used would need to be readied for use during each assay, but more specific information about the effect of chromatographs could be had. Control charts for each device could be constructed with the data generated. Comparisons of these charts would indicate whether one or more chromatographs had inconsistent or larger variation than the others or whether one or more of the chromatographs exhibited inconsistency or bias in its average reading.

If either of these two methods is adopted, it must be understood that a standard deviation estimated from the moving range charts may underestimate "measurement variation" since this estimate would not include variation caused by the differences between the devices.

Other studies of the large measurement variation might be indicated, depending on what sources technicians and chemists think most likely to create large measurement variation. Other sampling and subgrouping procedures, as discussed in Chapter 6, might be used to identify the effects of other suspected sources of variation. The ideas discussed in Chapter 10 on designed experiments might also be used at this juncture to investigate simultaneously the effect of numerous suspected causes of variation.

Using the methods for evaluating the magnitude of measurement variation described in the following section, it may be decided that the measurement variation is small enough for its intended purpose. Good practice would then require that standard operating practices for ongoing monitoring of the measurement process be put in place. Variation in the measurement process may not remain stable. New operators or changes in equipment, reagents, or procedures could cause changes in measurement variation. A standard sample should be analyzed frequently, if not with every assay, in order to continue monitoring short-term and long-term measurement variation. The practice of continuing to monitor the measurement variation is necessary in order to be able to rely on the quality of measurements made on production material. Monitoring the measurement process is not a passive activity; learning occurs and opportunities are identified for improved methodologies that might result in reduced variation or shorter cycle times.

Plans for monitoring the measurement process are incomplete if they do not include procedures for responding to signals of process changes on the moving range and individual charts. These procedures should be developed in light of the previously identified sources of variation thought to affect the measurement process. Table 9.3 provides an example of how the list of sources of variation affecting the measurement of MGC has been used to provide focus for the ongoing maintenance of the measurement process. (Table 9.3 is an abbreviated version of the table actually developed. The table used by the laboratory was four pages in length. A portion of this table has been reproduced for illustrative purposes.) Out-of-control conditions on the moving range and individuals chart have been linked with likely reasons for that condition. The intent of developing this table was to provide focus for responding to out-of-control signals. For example, if a moving range fell above the upper control limit, the conditions that might have caused this can be referenced and investigated. So one way in which the table is to be used is to promote a timely investigation of suspected causes. In addition, the table is to be viewed as a "work in progress." All possible reasons for process instability are surely not identified. When additional reasons are noticed, they should be added to the table. Although Table 9.3 provides a useful way of capturing the effects of known sources of variation, by itself it is incomplete. Additional documentation on how to correct any identified problems must be provided; this documentation would necessarily include who is responsible for the corrections and how such corrections must be made.

TABLE 9.3 *Possible Causes of Out-of-Control Conditions in the MGC Assay*

Information from Chart	Possible Causes
Moving range chart has point above UCL	1. Chromatograph repacked 2. Change in reagent 3. New standard 4. . . .
X chart has point outside of control limits	1. Chromatograph repacked 2. Change in reagent 3. New standard 4. . . .
X chart fails runs test	1. Standard stored improperly 2. Chromatograph needs repacking 3. Neutralized methanol? 4. . . .

Another key issue in our measurement study is that the standard used was known by the laboratory technicians to have a value of 96 units. A question about the validity of the estimated measurement variation was raised; a view was expressed that the only "correct" way to assess precision was to routinely have a production sample passed twice through the laboratory without the technicians being aware that the same sample was being measured. Such a procedure would provide a "quality control" check on the laboratory. However, this practice is at variance with the philosophy of studying a process to know about sources of variation in order to better manage the process. It is worth reiterating that management of processes, and in particular, measurement processes, consists of

- aligning the objectives of the measurement process with customers of that process,
- knowing the current process configuration and capability of attaining those objectives, and
- selecting and directing needed process changes and improvement.

Management has the information, resources, and incentive to meet these requirements. In meeting this responsibility, management will need help from diverse sources. Knowledgeable and experienced employees can monitor work, report, and document abnormalities, identify possible causes, suggest means for change and improvement, and help test and implement these improvements. Prerequisites for this help span many social and cultural factors, but at a minimum, employees must know how well they are doing, must know when something different occurs, and have ideas about why the event occurred. This knowledge requires information about ongoing results. Control charts meet that basic need; those charged with the operation of measurement processes must be evaluating stability and magnitude of process variation as a means of understanding the causal structure affecting this variation. "Blind checks" of the laboratory may be used as a quality inspection method or because such checks are required by some governing agency. However, this type of inspection procedure is not the subject of this book; the intention of this book is to describe means of studying and improving process operation. With this intention in mind, those who are charged with understanding the process flow, with knowing about critical sources of variation, and with evaluating the effects of that variation will need to estimate and examine process variation routinely.

The preceding analysis has only described the current operation of the measurement process. Whether the variation in this process is small enough will depend on what variation is required both for process improvement work as well as for judging the fitness of production material. The next section provides techniques for deciding on the adequacy of a measurement process.

9.3 THE EFFECT OF MEASUREMENT VARIATION ON PROCESS STUDY

If measurement variation is large relative to product/process variation, then understanding, and subsequently improving, process behavior is compromised because process abnormalities will be difficult to detect. Large measurement variation will also create difficulties in determining the capability of a process to meet customer requirements and perhaps result in disagreements between vendor and customer. Similarly, work on process change and improvement will be more difficult because determining the effects of common cause variables through ongoing study is more difficult. An examination of the formula that explains the relationship between measurement variation and the variation measured on a process illustrates the difficulties encountered. This formula is given by:

$$\sigma_m^2 = \sigma_p^2 + \sigma_e^2$$

where

σ_m^2 is the variance of values measured on a production process. $\bar{R}/d_2$ from process investigation estimates the square root of this quantity.

σ_p^2 is the variance of the production process itself. The square root of this quantity would describe the variability in results if there were no measurement error.

σ_e^2 is the variance of the measurement process.

The preceding equation states that the observed variance of values measured on output from a production process is not equal to the actual variance of the output, but is increased by the variance of the measurement process. Figure 9.6 provides a graphical illustration of how large measurement variation can hinder process study. Plot I of Figure 9.6 is a time plot of the density of a plastic part produced at a plant. In this first plot, the measurement variation is known to be small. Two sources of raw material are used for this plastic part. The letter A is used to plot the density of a part produced from raw material A and the letter B is used for a part from raw material B. The different behavior of the raw materials is readily apparent from plot I. Although both processes—that resulting from material A and that from B—appear to be stable, the average densities from the two materials are clearly different.

Stated another way, the differences in the two raw materials are a source of variation contributing to the variability in density of parts produced at the plant. By identifying the differences in raw material as an important source of variation, action is then possible for reducing the variation in the density of the plastic part. This action might be to use only one of the two types of raw material in the production of this part.

Plot II of Figure 9.6 is a similar plot of measured density of plastic parts, but the measurement process used to measure the densities was known to have large variation. From plot II, it can be seen that the differences in average densities from parts produced from the two different raw materials is masked by measurement variation; the difference in average density that was immediately apparent in plot I is not as easily seen in plot II. The ability to understand the sources of variation affecting product output is hindered by large measurement variation.

Since the magnitude of measurement variation impacts the ability to obtain

I   Time plot of product density with small measurement variation

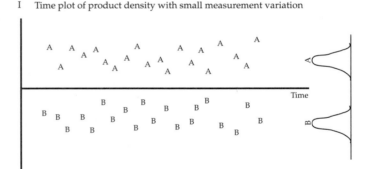

II   Time plot of product density with large measurement variation

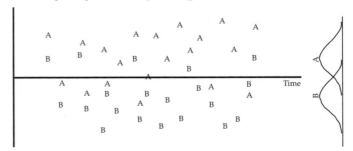

FIGURE 9.6   *Illustration of the Effect of Measurement Variation on Process Study*

useful information about a process, it is necessary to have some criteria for judging the adequacy of a measurement process. Two methods for assessing the impact of measurement variation on evaluating process performance are suggested in this book. The first method uses information generated from the ongoing study of both the measurement and production processes. The percent of variation in measured product output that is due to measurement variation can be determined to evaluate the adequacy of the measurement process; Section 9.3.1 illustrates this method. The second method uses a particular sampling and subgrouping scheme in conjunction with control charts to describe the ability of a measurement process to discriminate between different product output values; Section 9.3.2 contains an example of this technique. The impact of measurement variation on judging product is also a critical issue. Section 9.3.2 also describes means of evaluating the impact that measurement variation has on the ability to determine if a product is meeting requirements.

9.3.1   **Measurement Variation as a Percent of Measured Output Variation**

In Section 9.2, a study of a measurement process used to determine the amount of MGC in an herbicide was described. An examination of data on the measurement process led to the conclusion that the process was stable over time; the measurement standard deviation was estimated to be .62. The same measurement process, of course, is supplying data to production on the purity levels of every batch produced. Production data are routinely plotted and an analysis of the moving range and X charts constructed from these data show that the production process is in statistical control with an estimated process average of 94.4 and an estimated process standard deviation of 1.6. This standard deviation of 1.6 reflects both the variation in product purity levels as well as measurement variation. Using the notation developed earlier, this information on process and measurement variation could be summarized by:

$$\hat{\sigma}_e = .62 \qquad\qquad \hat{\sigma}_e^2 = .3844$$

$$\hat{\sigma}_m = 1.60 \qquad\qquad \hat{\sigma}_m^2 = 2.56$$

The variance of the measured purity, σ_m^2, is the sum of the variance of the actual product plus the measurement variance, σ_e^2. The size of measurement error relative to variability in measured results is found from:

$$\frac{\sigma_e^2}{\sigma_m^2} \times 100\% = \frac{.3833}{2.56} \times 100\% = 15.0\%$$

In summary, 15% of the variation observed in measured purity levels can be attributed to measurement variation. Is 15% too large a percentage? Although there is no one, correct answer to this question, a "rule of thumb" is that a value of 10% or smaller for the percentage of variation due to measurement is generally accepted as indicating that the measurement process will be adequate for studying process variation.

Since for the process under study, measurement variation accounted for 15% of the total variation observed, the adequacy of the measurement process should be questioned. Further study of the sources of variation that affect the measurement process could be undertaken. Methods for studying the measurement process with the intent of reducing measurement variation were discussed in Section 9.2.3. However, it might be that work to reduce measurement variation is unsuccessful or remains to be completed. In that case, laboratory personnel could reconfigure the process for making measurements by always performing multiple (two or more) assays on both the standards used to evaluate measurement variation as well as on every product sample. The reported levels of purity would then be the average of the multiple readings. If n multiple readings are obtained for each sample, then the measurement standard deviation should be reduced by a factor of $\sqrt{n}$. It will not, of course, be sufficient to assume that the measurement standard deviation is decreased by this amount; the results must be actually observed and evaluated. A reevaluation of the measurement process, which includes collecting multiple readings on every specimen of material, would need to be performed to verify the reduction.

9.3.2   Using the Components of Variance Technique to Evaluate the Discrimination of Measurement Processes

Large measurement variation, relative to process variation, makes it difficult to evaluate product and process characteristics and so inhibits the ability to improve process performance. When reliable information about a *consistent* measurement and production process is available, the percentage of observed product variation due to measurement variation can be determined. Another method for evaluating the effect of measurement variation on observed process variation would use the nested, hierarchical data collection strategies discussed in Chapter 8. This method might be used when information on measurement and production variation is not complete or when the stability of the production process is in question. The intent of collecting such data would be to separate the variation observed in measured product into that portion due to measurement variation and that due to variation created by the production process.

An example from the steel industry illustrates the use of nested sampling plans for separating measurement error from production process variability. In this specific example, time to complete a heat is a critical process result, one having significant economic implications. Investigation has revealed several reasons for finishing delays, one of the more important being in-process adjustments to obtain the correct chromium (Cr) percentage. Specifications for Cr percentage have a nominal of 21.3 with upper and lower specification limits of 20.80 and 21.80, respectively. Engineers attempt to achieve correct Cr levels for a heat by choosing an appropriate scrap and alloys mix. As the heat progresses, a molten mixture is obtained, a sample is drawn, and an assay of Cr content completed. If necessary, adjustments are made to attain the correct Cr level. Delays due to adjusting Cr content have a large, adverse impact on cycle time. In attempting to improve the process of selecting scrap and alloys as material inputs to a heat, past performance is reviewed. A cursory examination of production records reveals that variability between heats in Cr values is large. One of the reasons suggested for large variability in Cr values is measurement error. A review of the measurement process shows that under test circumstances the measurement method is consistently unbiased with apparent large variation; the measurement variation might be even larger under rated work conditions. An on-line study using the nested methodology to examine measurement under actual work conditions and to separate measurement from production variability is to be conducted. The primary questions to be addressed by this study follow:

1. How large is short-term measurement variation? Is it stable?
2. Can the measurement process discriminate among heats; in other words, can it distinguish Cr values from one heat to another?
3. How large is production variation and is it stable?
4. What percent does measurement contribute to total, observed variability?

The plan for using a nested (or hierarchical) study is provided below.

1. From each heat, randomly select a sample of the molten mixture. Do this four times a day for 10 successive days.
2. Split the sample into two portions. Have the lab make independent assessments of each portion. Because the samples are made homogeneous prior to splitting, differences between the two determinations on the same sample are thought to be primarily due to measurement error.
3. Construct a range chart using the multiple determination on each sample. Magnitude and stability of measurement error is evaluated from this chart.
4. Construct two charts from the heat averages, each being the average of two readings on the same sample. One of the *X*-bar charts has control limits based on the average range for measurement error. This chart is

used to assess the ability of the measurement process to discriminate between heats. A second X-bar chart has control limits based on the average moving range of the heat averages. A chart of the moving averages and this second X-bar chart permit assessment of the stability and magnitude of measured variability and level for Cr values.

The 40 subgroups of data collected according to this plan are provided in Table 9.4. Each of the 40 subgroups in Table 9.4 came from a different heat. Therefore, variation in the subgroup averages reflects variation due to different heats as well as variation due to the measurement process. The consequences of this subgrouping scheme are twofold. First, these data *do not* allow for an assessment of the stability of long-term measurement variation nor can the accuracy of the measurement process be determined. The second consequence is that the magnitude and stability of the subgroup averages *do* allow for an assessment of the effect the production process itself has on variation in Cr levels. The procedure for evaluating the effect of measurement variation and process variation on the total variation in measured Cr percentages is described in Figure 8.10 of Chapter 8. In the present example within-subgroup variation would be measurement variation and between-subgroup variation would be process variation.

A chart of the ranges from Table 9.4 appears in Figure 9.7. The range chart is in control, indicating that short-term measurement variation is consistent across time, conditions, people, and varying heat results Because sample portions were evaluated as they became available, deviations between the two readings for the same sample are explained by variations within and between devices, between setups, and within and between technicians. These variations in measurements on the "same" material, conducted under normal production conditions, are thought to represent measurement error realistically. An estimate of the measurement standard deviation is:

$$\hat{\sigma}_e = \frac{\overline{R}}{d_2} = \frac{.60}{1.128} = .53$$

The limits for the X-bar chart in Figure 9.7 have been calculated using the formulas:

$$UCL_{\overline{x}} = \overline{\overline{X}} + A_2\overline{R}$$

$$LCL_{\overline{x}} = \overline{\overline{X}} - A_2\overline{R}$$

TABLE 9.4 *Percent Chromium Found by Two Independent Determination in Samples from 40 Heats*

Sample	X_1	X_2	R	$\bar{X}$	Sample	X_1	X_2	R	$\bar{X}$
1	21.87	21.89	.02	21.884	21	19.52	18.53	.99	19.027
2	19.25	19.40	.25	19.327	22	21.39	19.82	1.57	20.603
3	21.40	20.98	.42	21.192	23	23.41	24.76	1.35	24.085
4	21.36	21.74	.38	21.552	24	20.35	19.76	.59	20.054
5	22.29	21.70	.59	21.995	25	21.25	19.95	1.30	20.598
6	18.97	18.90	.07	18.936	26	21.58	22.06	.48	21.822
7	19.79	19.82	.03	19.808	27	21.29	21.85	.56	21.566
8	20.97	21.43	.46	21.201	28	19.35	20.61	1.26	19.981
9	20.62	20.11	.51	20.368	29	24.33	25.12	.79	24.727
10	21.38	21.64	.26	21.508	30	23.03	23.33	.30	23.179
11	21.67	21.38	.29	21.526	31	22.63	22.18	.45	22.405
12	20.70	21.75	1.05	21.228	32	22.21	21.25	.96	21.730
13	22.84	23.00	.16	22.921	33	21.18	19.75	1.43	20.467
14	19.99	19.55	.44	19.771	34	21.33	22.05	.72	21.690
15	21.49	21.59	.10	21.538	35	19.89	19.75	.14	19.817
16	21.53	21.35	.18	21.443	36	20.59	20.31	.28	20.450
17	20.88	22.32	1.44	21.601	37	19.83	21.51	1.68	20.670
18	20.36	19.15	1.21	19.755	38	22.94	22.88	.06	22.910
19	20.10	20.48	.38	20.292	39	20.71	20.17	.54	20.441
20	22.28	22.66	.38	22.470	40	18.92	19.13	.21	19.027

The baseline variation against which the X-bars are being evaluated is measurement variation. These limits represent the measurement error effect; but heat averages vary because of measurement and process sources. The out-of-control X-bar chart indicates that process variations are larger than measurement variations. The conclusion is that the measurement process can indeed discriminate between different percentages of chromium with $n = 2$ observations per heat. On the other hand, if the X-bar chart had been in control, the conclusion would be that measurement variation is so large compared to process variation that differences in the percentages of Cr cannot be detected in the presence of this measurement variation.

Since the X-bar chart of Figure 9.7 is out of control, we can conclude that the process for measuring Cr can discriminate between varying percentages of Cr. An assessment of how well the measurement process discriminates will require knowledge of the magnitude of process variation (variation in Cr percentages.) As outlined in the flowchart of Figure 8.10, the variation in the X-bars can be used to provide information on the stability and magnitude of this process variation.

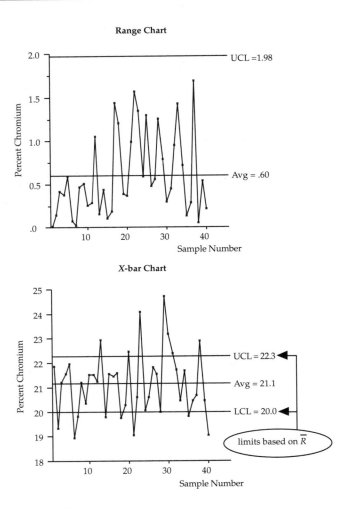

FIGURE 9.7   *Range and X-bar Chart for Percent Chromium*

The moving range of the X-bar values permits an assessment of short-term process variation. That chart, shown in Figure 9.8, is in control. Stability in the average for the production process is examined by treating heat averages as individual values and basing control limits for those values on the average moving range. Since both charts are in control, it is concluded that the variation in percentages of Cr levels is stable over the time of the study. An estimate of the magnitude of the variation in Cr levels is found by:

$$\hat{\sigma}_c^2 = \left(\frac{1.5}{1.128}\right)^2 - \frac{(.53)^2}{2} = 1.63$$

$$\hat{\sigma}_c = \sqrt{1.63} = 1.28$$

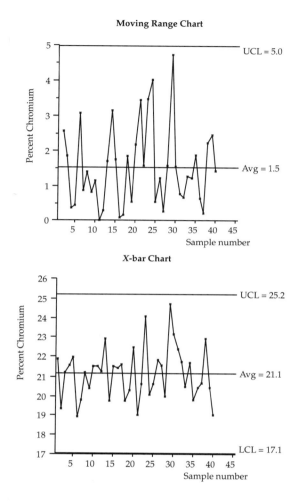

FIGURE 9.8 *Moving Range and X-bar Charts of Chromium Percentages*

The estimated standard deviation of 1.28 captures the variation in Cr levels that is attributable to the production process. With an estimated standard deviation for measurement of .53, the total variance and standard deviation of *measured* values of Cr would be:

$$\hat{\sigma}_m^2 = \sigma_c^2 + \sigma_e^2 = 1.63 + (.53)^2 = 1.92$$

$$\hat{\sigma}_m = \sqrt{1.92} = 1.39$$

If single measurements are made on the samples, an estimated standard deviation of these individual measurements would be 1.39. The percentage of total variation due to the measurement process is found from the ratio:

$$\frac{\hat{\sigma}_e^2}{\hat{\sigma}_m^2} \times 100\% = \frac{(.53)^2}{1.92} \times 100\% = 14.6\%$$

The X-bar chart of Figure 9.7 showed that the measurement process, with two readings per sample, could discriminate between the different percentages of Cr observed in production. The magnitudes of measurement variation and process variation were also determined. This information can be used to assess how well the measurement process discriminates—not just that it *does* discriminate. Furthermore, the impact that measurement variation has on assessing whether the product is capable of meeting requirements can also be judged from the size of measurement and process variation.

Using the measurement process to evaluate process variation requires a controlled measurement process. When this is the case, the percentage of total measured variation due to measurement variation provides a means of evaluating the ability of the measurement process to discriminate. As a general rule, the percent of variation attributable to measurement variation should be 10% or less. When measurement variation is this small as compared to process variation, data collected on the production process can be used to learn about sources of variation affecting the production process. Of course, any time a general rule is stated, there will be exceptions. For example, it may be the case that the ability to produce product with small levels of variation exceeds current capabilities to measure the output of the process. If measured variation in the production process is so small that work to understand sources contributing to that variation is not a priority issue, then monitoring the process with an assessment of product variation that includes a large component due to measurement may be acceptable.

The effect that the size of measurement variation has on determining whether product meets requirements is also an important assessment of the capability of a measurement process. For example, in the present instance of measuring Cr percentages, the requirement is that the percentage of Cr in a heat should fall between 20.80 and 21.80. In other words, an engineering tolerance of 1.00 was needed. The total variation of Cr percentage, which includes both process and measurement variation, was 1.39. The natural tolerance of the process would be 6(1.39) = 8.34, a number considerably larger than the engineering tolerance of 1.00. The process is not capable; delays are inevitable with the current process. Certainly, improvements in the ability to predict percentages of chromium are desired to avoid the large number of delays. At issue is how these improvements should be achieved.

One possible improvement would be to reduce the variation due to measurement. A reduction in the variation of the measurement process would surely reduce the total observed variation. Whether the size of the measurement variation is large enough that its reduction would have a real impact on the process delays can be assessed by comparing the natural tolerance of the measurement process to the engineering tolerance. With a standard deviation of .53, a natural tolerance of the measurement process could be defined by:

$$NT_e = 6\hat{\sigma}_e = 6(.53) = 3.18$$

This natural tolerance for measurement reveals the likely spread in a large number of readings on the same material. The impact of this spread is immediately appreciated by realizing that if the actual chromium percentage were at the nominal of 21.30, measurement on a sample would range across an interval from

$$21.30 - \frac{3.18}{2} = 19.71 \qquad \text{to} \qquad 21.30 + \frac{3.18}{2} = 22.89$$

and thus would be likely to indicate that a process adjustment is necessary. In this application, the measurement spread of 3.18 exceeds the engineering tolerance of 1.00. If the interval (20.80,21.80) is correct for process adjustment, then a more precise determination of Cr is needed; this measurement process will not serve to determine how to adjust the process. Reducing measurement variation might come from identifying sources of variation impacting the measurement process, as discussed in Section 9.2.3, or by making multiple measurements on Cr as described in the previous section. In this instance, multiple measurements may need to be adopted on a short-term basis until measurement variation can be otherwise reduced.

The estimated standard deviation for the production process, 1.28, can also be examined for the effect of the production process on the ability to manage Cr levels. The natural tolerance of the production process, exclusive of measurement variation, would be:

$$NT_c = 6\hat{\sigma}_c = 6(1.28) = 7.68$$

Clearly, the ability to project Cr levels adequately prior to a heat is insufficient to avoid delays in the process since the natural tolerance is so much larger than the engineering tolerance. Concurrent with work on the measurement process, work on knowing what sources are creating the large variation between heats and acting to reduce that variation must proceed.

A summary of what we have learned from the study plan developed and further information required to understand observed variation in chromium percentages follows.

1. Measurement and process variations are consistent across time and conditions.
2. Measurement variation is large: Measurement error accounts for 15% or observed variation in process results for chromium. The natural tolerance for measurement exceeds the engineering specifications (3.18 versus 1.00).
3. The natural tolerance of the production process, exclusive of measurement variation, is 7.68, a number considerably larger than the engineering tolerance of 1.00. The process for achieving correct Cr levels on first pass must be improved.
4. Although measurement variation is large, process work can proceed with the current measurement process, perhaps using duplicate readings, because the measurement process can differentiate heat-to-heat values of Cr.
5. The measurement process requires work: Because process work decreases that variation, at some point, the current measurement process will not be able to detect deviations from heat to heat. Work on the measurement process must also include an assessment of the accuracy of the measurement process. An assessment of the stability of the level of the measurement process and its accuracy is required because of the need to hold the Cr values to a nominal of 21.30.

9.4   COMPARISON OF MEASUREMENT PROCESSES

The preceding sections have dealt with methods for characterizing, evaluating, and improving measurement processes. Just as with production processes, it is often necessary to compare measurement processes. For example, if a change has been made to a measurement process, it will be necessary to evaluate the effect of the change on the results provided by the process. Or, if two measurement processes are in use, possibly at two different locations, knowledge of how the results from the two processes compare will be required. Comparison of measurement processes should address the following issues:

1. Do both processes have the same measurement variation? Are the variations stable across time? Are the measurement variations the same across the range of product to be tested?

2. Do both processes, on average, tend to provide results close to an accepted standard? Do they do this consistently? Do the two processes agree on average across a range of product values?

Any one study used to compare processes will probably not address all of these questions. Part of the process of designing a study to compare measurement processes will consist of determining the first questions to be addressed and how data should be collected to address these questions. The next example describes a situation where it was deemed necessary to compare measurement processes and what choices were made in designing a measurement study.

A customer and vendor were in disagreement on the measurement of an important quality characteristic. The vendor always reported an average and standard deviation for an outgoing lot of material. When the lot arrived at the customer's plant site, some parts were selected from the incoming material and an evaluation of the average and standard deviation of the lot was performed. One of the reasons for the disagreement was that the difference in the results reported by the vendor and those obtained by the customer were at times dramatic. Although the source of the reasons for the disagreement was not clear, the possibility that the source might be in part due to sampling differences or measurement methods was considered.

As a way to begin to narrow down possible reasons for the differences in the characteristics reported on the lots, managers from the two plant sites agreed to conduct a measurement study of the characteristic in question. They were particularly interested in determining whether the two measurement processes had similar levels of variation across a range of parts. Consequently, the study was performed on 20 parts selected from recently manufactured material with the 20 parts chosen by the customer group to reflect the range of measurements observed by production. These 20 parts were first passed through the measurement process at the vendor's facility. These same 20 parts were measured a second time on the following day. After the two measurements were made on the 20 parts, the parts were transported to the customer's facility and passed through the customer's inspection procedure on each of two successive days. (The characteristic of interest is not changed by test, transport, or time.) The results of this measurement study are reported in Table 9.5. This table also summarizes the range of the measurements of the same part for both the vendor's and customer's measurements as well as the average measurement obtained by both the vendor and customer. Before looking at a statistical analysis of the data, it is instructive to consider what can be learned about the two measurement processes from these data.

The ranges reported on the vendor's measurements capture information about the precision of the vendor's measurement process; a similar statement can be made about the ranges calculated from the customer's data. A day's separation allows for some expansion of the sources of variation in measurement, providing a more realistic representation of measurement error than that which would be obtained by quick, repeated measurements in a short interval of time. Each of the ranges describes the difference observed from one day to the next when the same part is measured. By charting these ranges, the stability of measurement variation across the range of part dimensions selected by the customer group can be evaluated. If different size parts resulted in different measurement variation, then this should be picked up as a special cause on the range chart. The range charts allow for a check on the stability of the measurement processes across parts, but not across time. Care will need to be exercised in applying these results, since there is no basis by which to say that measurement variation is stable across time and conditions. Further data will need to be collected to assess stability across time. The advantage of this approach is that the stability of the two processes across a range of dimensions *can* be evaluated. If both processes deliver consistent variation across these values, then the centerlines of the two range charts can be compared to determine if the measurement variation reflected in the ranges is approximately the same for the two groups. If they are, then follow-on data will need to be collected to see if both processes deliver stable variation across time.

Figure 9.9 contains the two range charts constructed from these data. The horizontal axis on both charts has been labeled with the word "part"; this is consistent with the earlier statement that these range charts are a check on the consistency across parts, not across time. Both of the range charts in Figure 9.9 are in control; both the vendor's and customer's measurement processes appear to deliver the same level of variation across the 20 parts used in the study. However, even though both processes are consistent, differences in the measurement processes are immediately apparent by examining the two charts. The average range for the vendor's measurement process is considerably larger than that of the customer's. The larger average range means that for any one part, the vendor's measurement process is likely to report a much larger or smaller number than the customer's process; this may offer a partial explanation for the disagreement and uncertainties between the two as to the contents of shipped lots. With the understanding that the within-subgroup variation represents day 1 to day 2 differences in the measurement process, the estimated day-to-day standard deviation for the vendor and customer are estimated to be:

$$\hat{\sigma}_v = \frac{4.7}{1.128} = 4.27$$

$$\hat{\sigma}_c = \frac{1.4}{1.128} = 1.24$$

The observed variation in the vendor's measurement process is about three times that of the customer's.

Before looking at the two X-bar charts, the sources of variation affecting the X-bars should be considered. The averages for both charts are computed from two measurements made on each of the 20 parts. Although some of the differences in these twenty averages would be due to the variation from the measurement process, the variability in averages also reflects the variation in the 20 parts. In fact, if the measurement process used to measure the parts had negligible variation, then all the variability in the averages would be due to part-to-part differences, a production process characteristic.

TABLE 9.5 *Comparison of Two Measurement Processes*

	Vendor's Process				Customer's Process		
Part	Measurements		Range	Average	Measurements		Range	Average
1	28.5	32.9	4.4	30.70	34.7	36.1	1.4	35.40
2	32.5	28.1	4.4	30.30	32.3	30.7	1.6	31.50
3	28.0	22.8	5.2	25.40	29.9	30.8	.9	30.35
4	41.4	50.5	9.1	45.95	43.9	42.7	1.2	43.30
5	54.6	54.7	.1	54.65	47.5	50.5	3.0	49.00
6	37.5	46.8	9.3	42.15	45.7	43.5	2.2	44.60
7	36.8	38.3	1.5	37.55	44.8	44.1	.7	44.45
8	32.4	40.3	7.9	36.35	36.7	36.5	.2	36.60
9	41.0	40.4	.6	40.70	41.1	41.3	.2	41.20
10	37.0	40.4	3.4	38.70	33.9	36.3	2.4	35.10
11	53.5	45.4	8.1	49.45	40.7	42.0	1.3	41.35
12	35.3	34.5	.8	34.90	36.3	36.3	.0	36.30
13	34.2	38.7	4.5	36.45	39.7	38.7	1.0	39.20
14	49.9	46.2	3.7	48.05	45.0	46.5	1.5	45.75
15	46.4	39.7	6.7	43.05	48.9	45.9	3.0	47.40
16	37.0	37.3	.3	37.15	36.3	37.1	.8	36.70
17	46.9	45.9	1.0	46.40	47.0	46.5	.5	46.75
18	39.8	25.9	13.9	32.85	31.3	30.0	1.3	30.65
19	46.3	42.3	4.0	44.30	46.4	43.4	3.0	44.90
20	35.5	41.3	5.8	38.40	41.2	40.0	1.2	40.60

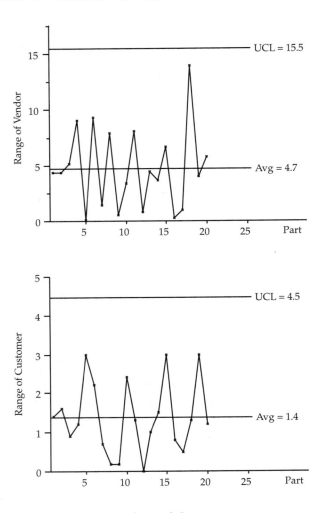

FIGURE 9.9   *Range Charts for Vendor and Customer*

An examination of the respective average charts for the vendor and customer provide a way of evaluating the effect the additional measurement variation from the vendor's measurement process has on the ability to evaluate product characteristics. In constructing the X-bar charts of Figure 9.10, the average range from the respective range charts has been used to calculate the upper and lower control limits. The sources of variation included in the determination of the control limits for the X-bar charts are due to measurement variation; these limits do not include any allowance for part-to-part differences.

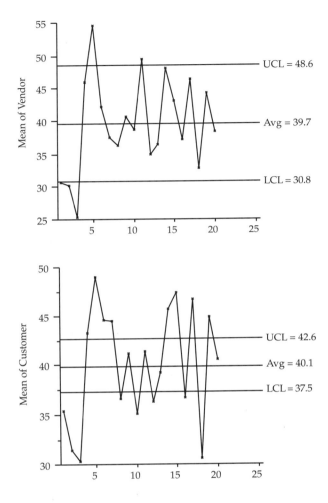

FIGURE 9.10 *X-bar Charts for Vendor and Customer*

As discussed in the previous section, using measurement variation to determine the limits for the X-bar charts is not a deficiency in the study. The designers of the data collection plan wanted to know if the respective measurement processes had the ability to discriminate among parts and if the measurement processes tended to read to the same averages. Because the control limits on the X-bar chart only reflect measurement variation and the X-bars are subject to part-to-part variation as well as measurement variation, it would be expected that the X-bar charts would be out of control. An examination of the two X-bar charts shows that both charts display average values outside of the reported control limits. Both the vendor's and customer's measurement processes are capable of discriminating between parts; measurement variation does not hide part-to-part variation.

Again, it is useful to note the differences seen in the two *X*-bar charts. The *X*-bar chart from the vendor's study does not have as many points outside the control limits as that of the customer; nor are the points on the vendor's chart as far outside the control limits. This qualitative difference can be summarized by concluding that, because of the smaller variation experienced by the customer's measurement process, this process has a greater ability to discriminate among parts than the vendor's measurement process. As previously stated, the parts selected for inclusion in the study covered the range of values observed by the customer. The vendor's measurement process only noted that five of these parts exhibited more variation than could be explained by measurement variation. On the other hand, the customer's *X*-bar chart had 16 of the 20 parts falling outside of the limits defined by measurement variation. The consequence of the difference in the magnitudes of measurement variation is that when samples of actual production parts are measured only once and a value reported, vendor and customer are likely to experience differences in reported values that appear significant. In the vendor's shop, parts outside of specifications may be measured as falling within specifications and parts that are within specifications may be scrapped or unnecessarily reworked.

The preceding study provides a clear direction for further work. The vendor will need to reduce measurement variation if customer and vendor are to agree on measurement of shipped parts. The vendor may find it possible to improve significantly the measurement process by studying procedures and practices at the customer's site; insight may be gained from the customer's process if the two measurement processes are similar.

9.5   GAUGE R AND R STUDIES

The studies of measurement processes in this chapter have focused on designing measurement studies. A one-time study of a measurement process is not sufficient to ensure the ongoing quality of the measurement process. As discussed in Section 9.2.3, it is important that an ongoing evaluation of a measurement process be put in place. As materials, personnel, techniques, etc., change over time, a measurement process will change as well. An ongoing practice of monitoring a measurement process needs to be put in place if the measurements are to be relied on to report on the behavior of product and process outcomes.

A technique in widespread use for describing the quality of a measurement process is a gauge R and R (repeatability and reproducibility) study. The information provided by R and R studies is useful in understanding the short-term impact that some selected sources of variation may have on a measurement process. But, as with many other measurement studies, the gauge R and R methods do not provide any information about the behavior of a measurement process over time. Ongoing monitoring of a measurement process must be considered. These ideas of the insufficiency of gauge R and R studies are further developed in the following example. This example describes how one gauge R and R study was performed, what is typically learned from this type of study, and what remains to be determined.

The example deals with the use of a gauge R and R study to evaluate a newly purchased device for measuring viscosity. A worksheet completed by a group in the analytical laboratory where the device was installed is reproduced in Figure 9.11. This worksheet is typical of prepared forms available for studies of this type. The data in the worksheet were gathered by having each of three operators make a series of duplicate readings on a material known by best current methods to have a viscosity of 75. A set of 10 duplicate readings by each operator were completed in a two-hour period on August 10. Ranges for the three sets of duplicate readings have also been recorded on the worksheet. A range chart constructed from these 30 ranges appears in Figure 9.12. Each of the ranges captures short-term variation within operator. The fact that the range chart is in control indicates that operators are internally consistent and consistent with each other in the variation observed short term in the measurement process. Caution must be exercised in how this conclusion is used. The brief two-hour time period of the study provides no real indication of whether this component of variation, within operator, would be the same across the variety of circumstances and conditions encountered over time. At best, the rather weak conclusion is that during initial evaluation of the gauge, no problems regarding within-operator variation are noted and all three operators have the same levels of within-operator variation. The value of the centerline of the range chart is used to provide an estimate of the within-operator (wo) variation:

$$\hat{\sigma}_{wo} = \frac{.387}{1.128} = .343$$

The X-bar chart in Figure 9.12 has been constructed from the averages of the 30 subgroups used to form the range chart. The centerline value from the range chart, .387, has been used to calculate the limits for the X-bar chart. All

GAUGE REPEATABILITY AND REPRODUCIBILITY DATA

Measurement Instrument	ESRAJ	Date 8/12/93
Characteristic	Viscosity	Performed by CP & SA
Specification limits		

Condition	Operator A				Operator B				Operator C			
Sample	1st Reading	2nd Reading	3rd Reading	Range	1st Reading	2nd Reading	3rd Reading	Range	1st Reading	2nd Reading	3rd Reading	Range
1	74.3	74.0		.3	74.7	74.7		.0	74.9	74.5		.4
2	75.5	74.5		1.0	74.4	74.2		.2	75.0	75.1		.1
3	74.1	75.0		.9	74.4	74.5		.1	74.4	74.3		.1
4	75.1	75.4		.3	74.8	75.8		1.0	74.2	74.0		.2
5	75.2	74.9		.3	74.2	74.6		.4	74.8	74.6		.2
6	74.6	74.3		.3	74.5	74.4		.1	74.6	74.5		.1
7	74.1	75.2		1.1	74.2	74.1		.1	74.5	75.5		1.0
8	74.7	74.7		.0	75.2	74.9		.3	75.1	74.7		.4
9	75.6	74.8		.8	74.6	75.0		.4	74.6	74.6		.0
10	74.5	74.9		.6	74.7	75.3		.6	74.6	75.1		.5
Average	74.77	74.77		.56	74.57	74.75		.32	74.67	74.69		.30

X_1 X_2 X_3 R_A X_1 X_2 X_3 R_B X_1 X_2 X_3 R_C

$\bar{X}_1$	
$\bar{X}_2$	
$\bar{X}_3$	
$\bar{X}_A$	

$\bar{X}_1$	
$\bar{X}_2$	
$\bar{X}_3$	
$\bar{X}_B$	

$\bar{X}_1$	
$\bar{X}_2$	
$\bar{X}_3$	
$\bar{X}_C$	

$\bar{X}_A$	
$\bar{X}_B$	
$\bar{X}_C$	
Max $\bar{X}$	
Min $\bar{X}$	
Range $\bar{X}$	

$\bar{R}_A$	.56
$\bar{R}_B$	.32
$\bar{R}_C$	.30
$\bar{\bar{R}}$	.393

FIGURE 9.11 *R and R Worksheet*

of the averages fall within these limits. The conclusion that can be reached is that, over the two-hour time period examined, the operators obtain consistent averages. No operator finds a lower or higher average at some times and the operators appear to measure to about the same average. The second conclusion could be stated as "any between-operator variation is small when compared to within-operator variation." The weak nature of this conclusion must be understood. The gauge R and R study has not identified any initial problems in between-operator variability. However, the long-term stability of the measurement process has not been determined.

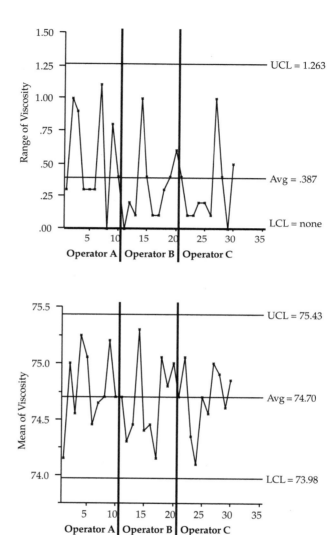

FIGURE 9.12 *Range and X-bar Chart for Gauge R and R Study*

The cause and effect diagram in Figure 9.13 captures some important sources of variation to be considered in the measurement of viscosity. The major headings of this cause and effect diagram correspond to the characterizations of sources of variation in a measurement process that were introduced in Section 9.2. Note that, of these five sources, only two, equipment and people, are typically examined in a gauge R and R study. Furthermore, in such a study, the examination of these two sources of variation is typically performed in a short interval of time and often under conditions that are not representative of the work environment. Although problems with the consistency or magnitude of these sources of variation might be detected in such a study, the variation due to changes over time in these sources of variation, will not be discovered in a typical gauge R and R study.

At the site where the new gauge was installed, other studies were initiated after the initial gauge R and R study was completed. To examine the effects of sources of variation over time, two viscosity checks of the standard with known viscosity of 75 were performed at two different times (morning and afternoon) each day for 31 days. During the 31 days studied the same measuring device was used and the same technician performed all 62 measurements. Three different batches of solvent were used during the 31-day period.

Table 9.6 contains the 31 days of data. Ranges and averages for each of these days have also been recorded. The magnitude of the ranges for each of the 31 days reflects within-day variation. From the information on the cause and effect diagram, these within-day variations may be due to short-term (within day) drifts in the gauge, short-term within-operator variation, temperature variations that occur within a day or to variation within the batch of reagent in use on a particular day. The X-bars will capture day-to-day variations that occur in the measurement process. These variations may be due to differences in the manner in which the operator performs the measurement process on different days or to temperature variations from one day to the next, or possibly because a batch of reagent has changed from one day to the next.

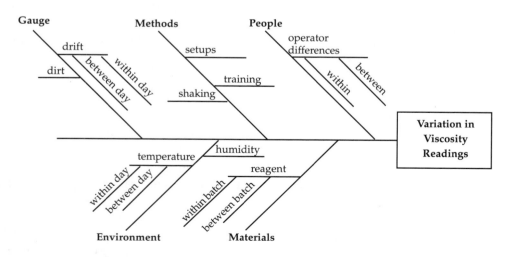

FIGURE 9.13 *Cause and Effect Diagram for the Viscosity Measurement Process*

The control charts for these data are contained in Figure 9.14. The range chart indicates that those sources creating within-day variations are consistent across the 31 days studied. However, the X-bar chart fails the Rule of Seven runs test, which indicates a distinct change in the average viscosity during the 31-day period. The fact that there appears to be three distinct averages is consistent with the fact that three different batches of solvent have been used during the course of the 31 days. The timing of the change in the batches would provide a more solid indication that the average viscosity reading is affected by the change in reagent batches.

Three key points are provided by this second study of the viscosity measurement process:

Point 1

The initial gauge R and R study did not provide information on the stability of the measurement process across time. This information can only to be obtained by collecting data on the process over time.

Point 2

The formation of subgroups for a time-oriented measurement should rely on knowledge of the sources of variation that might affect the measurement process, sources that are typically captured on a cause and effect diagram.

Point 3

As with any on-line study of a process, the correlation of observed behavior on a control chart with knowledge of changes in process parameters provides direction for further investigation of the effect such process parameters may have on process outcomes. As is further discussed in Chapter 10, the suspicion that such causal relationships may exist will often require the use of a designed experiment to verify the suspected relationship.

9.6　RECOMMENDATIONS FOR ENSURING GOOD MEASUREMENTS

The ability to know what is being delivered by a production process relies heavily on knowing that good measurements of the characteristic of interest are being used. "Good" measurements are those provided by a measurement process that consistently over time provides accurate and precise numbers. Determining the characteristics of a measurement process should be done just as for any other process. In other words, the following elements of process study should be considered as part of the plan for studying the quality of a measurement process.

TABLE 9.6 *Viscosity Check*

Day	X_1	X_2	Range	$\bar{X}$
1	75.2	76.1	.9	75.65
2	73.9	76.6	2.7	75.25
3	76.5	75.4	1.1	75.95
4	77.1	74.0	3.1	75.55
5	79.4	78.1	1.3	78.75
6	76.0	79.8	3.8	77.90
7	77.3	75.0	2.3	76.15
8	78.7	80.3	1.6	79.50
9	78.4	79.5	1.1	78.95
10	77.2	83.1	5.9	80.15
11	82.7	80.4	2.3	81.55
12	82.3	81.1	1.2	81.70
13	80.6	77.6	3.0	79.10
14	81.2	82.0	.8	81.60
15	81.5	81.0	.5	81.25
16	79.1	84.5	5.4	81.80
17	80.8	80.8	.0	80.80
18	78.3	80.0	1.7	79.15
19	81.3	81.4	.1	81.35
20	79.2	79.4	.2	79.30
21	80.8	78.2	2.6	79.50
22	80.0	75.2	4.8	77.60
23	79.1	79.6	.5	79.35
24	77.7	82.7	5.0	80.20
25	76.7	75.1	1.6	75.90
26	75.4	75.6	.2	75.50
27	75.5	76.5	1.0	76.00
28	77.7	75.1	2.6	76.40
29	72.1	78.4	6.3	75.25
30	73.4	76.1	2.7	74.75
31	75.9	75.0	.9	75.45
Sum			67.2	2427.30

Element 1: The appropriateness of the measurement for judging the characteristic of interest should be known.

The statistical methods discussed in the present chapter have addressed the study of a given measurement process. What is not determined by these methods is whether the characteristics being measured are ones that are

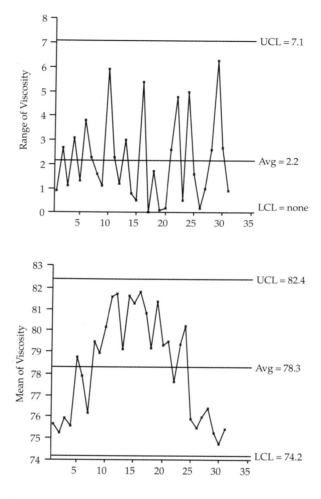

FIGURE 9.14 *Range and X-bar Charts for Viscosity*

important to the customer or ones with known impacts on characteristics valued by the customer. However, it is one of the first issues that should be addressed prior to beginning a study of a measurement process. Too often measurements are taken on a characteristic because it has been past practice to do so or because it is suspected, but not known, that good measurements at one stage of a process are indicative of the quality of the final result of the process. An example of this practice occurred in the manufacture of pulp for making paper. A characteristic measured on the pulp by the paper company was "freeness." Extensive effort went into taking this measurement and efforts were continually made to see that operators were consistent in their measurement practices. Yet, neither the operators nor the managers of this process stage knew if, and if so why, freeness was an important determinant of the final quality of the paper produced. Freeness was simply something that was always measured at this stage of the process. Without understanding why it was measured, there was no understanding of what kinds of variation in freeness and consequently in the measurement of freeness were appropriate. Prior to an investigation of a measurement process it is recommended that answers to the following questions be known:

1. What is the engineering, physical, or chemical basis on which the measurement method used is known to provide information on the characteristic studied? For example, in the study of the amount of MGC found in an herbicide, this question asks the chemical basis by which the assay is known to deliver information on this amount. The answer to this question cannot, of course, be provided by a statistical study.
2. Why is the measurement being taken? What physical properties of the process or product are being measured?
3. What levels of variation are required by the measurement process? An answer to this question will require knowing the required tolerances of the characteristic being measured and the amount of variation imparted to the characteristic by production.

Element 2: A flowchart of the measurement process should be in place.

A flowchart of a measurement process serves several purposes. It ensures that there is a common understanding of the method by which the measurement is made and may point out discrepancies between different groups performing the process. It also indicates critical points in the process that may affect measurement variation. But possibly the most important aspect of the flowchart of the measurement process is that the boundaries of the measurement process begin to describe what might be meant by measurement variation. For example, in the study of Section 9.2 on the herbicide, the intent was to learn how measurement contributed to the variation of amounts of MGC reported by the laboratory to the production floor. The flowchart of the measurement process should begin with the delivery of the sample to the lab and end with the delivery of a result. The effect of, say, different operators, devices, reagents, etc., would all be included in the boundaries defined by the measurement process.

Element 3: A list of possible sources of variation and their impact on the measurement process should be completed.

The list of possible sources of variation are often displayed on a cause and effect diagram. In any case, it should be understood that the intent of developing the list or the diagram is to begin learning about which effects are important and how they affect variation in measurements. The effects of people, materials, equipment, methods, and environment that may occur from the start of the process to the end form one means of sorting the effects on a cause and effect diagram.

> **Element 4:** *Plans for appropriate, ongoing evaluation of the measurement process should be in place.*

As is true for any process, measurement processes are subject to change across time and conditions. Consequently, it is not possible to install a measurement method without making plans for evaluating its behavior over time. The nature and frequency of these evaluations will depend in large part on the knowledge gained about the effect of sources of variation on process variation. For example, in the study of the process for measuring viscosity, it was suspected that changes in reagents might affect the results of the measurement process. In this instance, it would be necessary to take measurements, at a minimum, more frequently than changes in reagents occur so that possible effects of these changes could be evaluated.

The practice of a one-time application of a gauge R and R study has been described and the insufficiency of this approach for ensuring the quality of measurements has been explained. Although the practice may not be referred to as an R and R study, one-time evaluations (also called validations or qualifications) of measurement processes are a common phenomenon. But they do not serve to ensure that the output of a measurement process will be adequate across time and circumstances.

> **Element 5:** *An individual responsible for ensuring the quality of the measurement process must be identified.*

The response that everyone who uses a measurement process should be responsible for ensuring the quality of the process is the same as saying that no one is responsible. Different shifts using the same measurement methods will likely provide different results if attention is not paid to the consistency across groups. Further, changes in materials, equipment, and environment will require ongoing investigation to ensure the adequacy of the measurement process. The person identified as responsible for ongoing management of the measurement process must be equipped with the knowledge and authority to maintain and improve the measurement process and must know that it is their responsibility to do so.

9.7   APPLICATION PROBLEMS

9.7.1 For a particular measurement process, there is one device and four technicians. Design a study whose objective is to examine within technician–instrument variability, its magnitude, and stability.

a. Would you recommend using the same material during the study? Is that necessary?
b. Suppose it were not possible to use the same material during the course of the investigation. Would the consistency of the measurement process's ability to read to the same average be an issue? Why or why not?
c. How do potential drifts or shifts in average reported readings *on the same material* affect the study of within technician–instrument variability?
d. How would less (or more) precise technician methods be identified?

9.7.2 We want to study a measurement process involving two technicians. Two instruments are available; either of the instruments is likely to be used by either technician. Our objectives are to determine if short-term variations are about the same and to determine if both read to the same average.

In planning the study, one of the technicians suggests that each measure the same material 30 times over the next two or three days and that the average and standard deviation be calculated and compared for each technician. The technician's reasoning is that because material is the same, then any difference in the standard deviations would suggest difficulties about precision and any difference in averages would reveal any difficulties in attaining accuracy.

a. What are the advantages and disadvantages of this plan?
b. Do you consider the objectives to be well stated and relevant? Would you expand on these?
c. How would you design the investigation? (Suppose you have to quantify how much the deviations between technicians adds to measurement variation.)

9.7.3 To ensure the ongoing quality of results delivered by a measurement process, an analytical laboratory measures the purity of a standard each time an assay is run on production material. The measured values of the standard for the most recent five assays are provided in Table 9.7. We know the following about the past evaluation of the measurement process:

- Previous moving ranges of the purity of the standard have been in control with $\overline{MR} = .70$.
- The standard used is an internally developed one, so the purity of the standard cannot be considered "known." However, the purity of the standard is thought to remain stable across time. In the past the individual chart of determinations on this standard have been in control with a centerline of 95.8.

Using the data provided in Table 9.7, answer the following questions:

a. Has the short-term variation of the measurement process used to check for purity remained consistent?
b. Has the measurement process maintained a consistent average?
c. What, if anything, is known about the accuracy of the measurement process?
d. If appropriate, estimate the measurement variation for the measurement process used to evaluate purity.

TABLE 9.7 *Purity Measurements of Standard*

Determination	Purity Reading
1	96.2
2	96.5
3	95.3
4	96.0
5	95.5

10 The Role of Designed Experiments in Process Management

Previous chapters in this text have discussed the use of statistical methods, primarily control charts, for studying processes with the intent of providing consistent output and then of improving that which is provided and how it is provided. In some cases, the collection of data from an ongoing process is a passive activity; data are collected at frequent intervals in order to characterize the stability and magnitude of the variation and average of a process characteristic. At other times or in other situations, a more aggressive approach is taken whereby a planned subgrouping strategy leads to a better understanding of sources of variation contributing to total process variation. In still other situations, a process parameter is changed, data are collected after the change, and these data are charted in order to determine what effect, if any, the process change has had. An example in which these approaches were used was provided by the pot pie example of Chapter 3.

1. Initially data were taken after the formation of the bottom crust line; this information served as a baseline for describing current operation, but provided little information on the reasons for the high level of nonconforming crusts.
2. Next, a subgrouping strategy that separated the "end" and "middle" crusts showed distinct differences in the levels of nonconformities for the two groups.
3. Finally, an experiment was performed in which the recycling of dough was stopped in order to evaluate the impact on end nonconformities.

In the last two data collection activities, data were collected on the process to study the effect, if any, of varying process factors or characteristics (tin position and recycling) on an output characteristic. It is in just such situations that process investigators may find the ideas of designed experiments useful.

In the pot pie study, effective study and process improvement were achieved without the use of designed experiments. So it is instructive at this point to describe why, in similar situations, designed experiments should be considered as a component of process study and improvement.

10.1   DEFINITION OF DESIGNED EXPERIMENTS

Design of experiments is the practice of deliberately changing process variables in specific, patterned ways in order to evaluate the effects these changes have on one or more response variables. An effect is the average measured change, increase or decrease, in a response variable associated with changes in a process factor. An advantage of using designed experiments is that the effect of many factors can be studied simultaneously. Simultaneous study of factors offers several advantages, two of which follow:

1. Time can be saved in investigating the influence of many variables. In the pot pie study, tin position was studied in one set of data collected and recycling in another set. Designed experiments offer an approach for simultaneously studying the effect of numerous factors, maybe as many as seven or eight, in order to more rapidly provide information concerning the selection of improved or preferred raw materials and process parameter values.

2. The simultaneous study of the effect of two or more factors allows the investigator to understand complex relationships between factors and the output characteristic of interest. As an example, large variations in incoming density of a raw material might not affect final particle size of a batch process unless the crystallization phase proceeded too rapidly. Consequently, studying the effect of variation in density while holding cycle time constant may not provide adequate process information. (This relationship between factors is called an "interaction" and is more fully discussed in Section 10.3.) Designed experiments can aid the process investigator in sorting out complex relationships among process settings, parameters, and output characteristics.

These ideas provide a brief overview of the advantages of using designed experiments. More definitive examples of the power of these techniques will be discussed in connection with the examples provided in the following sections. However, the material in the following sections is in no way a complete discussion of the topic of designed experiments; a list of useful works on this subject is provided at the end of this chapter. Instead, the material in this chapter is presented for the purpose of describing the integration of designed experiments with a strategy for process study and improvement.

In the design phase of a product, statistically designed experiments can be used to select raw materials, set specifications, determine machine operating conditions, and recommend parameter settings. However, once a process exists for producing a product, designed experiments must take on a very different role. Methods, materials, work practices, machine settings, and a host of other operating conditions for the process are already in place. At issue then is the selection of improved methods, practices, and parameters to improve what is delivered by the process. Consequently, designed experiments for on-line investigations will be most effective when coupled with knowledge of current process behavior. The graphic in Figure 10.1 illustrates the manner in which designed experiments fit into a "chain" of process investigation. Knowledge of which process output characteristics are critical to providing customer value as well as knowledge of past experience in managing not only those characteristics but the factors that affect them should be in place. Designed experiments then offer an option for rapidly exploring process responses to changes in process parameters, settings, or materials. The more complete the available process information is, the more efficient an experiment will be in providing additional process knowledge. Since designed experiments provide relatively quick answers to specific process questions, the limited time frame within which the experiment is run requires that the validity of the experimental results be verified across time and circumstances. Data must be collected on the process after implementing a change indicated by an experiment to ensure that the change not only provides the desired improvement, but that the improvement can be maintained over time. To further define the elements of designed experiments and their role in process study, the results of an experiment designed to achieve more uniform pore size in membrane manufacture is discussed in the following section.

10.2   ELEMENTS OF DESIGNED EXPERIMENTS

Work conducted on a process for making porous nitrocellulose membranes was described in Chapter 8. For that product, the average variability in pore size greatly affected performance. Work toward achieving greater uniformity of pore size could have taken several paths. Demands could have been made of the vendor to provide more uniform material. Supervisors and other managers could demand more consistency over time and runs in any or all of identified methods and practices. Technical people could have been assigned to investigate and rebuild the entire process for making laminas. The sampling study that was used was not designed to make anything better; it was designed to reveal where most of the variation in pore size originated so that efforts could be focused on the largest contributors to variation. The investigation showed that most (67%) of the variation in pore sizes occurred within a lamina. Stated

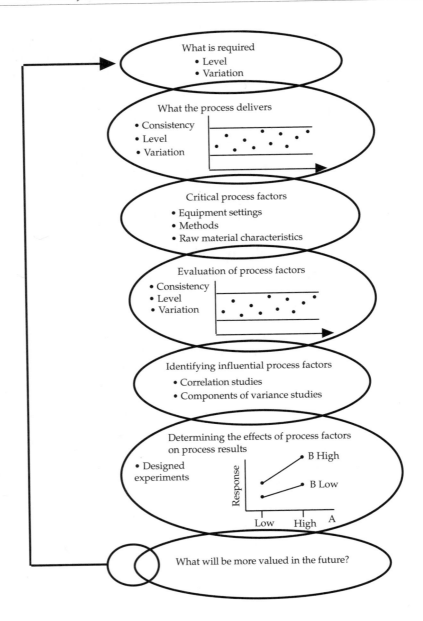

FIGURE 10.1 *Chain of Investigation*

differently, the behavior of the membrane process was such that laminas made across time and from different raw materials created membranes consistently alike, but within laminas there were many different pore sizes. Consequently, the process for producing the nitrocellulose membranes became a focus for study. In particular, attention was directed toward the process causes that resulted in large variation within a lamina.

A closer look at the process showed that membranes are manufactured in a batch process where a polymer and a copolymer are dissolved in a solvent at an elevated temperature. As the temperature is reduced, the polymer precipitates out of the solution to form a thin film membrane (lamina). Pore size is affected by these variables:

1. *Precipitation rate:* Precipitation rate, in turn, is controlled by solvent type and by temperature-pressure combinations.
2. *Molecule length:* Molecule length is primarily affected by the copolymer and is secondarily affected by an elastomer that is added to strengthen the membrane.

Atmospheric conditions, particularly moisture percentage, affect both precipitation rate and molecule length. Drying agents, used to recover solvents, can be used to control moisture. The graphic in Figure 10.2 illustrates the relationship between process parameters (or factors) and pore size of laminas.

Membrane Process

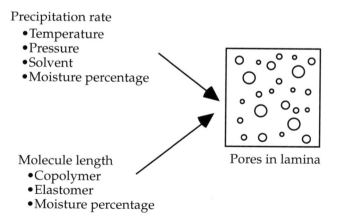

Precipitation rate
  • Temperature
  • Pressure
  • Solvent
  • Moisture percentage

Molecule length
  • Copolymer
  • Elastomer
  • Moisture percentage

Pores in lamina

FIGURE 10.2 *Relationship Between Process Parameters and Pore Size*

10.2.1   Defining Experimental Objectives

The objectives of an experiment will be determined by knowledge of

- what other or better results may be needed from the process in order to enhance customer value,
- current process behavior, and
- how to make optimal use of available resources to design an experiment that will provide a path for achieving those values.

The amount of available process knowledge will influence the objectives of an experiment; the more that is understood about a process, the more specific will be the statement of experimental objective(s). With respect to the membrane process, we wanted to increase uniformity of pore size in membranes. Since the major contributor to variation was within-lamina variation, a designed experiment whose objective was reducing pore size variation within a lamina was run.

Other objectives for a designed experiment will be determined in other situations. Some of these objectives might be to:

1. Find process parameters that can be used to move, correct, or control the average of a product characteristic or correct, reduce, or control variability of a product characteristic.
2. Determine the required tolerance range for critical process parameters.
3. Find which variables, and in what combinations, tend to be the most influential in their effect on the average or variability of a product characteristic.
4. Test or develop a theory regarding the relationship between process parameters and a product characteristic.
5. Determine those process parameters that might need to be tightly controlled.

Defining the experimental objective requires consideration of the following:

1. *Information about the past behavior of that characteristic under investigation is required.* What has been the experience in maintaining an appropriate level and variation of the characteristic? If the knowledge is not resident in the organization to run the process at stable variation and average, then consistent application of results of experiments designed to provide knowledge for moving the average or reducing variability are likely to be disappointing. If the process cannot be run consistently before the experiment, it is doubtful whether the knowledge gained from such an experiment can be put to good use.
2. *Information on the behavior of causes thought to affect the characteristic of interest is also required.* In situations where a process is subject to frequent or

infrequent upsets, it might be useful to run an experiment whose objective is to learn what factors are responsible for erratic process behavior. By identifying influential variables, it may be possible to develop control strategies that result in improved process consistency. In such a situation, speculations about important sources of variation are obviously required. Further process work prior to experimentation is also indicated. Information concerning the past practices in managing suspected causes will be needed. For example, maybe it is thought that variation in incoming raw materials causes process upsets. If so, what is known about the properties of the raw materials that can cause process upsets? If specific factors of the raw material have been identified, what information about the level and variation over time of these factors is available from charting these factors?

10.2.2   Selecting the Response Variable(s)

The response variable(s) of a designed experiment is the process characteristic(s) to be measured. The selection of the response variable is based on the major objectives of the study. For example, in the membrane process the intention was to explore the effect of process parameters on within-lamina variation; a measure of that variation is the required response variable. Five squares were randomly chosen from each lamina manufactured in the experiment with the average pore size determined for each of the five grids. The response in the experiment was the standard deviation (STD) of those five numbers.

Specifying the method for calculating the response variable, STD, is by itself an insufficient description of the response variable. This description states *what* is measured, but not *how* and *when*. Once the variable to be measured has been determined, it is necessary to know *how* the measurement method for evaluating the response is to be used and whether this measurement process is adequate. As with any process study, the process for making the measurement must have been assessed for stability, precision, and accuracy. In addition, *when* the process response is determined in an experiment is also critical. In the course of a designed experiment, process settings are adjusted and their effect on the response determined. At issue in deciding when to measure a response is the length of time it would take for the change in process settings to have their full impact on the response variable. Previously developed control charts and/or knowledge of process dynamics will be required in order to know the amount of time that must elapse after a change in process settings before the response variable of interest should be measured.

10.2.3   Determining Experimental Factors

Experimental factors are those causes thought to affect the response variable in a designed experiment. In identifying the factors to be studied in an experiment, the experimenter is projecting what factors might be important. Whether any knowledge is gained from the experiment will rest heavily on whether the projected factors do indeed have an effect on the response variable. Thus, the choice of process factors must rely on sound process knowledge, such as that which has been generated from the use of control charts to monitor, control, and evaluate process changes. For example, the preceding work on the membrane process determined that the major source of variation in pore size of laminas occurred within laminas. Consequently, factors whose effects tend to be active in the short term (within lamina) become candidates for exploration in a designed experiment. On the other hand, factors that affect, say, variation between cars of material might reasonably be excluded from study since the experimental objective is to identify factor settings that reduce within-lamina variation. For the membrane experiment, the factors to be studied because of their suspected effect on the within-lamina variation were

> temperature,
>
> pressure,
>
> copolymer,
>
> solvent,
>
> elastomer, and
>
> moisture.

10.2.4   Types of Factors

A useful way of classifying experimental factors is whether they are quantitative or qualitative. Quantitative factors are those whose values or levels can be associated with points on the real number line. In the preceding list of factors for the membrane study, temperature, pressure, and moisture are quantitative factors. Values for quantitative factors can be ordered from smallest to largest. Changes in the response due to changes in a quantitative factor can be meaningfully described by a signed magnitude; for example, it could be stated that as temperature increases, the response variable increases (or maybe decreases). Copolymer, solvent, and elastomer are called qualitative variables. The levels of these variables reflect a characteristic used to type or classify the variable. The levels of a qualitative variable cannot be ordered in a meaningful way; designating one level as "low" and another as "high" is artificial and not mathematically meaningful. For example, the level of copolymer does not refer to an amount of copolymer used but to what type, kind, or brand of copolymer. Changing from one copolymer to another may cause a change in the uniformity of pore size, and the magnitude of that change is important. But the sign attached to the change is not meaningful; it is merely a result of the particular assignment of types to levels.

10.2.5 Determining Factor Levels

Factor levels are the settings, amounts, or type of factors to be studied in an experiment. For example, temperature is one of the quantitative factors whose effect on pore size is to be evaluated; the information required is, as temperature changes, what changes, if any, occur in uniformity of pore size. The choice of the factors along with the levels at which they are to be investigated are the main determinants of the quality of information obtained from an experiment. In deciding on the levels of temperature to be investigated, it is possible that the range in temperature levels selected for study is so small that the resulting effect on pore size variation is also too small to be detected or to small to be of practical use. By the same token, two levels may be so widely separated that the observed effect is very large. Interpolating between these values would have to be done in an attempt to identify an appropriate setting for obtaining the needed effect on the response. But if the relationship between the response and the factor is not linear, this interpolation will not provide a useful choice of an operating level for that factor. Consequently, the choice of the levels of the factors requires careful consideration. Again, process knowledge, both in terms of preceding data collection and analysis, as well as understanding the physics, chemistry, or engineering of the process is necessary to guide an appropriate selection of the range of values for factor levels. Process engineers and operators, both of whom should help plan the experiment, are likely to have helpful information on the required length of an interval to achieve a change in the response. Previous experiences in adjusting the process or engaging in correction activities might provide useful information on the magnitude of a change in a factor that will produce a measurable effect on the response.

In addition to deciding on the range of process levels, the number of levels to be investigated for each factor must be determined. For qualitative variables, this decision entails choosing the number of types or classes of a factor. For quantitative factors, the decision of the number of levels will, in part, rely on the anticipated behavior of the response as levels of the factor are changed. The graphs in Figure 10.3 provide one aspect to be considered in deciding on the number of levels. In those graphs, values of the response variable, STD, are plotted on the vertical axis; values on the horizontal axis represent values of an independent variable, factor A. Scenario 1 depicts a linear response between the factor named A and the response variable STD. The response in scenario 2 is nonlinear; as the level of factor A increases, the amount of increase in STD changes. Of course, the information about the nature of the change in STD with changing levels of A cannot be so clearly described before an experiment is conducted; it is the objective of the experiment to gain information about this response. Maybe the decision is made to investigate the effect of A on pore size by evaluating STD when the process is set at two levels of A. The experiment is run, STD is determined, and a plot is generated from the resulting data. The resulting plot might look like the one labeled "Experiment One" in Figure 10.4. From this plot it can be seen that STD increases with increasing A. But whether this increase is linear cannot be determined with the results from two levels of A. To know whether a response is nonlinear, at least three levels of A would need to be investigated. Three levels of Factor A have been studied, plotted, and labeled "Experiment Two" in Figure 10.4. When the three levels are studied, a nonlinear effect of A on STD is observed.

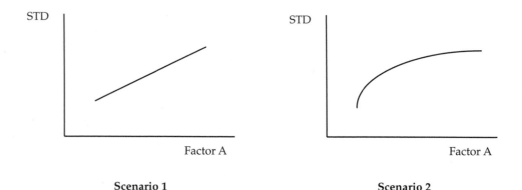

FIGURE 10.3 *Possible Effects of Factor A on Variation in Pore Size*

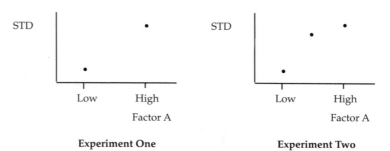

FIGURE 10.4 *Experimental Results for Investigation of Two Levels of Factor A*

More than two levels of a factor are needed to explore possible curvature in a response. But if three levels of factors are studied the size of the experiment increases; the magnitude of the increase gets larger as the number of factors increases. (A more precise description of the size of an experiment is given in Section 10.3.) Consequently, when investigating a large number of factors (more than two or three) to determine their effects on the response variable, typically two levels of each variable are initially investigated. Once the list of factors has been narrowed to those that do have an experimentally verified effect on the response, follow-up experiments may be performed with this fewer number of factors to begin to explore curvature.

In the membrane example, six different factors were to be studied. Since there were initially so many factors, two levels of each were to be investigated. The factors and their levels are given in Table 10.1. In that table, the column labeled "Code" introduces the common practice of using "coded" values to represent the values of the independent variables. For quantitative variables, −1 corresponds to the smaller of the two values and +1 to the larger.

10.2.6 Summary for the Membrane Experiment

The next step in designing the membrane experiment was to decide in what combination and in what order the levels of the process factors should be set throughout the experiment. A complete description of the basis on which that decision was made would require some additional learnings about the basic mechanics of designed experiments. These mechanics are discussed in the next section. However, a brief summary of what was learned from the membrane experiment is provided in the following paragraph.

TABLE 10.1 *Factors and Factor Levels for the Membrane Experiment*

Name	Symbol	Value	Code
Temperature	T	High	+1
		Low	−1
Pressure	P	High	+1
		Low	−1
Copolymer	C	A	+1
		B	−1
Solvent	S	A	+1
		B	−1
Elastomer	E	A	+1
		B	−1
Moisture	M	5%	−1
		7%	+1

The group exploring the causes of within-lamina variation in pore size found that the type of copolymer and solvent were critical to the resulting variation in pore size. Type B copolymer used in conjunction with type A solvent resulted in less variation than any other combination of copolymer and solvent. This finding provided information for selecting materials for trial runs of a revised process. In the same experiment, it was also discovered that temperature and elastomer had an effect on pore size variation. Variation in pore size was smaller when type A elastomer was used and when temperature was held at the highest level. Again, these findings suggested a material type and a factor setting to be used in the revised process. Trial runs for the adjusted process were then run to confirm that reduced variation in pore size occurred.

10.3   BASIC MECHANICS OF DESIGNED EXPERIMENTS

In Chapter 5 a description of a process for making particle board was provided in the flowchart of Figure 5.7. The description and use of the basic tools of designed experiments are illustrated here with an example from this same process. The data provided below and the description of the work and knowledge of the personnel at the plant site are fictitious. However, the problem situation and learning gained from a similar experiment are real.

One of the critical characteristics of final board quality is density; this critical characteristic was also one that had been difficult to manage to correct levels. At times, for reasons not well understood, board density would decrease below required levels. On these occasions it was common practice for operators to increase the amount of resin at blending in order to bring density levels, measured after the press cycle, back on target. The amount of the resin increase was based on the judgment of the operators and supervisors working in this area. Because blending of the boards occurred at an early stage of the process, the effectiveness of the adjustment could not be immediately evaluated. Approximately one hour had to pass before the density level could be checked for the adequacy of the adjustment. The results of this check did not always indicate that the adjustment to resin had been effective in returning density to an appropriate level; sometimes the density would be too large, indicating too much resin had been added. At other times, the increase in resin was insufficient to bring density up to the required level. In either of these last two cases, additional adjustments to resin were made. It should be noted that informal experimentation was routinely occurring at this plant site. A change was made; the effectiveness of the change was evaluated; and further "experimentation" was done (i.e., additional changes to resin were made) until the density was at its appropriate level.

The management of the plant site decided to adopt a different approach for using resin to manage board density. This approach was to begin to better define the relationship between resin *and other factors* that affect board density. They knew that resin alone was not the sole determiner of final board density. Some process variable or variables were clearly causing unplanned decreases in density and something was impacting the effect the resin had on changing the level of board density. It was with the intention of better understanding these complex relationships between resin, other process factors, and final board density that a designed experiment was planned.

The implication of running a designed experiment is that process variables will be deliberately adjusted in order to observe the impact on a critical characteristic, board density in this case. The idea of perturbing process variables with the intention of studying process behavior required, as is often the case, a fundamental change in the way in which the plant site viewed appropriate means for process improvement. The passive activities of charting process data had become an acceptable practice. But deliberately "upsetting" the process for the purpose of process study did not immediately gain acceptance among plant management. Yet, informal experimentation routinely occurred at the plant. By reporting the number of changes in resin and the number of times that these changes were ineffective, acceptance of a more planned, deliberate approach for understanding the management of board density was achieved.

Another critical factor thought to affect final board density as well as the impact of resin on density was the moisture content of the material entering the blending stage of the process. Consequently, a designed experiment that examined the effect of these two factors on board density was performed. The intention of the experiment was to adjust both factors, amount of resin and moisture content, and observe the effects these changes had on density. So two different amounts, or levels, of resin and of moisture were identified. The selected levels of the two variables were based on speculations of operating personnel; the guiding principle in the choice of the levels was that changing from one level to the other should result in a measurable impact on the process. The levels chosen are given in Table 10.2. A "–1" is used to identify the low level of a factor and "+1" to identify a high level. The advantage of this notation will be demonstrated in the subsequent analysis of data generated from the experiment.

TABLE 10.2 *Factor Levels for Density Experiment*

Factor	Low level −1	High level +1
Moisture	4%	7%
Resin	6%	8%

10.3.1 Completely Randomized Designs

The values in Table 10.2 identify the levels of the moisture and resin factors examined in the experiment. The next decision made was how these levels should be adjusted throughout the course of the experiment. The decision was made to examine all possible combinations of factor settings. In an experiment with two factors, each to be explored at two levels, the number of distinct combinations of factor levels is determined by:

(number of levels of first factor) × (number of levels of second factor)

OR

(2 levels of first factor) × (2 levels of second factor) = 4 combinations

OR

$(2 \text{ levels})^{2 \text{ factors}}$ = 4 combinations

The four experimental conditions listed in Table 10.3 were investigated. This table indicates four different processing conditions for which information on board density was to be collected. For example, for experimental condition 3, moisture was to be set at its high level (7%) and resin at its low level (6%). After the process stabilized at these values, board density was to be measured.

The measurement of board density was intended to reflect the effect that particular settings of moisture and resin had on board density. However, it was also understood that board density would be subject to variations caused by many other factors besides the two under investigation. For example, the thickness of the board, which is primarily managed by press settings, will clearly have an effect on density since a "tighter" press will cause a denser board. At the time the current work was performed, the thickness settings on the press were being maintained at a near constant value. With constant thickness settings, a decreasing amount of resin should cause the board to increase in thickness (or spring back) after going through the press, thus resulting in a thicker and less dense board.

TABLE 10.3 *List of Experimental Conditions*

Experimental Condition	Moisture Setting	Resin Setting
1	−1	−1
2	−1	+1
3	+1	−1
4	+1	+1

Other factors also influence process results; the level and variation of these other factors and their effect on density may or may not be as well understood as were those of thickness settings. In either case, how to run the experiment in light of the fact that other variables will be creating variation in the response must be addressed. In particular, it will be necessary to consider the following three questions:

1. *How many times should the set of four conditions given earlier be replicated in order to be confident that the effect of altering moisture and resin is understood?* Even without changing the settings of moisture and resin, the observed values of density will vary. If this variation in density values is large, then when moisture and resin *are* adjusted, the effects of this adjustment will be more difficult to detect. Consequently, a number of readings of density at a certain choice of moisture and resin settings may be required in order to be confident of the change in density that is or is not caused by these factors.
2. *In what order should the replications of the basic four-run design be performed?* The choice of the order of the experimental runs will depend on the manner in which factors not included in the experiment are known or suspected to affect density as well as on the time and cost of running the experiment.
3. *Since other factors may have an impact on variation in board density, how should these factors be managed during the experiment?* The decision to hold the press settings that control thickness at a constant level was made prior to conducting the experiment. Similarly, how to manage other influential factors must also be considered. One choice would be to tightly control these factors throughout the experiment. This tactic would be useful if there were concerns that changes due to other variables would disguise the relationship between resin, moisture, and density. Another tactic would permit those variables to be maintained at typical production tolerances experienced at rated production. This tactic would be useful if experimenters thought that resin and moisture effects would be detectable in the face of this variation and if knowledge regarding consistency of effects across these various conditions were to be used for future rules for maintaining and recovering control.

The group planning the experiment decided to replicate the basic four-run design five times. With five replications there would be a total of $4 \times 5 = 20$ different runs in the experiment. Once the experiment was run, there were five responses at each of the four combinations from which an estimate of the variability and the levels of the response for those factor settings was calculated. Deciding whether five replications is adequate for an experiment is equivalent to deciding whether five numbers at a particular combination of factor settings will provide adequate information about the process at that setting. If a process has large inherent variation (i.e., the value of $\bar{R}/d_2$ obtained from previous study of the response variable is large), it will be more difficult to detect a change in the process level caused by changing factor levels. On the other hand, selecting a very large number of replications might be quite expensive and only serve to show that a statistically significant difference in level occurs with changing factor levels even though this difference is of little practical significance.

The 20 different experimental runs performed are listed in Table 10.4. The runs have not been listed in the order in which they were actually performed in the experiment. Column 2 of Table 10.4 contains that ordering. The ordering used was determined by randomizing the order of the 20 runs. Such a randomization could be found by writing each of the 20 runs down on a separate slip of paper, mixing the slips of paper together, and then drawing slips of paper, one at a time, from the mixture to determine the order of the 20 runs. Because all 20 runs have been randomized, the experiment is said to be a *completely randomized design*. The reasoning behind using such a randomization scheme can be seen by noting the pattern of runs listed in Table 10.4. The first 10 runs listed correspond to low moisture levels and the last 10 runs to the higher moisture level.

When the experiment is run, attempts might be made to hold constant other process settings or parameters not included as factors in the experiment. However, since knowledge of those items that influence density is surely imperfect, it is possible that some parameter may change in the course of the experiment without the knowledge or planning of those responsible for the execution of the experiment. If this factor or factors had the effect of gradually increasing density, the observed increase in the recorded values of density might be falsely attributed to the change made in moisture levels. As another example, suppose that during the course of the experiment, the group running the experiment learned how to better adjust factor settings. As the experiment proceeded, there would be decreased variation in the results. This decrease might be falsely attributed to changing moisture if the runs were not randomized. Or suppose that an uncontrolled factor had the effect of increasing density at some times and not at others. If the factor was active only during the time when runs of one of the four factor combinations were being made, again the effect of that factor combination might be falsely interpreted. Because of these possibilities, runs are randomized. By randomizing the order of the runs, each of the four experimental conditions will be just as likely to be affected by additional variation created by uncontrolled factors.

The choice of a completely randomized design is only one of many options available to the experimenter. Note that by choosing to use such a randomization scheme, moisture and resin levels will need to be changed 20 times in the course of the experiment. In some experimental situations, performing so many changes may result in an experiment that is prohibitively expensive or time consuming and other options for running the experiment must be considered. A brief description of other randomization schemes is provided in Section 10.5. At this point suffice it to say that the succeeding analysis of the experimental data from the density experiment is based on the assumption that runs were performed in a completely randomized manner.

TABLE 10.4 *Experimental Runs for Density Experiment*

Run Number	Run Order	Moisture Level	Resin Level	Density
1	7	−1	−1	23.2
2	16	−1	−1	22.4
3	3	−1	−1	21.8
4	11	−1	−1	22.9
5	19	−1	−1	22.3
6	12	−1	+1	28.2
7	1	−1	+1	27.7
8	15	−1	+1	28.3
9	8	−1	+1	27.7
10	2	−1	+1	29.1
11	18	+1	−1	18.8
12	10	+1	−1	19.2
13	4	+1	−1	17.6
14	20	+1	−1	18.8
15	6	+1	−1	18.1
16	13	+1	+1	20.1
17	5	+1	+1	18.6
18	17	+1	+1	19.2
19	9	+1	+1	18.7
20	14	+1	+1	19.5

The choice of a completely randomized design was based on a consideration of the possibility of other unknown and uncontrolled factors affecting the process. Of course, in most experimental situations other factors will also be identified as having a possible effect on density. As previously discussed, reasonable attempts are often made to hold these factors constant. When this strategy is followed, it must be understood that information gained from the experiment is only known for those conditions that have been maintained throughout the experiment. For example, it may be thought that milling practices, which set fiber length and width, have an effect on board density, so throughout the course of the experiment attempts were made to hold the milling of wood within certain parameters. After completing the experiment, it must be understood that what is learned about the effects of moisture and resin on density is only valid when milling is performed as in the experiment. Should milling practices be altered, then the effect of moisture and resin on density might also be different.

10.3.2   Graphical Summaries of Experimental Results

The fifth column of Table 10.4 contains the (coded) density of a board created by the process when moisture and resin were held at the selected values. A summary of these numbers is provided in Table 10.5. The objective in conducting the experiment was to determine how much, if any, the density values changed with changes in moisture and resin. The column in Table 10.5 labeled "Average" provides a report on this information. However, before examining these averages, it is first necessary to consider the information provided by the standard deviations. The first standard deviation in the table, .5450, is the standard deviation of the five density values obtained when moisture and resin were both at their low levels. This standard deviation provides some information about the amount of variability that would be observed in the process if it were run at the low levels of moisture and resin. Of interest, then, is whether the data from the experiment indicate that the variation at the four combinations of factor settings is of a similar or different amount. If, for example, it were determined that one of the combinations of settings resulted in significantly larger levels of variation in density, it might be decided that it was necessary to avoid this particular combination of settings in process operation. Additionally, further work to understand why variation was observed to be larger would be indicated. In the present situation, the standard deviations all appear to be of roughly the same magnitude. An s chart provides a simple method of assessing the relative magnitudes of the four standard deviations. The average standard deviation is:

$$\bar{s} = \frac{.5450 + .5745 + .6403 + .6140}{4} = .6002$$

TABLE 10.5 *Summary of Density Experiment*

Moisture	Resin	Average	Standard Deviation	Label
−1	−1	22.52	.5450	Y1
−1	+1	28.20	.5745	Y2
+1	−1	18.50	.6403	Y3
+1	+1	19.22	.6140	Y4

and the UCL is:

$$UCL_s = B_4 \bar{s} = 2.089(.6002) = 1.2538$$

No lower control limit is calculated since the standard deviations are based on samples of size five. All four of the standard deviations in Table 10.5 are well below the UCL, justifying the earlier statement that the four standard deviations are of roughly the same magnitude.

Attention now turns to analyzing the effect of changing moisture and resin on the level of density. A check of the four averages in Table 10.5 reveals that the highest average density, 28.20, occurs when moisture is at its low level and resin at its high level. The lowest average density, 18.50, occurs when moisture is high and resin low, although the average density when both moisture and resin are high is not much larger, 19.22. These descriptions of the experimental results are more easily understood by plotting the averages using an interaction plot. Figure 10.5 is a completed interaction plot for the experimental results.

An interaction plot is drawn by plotting values of the response variable on the vertical axis and marking a scale for one of the factors, preferably a quantitative one, on the horizontal axis. A separate line, or curve, for each level of the other factor is then drawn. Figure 10.5 shows average density, on the vertical axis, ranging from 15 to about 30. Values, low and high, for resin amount are marked on the horizontal scale, with a distinct, separate line drawn to indicate results for the two moisture levels. The letter Y denotes the average density across a number of runs. The number attached to the symbol indicates the process condition used and is consistent with the results reported in Table 10.5.

The interaction plot provides powerful insights into the effect of moisture and resin on density. As moisture content increases, density decreases, and as resin amount increases, density increases. Significantly, the amount of increase in density as resin increases is different depending on the amount of moisture. At high moisture levels, the increase in average density as resin increases from low to high (from 4% to 7%) is only 19.22 – 18.50 = .72. However, if moisture is at its low level, the increase in average density with increasing resin is 28.20 – 22.52 = 5.68. The phenomenon that the effect of a resin increase varies with changing moisture levels is an example of a *two-way interaction* between moisture and resin. The presence of an interaction between two variables indicates that the effect due to one is dependent on the operating value of the other factor.

The information that there is an interaction between resin and moisture should be considered in the light of previous practice for managing the process. Prior to the experiment, resin was added to the process in order to obtain higher density levels; but the amount of resin to be added was not well known. The fact that the effect of increased resin changes with changing moisture helps explain this earlier experience. If the moisture of the incoming material were high, then adding more resin would not have as great of an impact than if moisture content had been low. Further, the possibility of interaction effects means that experimenting by changing just one factor at a time (as had been done informally at the plant site) is often inadequate for understanding the effect process factors have on a response. In the present situation, if density is to be managed through resin use, then the amount of resin used must be based on the level of moisture chosen. Further experimentation at other levels of moisture or resin may be desired to better map out the simultaneous effects of these factors on density.

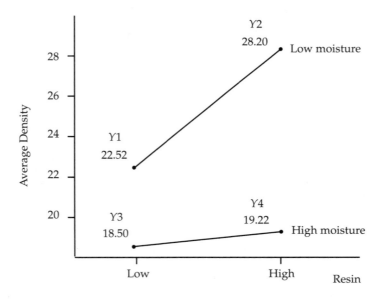

FIGURE 10.5 *Interaction Plot for Density*

10.3.3   Numerical Summaries of Experimental Results

Information obtained from the interaction plot can also be reported numerically. Quantification of effects is useful for understanding the relative magnitudes of the effects of moisture and resin increases on density level. These three pieces of information were obtained from the interaction plot:

1. An increase in moisture results in a decrease in density.
2. An increase in resin results in an increase in density.
3. The effect of a resin increase is different at different moisture levels.

The effect due to increasing moisture is noted on the graph of Figure 10.5 by the fact that the two points where moisture is at its low level (labeled $Y1$ and $Y2$) have a higher average density than the two points where moisture is at its high level (labeled $Y3$ and $Y4$.) The average density observed when moisture is at its low level can be found by averaging the two averages, $Y1$ and $Y2$. The resulting number:

$$\frac{Y1 + Y2}{2} = \frac{22.52 + 28.20}{2} = 25.36$$

is the average density for low levels of moisture. Similarly, the average density for high moisture levels is found by averaging $Y3$ and $Y4$ to obtain 18.86. The difference in the average density at high levels of moisture and low levels of moisture is called the *main effect* due to moisture. The main effect is thus found by:

$$\text{moisture main effect} = \frac{Y3 + Y4}{2} - \frac{Y1 + Y2}{2} = 18.86 - 25.36 = -6.50$$

The main effect for moisture is negative, indicating that as moisture increased from a low of 4% to a high of 7%, density *decreased* by 6.50 units. Because density at, say, the low level of moisture is averaged over both the low and high levels of resin, calculation of the main effect for moisture ignores the interaction effect observed between moisture and resin. This suggests that the experimenter should be cautious about interpreting main effects without using interaction plots.

Plots for main effects are often provided and can be informative; an example for moisture is shown in Figure 10.6. Main effect plots, like reports on main effects, can be misleading in the presence of a strong interaction. Further insight into the misleading nature of the main effects plot can be developed by using the interaction graph of Figure 10.7 to compute the moisture main effect.

A similar calculation of the resin main effect is performed by subtracting the average density determined at low levels of resin from the average density found at high resin levels. The calculation of this main effect is found by:

$$\text{resin main effect} = \frac{Y2 + Y4}{2} - \frac{Y1 + Y3}{2} = \frac{28.20 + 19.22}{2} - \frac{22.52 + 18.50}{2} = 23.71 - 20.51 = 3.20$$

The graph in Figure 10.8 illustrates the computation of the resin main effect. Again, this graph reveals the need to know about interaction before interpreting main effects.

The previously introduced +1 and −1 notation for low and high levels of a factor provides a useful schematic for calculating factor main effects. Table 10.6 contains the same information about the levels of moisture (M) and resin (R) found in Table 10.5. The columns of +1's and −1's in this table were used to describe the manner in which the data were collected, but they can also be used to describe the calculation of the main effects. The main effect for either moisture or resin is calculated as follows:

Step 1

Average the response (average density) corresponding to the +1 values.

Step 2

Average the response corresponding to the −1 values.

Step 3

Subtract the average at the −1's from the average at the +1's.

For the main effect of moisture the average density at the +1 values of moisture is:

$$\frac{Y3 + Y4}{2} = \frac{18.50 + 19.22}{2} = 18.86 \; \left(\text{Step 1}\right)$$

TABLE 10.6 *Summary of Density Experiment*

M	R	Average Density	Label
−1	−1	22.52	Y1
−1	+1	28.20	Y2
+1	−1	18.50	Y3
+1	+1	19.22	Y4

and the average at the –1 values is:

$$\frac{Y1 + Y2}{2} = \frac{22.52 + 28.20}{2} = 25.36 \quad (\text{Step } 2)$$

so the main effect for moisture, as previously determined, is:

$$\frac{Y3 + Y4}{2} - \frac{Y1 + Y2}{2} = 18.86 - 25.36 = -6.50 \quad (\text{Step } 3)$$

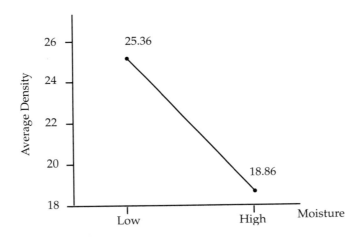

FIGURE 10.6 *Main Effect Plot for Moisture*

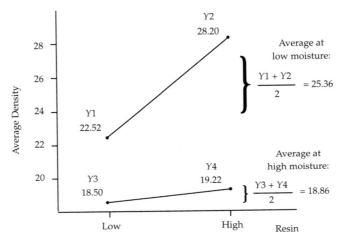

FIGURE 10.7 *Computation of Moisture Main Effect*

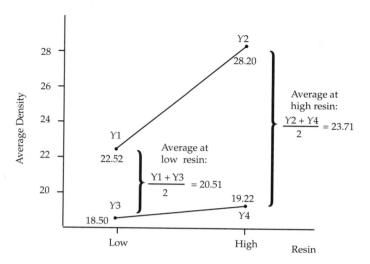

FIGURE 10.8 *Illustration of Resin Main Effect*

Having described the main effects of moisture and resin, it is necessary to summarize the different effect due to increasing moisture for the two different levels of resin. This effect is called the *interaction effect*. (The interaction effect could, equivalently, be described as the effect due to increasing resin for the two different levels of moisture.) An examination of Figure 10.9 shows that when resin is maintained at its low level, density decreases 22.52 − 18.50 = 4.02 units as moisture increases from its low to high level. But when resin is at its high level, the decrease in density with increasing moisture is 28.20 − 19.22 = 8.98. The different size of these decreases is what has been referred to as an interaction effect. The magnitude of this effect is calculated by:

$$\text{interaction effect} = \frac{-8.98 - (-4.02)}{2} = -2.48$$

The −1 and +1 notation used for calculating the main effects can also serve as a means for describing the calculation of the interaction effect. Table 10.7 contains some numerical results required for calculating the interaction effect. The column of −1's and +1's under M (for moisture) was used to calculate the main effect for moisture by subtracting the average density when M is −1 (25.36) from the average density when M is +1 (18.86). The column under $M \times R$ contains +1's and −1's that can be used in a similar fashion to determine the interaction effect. The column under $M \times R$ is found by multiplying the entry in column M by the corresponding entry in column R. For example, the first entry in both column M and column R is a −1. Since (−1) × (−1) = +1, the first entry in column $M \times R$ is +1.

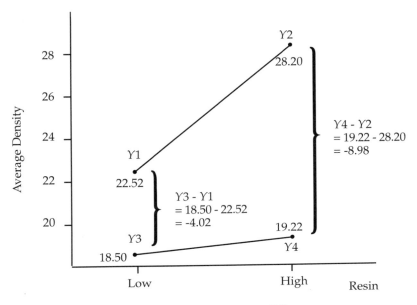

FIGURE 10.9 *Illustration of the Interaction Effect*

Using the $M \times R$ column, the interaction effect is determined to be:

$$\frac{Y1 + Y4}{2} - \frac{Y2 + Y3}{2} = 20.87 - 23.35 = -2.48$$

This is the same calculation used previously to estimate the interaction effect, since:

$$\frac{(Y4 - Y2) - (Y3 - Y1)}{2} = \frac{Y1 + Y4}{2} - \frac{Y2 + Y3}{2}$$

TABLE 10.7 *Calculation of Interaction and Main Effect Estimates*

	M	R	Average $M \times R$	Density	Label
	−1	−1	+1	22.52	$Y1$
	−1	+1	−1	28.20	$Y2$
	+1	−1	−1	18.50	$Y3$
	+1	+1	+1	19.22	$Y4$
Ave. density at +1's	18.86	23.71	20.87		
Ave. density at −1's	25.36	20.51	23.35		
Difference = effect estimate	−6.50	3.20	−2.48		

The numerical summaries of the experimental results, the moisture and resin main effects and the interaction effect, require interpretation. An interpretation of these numbers should first consider the size of the interaction effect. Since this number, −2.48, is of a similar order of magnitude to the other two effects, an interaction between moisture and resin is indicated. An interaction means that the effect of resin changes are dependent on the current operating level of moisture. In the presence of a strong interaction, reported main effects can be deceptive. Given that there is a sizable interaction effect, it is not meaningful to describe the way in which density decreases with increasing moisture without knowing or specifying the level of resin. The main effect for moisture was found to be −6.50, which at first glance might seem to indicate a decrease in density of 6.50 units as moisture moves from a low to a high level. However, this change of −6.50 is not an adequate summary of the effect of moisture at either level of resin. At low levels of resin the change in density is − 4.02 units, but the change in density at high levels of resin is −8.98 units. Thus the value of −6.50 underestimates the change in one case and overestimates it in the other.

In the preceding paragraph, the size of the interaction effect, −2.48, was judged to be large by comparing it to the size of the other two effects. However, it should be considered that the size of all three effects, although of a similar magnitude, might be due to the variation present in the process. In other words, even if there is no effect on density caused by changing moisture or resin levels, effect estimates of size 6.50 or smaller could be observed just by chance variation. Consequently, it is useful to compare the variation created by changing moisture and resin (that variation captured by the effect estimates) with the variation observed in the process when moisture and resin are held constant (as captured by the standard deviations of Table 10.5). This kind of an evaluation is usually performed with the aid of an ANOVA table. The construction and use of such a table is described in the following section.

10.3.4   Using ANOVA Tables to Summarize Experimental Results

ANalysis Of VAriance (ANOVA) tables provide a commonly used method of summarizing and evaluating experimental data. They provide information on the magnitude as well as the statistical significance of factor effects. However, ANOVA tables do not provide information about the nature of the factor effects. An ANOVA table only indicates whether changing a factor's setting caused some response, but not whether an increase or a decrease in a response variable resulted from the change in the factor setting. Consequently, ANOVA tables should be used in conjunction with the graphing techniques previously discussed in Section 10.3.2.

To illustrate the construction and analysis of an ANOVA table, the data from the experiment on density have been summarized in an ANOVA table. Although, for instructional purposes, the calculation of each element of the table will be illustrated, statistical software is typically used to perform these calculations. The ANOVA table of Table 10.8 was constructed using the software package JMP. The elements (or columns) of an ANOVA table are

Sources of variation (Source),
degrees of freedom (df),
Sums of Squares (SS),
Mean Squares (MS),
F values (F), and
p values.

A summary of what information about experimental data is captured by these elements is given next.

Source
The column in the ANOVA table labeled "Source" contains a list of the sources of variation that contribute to the variation observed in the experimental data. As shown in Table 10.8, the main effects and the two-way interaction effects are listed first. The source labeled "Residual" refers to the inherent, unexplained variation in the process under study. "Total" refers to the variation observed in all experimental values.

SS
The sums of squares reflect the variation that can be attributed to each source listed in the ANOVA table. For a two-level design, the sums of squares for main and interaction terms can be calculated from the effect estimates for these sources. A large positive or small negative effect will result in a large sum of squares. Table 10.9 shows the effect estimates for the experiment on board density. The sums of squares for each experimental effect can be determined from these values by:

$$\text{SS (effect)} = \frac{kn}{4} \left(\text{effect}\right)^2$$

where k is the number of trials or experimental conditions and n is the number of replicates in the experiment. For example, the sum of squares for the moisture main effect is:

$$\text{SS (moisture)} = \frac{4 \times 5}{4} \left(-6.50\right)^2 = 211.25$$

The sums of squares for the other experimental effects are calculated in Table 10.9. It should be noted that, unlike the effect estimates, the sums of squares do not reflect the nature, positive or negative, of the effect.

As the name implies, the total sum of squares represents the variation observed in all of the experimental data values. The total sum of squares can be calculated from the formula:

$$\text{total SS} = \sum_{i=1}^{20} (Y_i - \bar{Y})^2$$

The values of Y_i represent the responses from individual runs and $\bar{Y}$ is the average response across all runs. The total sum of squares reflects the total dispersion for all experimental runs. This dispersion is due to the effects of perturbing the factors and to inherent process variation.

The residual sum of squares is the fraction of the total variation that is not explained by varying the factors. Thus, the residual sum of squares can be calculated by

residual SS = (total SS) – (SS for M) – (SS for R) – (SS for $M \times R$)
= 298.858 – 211.250 – 51.200 – 30.752 = 5.656

df
The degrees of freedom reflect the amount of information from the experiment that was used to calculate the sums of squares. Note for the two-level design that the degrees of freedom is 1 for each factor effect. This reflects the fact that only two averages are compared to determine effect estimates, so

df = (number of means) – 1 = 2 – 1 = 1

Since the total sum of squares was determined from the variation of 20 numbers, the total degrees of freedom in the experiment is 20 – 1 = 19. The remaining degrees of freedom (19 – 1 – 1 – 1 = 16) is attributed to the residual sum of squares.

Mean squares
The mean squares are the sums of squares divided by their respective degrees of freedom. The mean square residual (also called the mean squared error, MSE) is found to be .3535. It is instructive to note that this term could also be calculated from the four standard deviations in Table 10.5. The standard deviations of Table 10.5 reflected the amount of variation observed at each of the four factor combinations. Each one of these standard deviations captured process variation observed when the factors were held at a particular combination of

factors. For this reason, these variations are thought to reflect "inherent process variation." The mean squared error is calculated from these numbers by squaring each one to obtain a variance, and then averaging the four variances. In other words, the mean squared error, also referred to as s_p^2, is found by:

$$S_p^2 = \frac{(.5450)^2 + (.5745)^2 + (.6403)^2 + (.6140)^2}{4} = .3535$$

In control chart work, $\bar{R}/d_2$ is used as an estimate of the process standard deviation for a controlled process. The square root of MSE is often thought of as an estimated standard deviation, but it is likely to be quite a bit smaller than $\bar{R}/d_2$. Since experiments are often conducted under tightly controlled conditions, the amount of "inherent process variation" observed in an experiment is likely to be considerably smaller than what is observed under normal operating conditions.

F Values

The F values are the mean squares for effects divided by the mean square error. In each case, the amount of variation that occurs because of adjusting a factor (as captured by its mean square) is compared to the amount of variation observed when the factors were held constant (mean squared error). A large F value would therefore indicate an important effect.

A statistical decision about whether an F value is large enough to indicate an important effect is based on considering how big F might be if there were no effects due to the experimental factors. The F tables in Appendix D describe values of F that would be observed if no effect were present. The generation of these tables is based on the statistical knowledge of how ratios of mean squares behave. For example, the table labeled with an α value of .05 contains critical values of F for different combinations of numerator and denominator degrees of freedom. The F value used to evaluate the significance of the interaction effect was found by dividing the interaction mean square (with 1 degree of freedom) by the mean square for error (with 16 degrees of freedom). The F table with an α value of .05 contains the critical value of 4.49 for these degrees of freedom. The interpretation of the value 4.49 is that if there were no interaction effect, then there would only be a 5% chance of observing an F value as large as 4.49. The actual, observed F value is much larger than 4.49 (the observed F value was 86.990!). Since the observed F value is at variance with the assumption of no interaction effect, we can conclude that there is an interaction effect.

p Values

When using most software packages, the physical use of the *F* tables can be avoided since software packages typically perform the kind of comparison needed to make a decision about the significance of an effect. The *p* values listed in an ANOVA table are a calculation of the chance (or probability) of observing an *F* value as large as the one contained in the table if there is no effect due to that source of variation. The *p* value in Table 10.8 for the interaction effect is .0000. The interpretation of this *p* value is that the probability of seeing an *F* value as large as 86.990 if there were truly no interaction effect is less than .00005. (JMP rounds these numbers to the nearest ten thousandth.) Since this probability is so small, the conclusion is, again, that there is a significant interaction effect. As a rule of thumb, the numbers in the *p*-value column are compared to .05 to decide on the significance of an effect. Small *p* values (less than .05) indicate a significant effect. Values larger than .05 indicate that the magnitude of *F* can be explained by chance variation.

The completed ANOVA in Table 10.8 indicated that the two-way interaction between resin and moisture is highly significant. Of course, this conclusion is based on a comparison of the variation due to the interaction effect with a measure of inherent process variation. The ANOVA table indicates that moisture and resin have a significant impact on density levels, but the table does not describe the nature of the response. Plots such as the one in Figure 10.5 are also needed to describe the way density changes with changing moisture and resin.

Another important point about the ANOVA table is that the data used to construct the table were, as is almost always the case, gathered in a fairly short time frame and under a certain set of conditions. Whether the results found in this (or any other industrial experiment) will hold in practice and under a variety of conditions will need to be verified with validation runs conducted under diverse, practical operating conditions. *F* values and *p* values are based on exact mathematical models, which are not exactly achieved in processes. These values, and the mean squares from which they are determined, will vary from one experiment to another one on the same process. Although the *F* values and *p* values provide support for identifying significant effects in an experiment, they should not be interpreted as exactly describing process results. For example, a large *F* value does not guarantee that experimental results can be replicated at another time, possibly under different conditions.

TABLE 10.8 *ANOVA Table for the Density Experiment*

Source	df	SS	MS	F	p Value
M	1	211.250	211.2500	597.600	.0000
R	1	51.200	51.2000	144.840	.0000
M $\times$ R	1	30.752	30.7520	86.990	.0000
Residual	16	5.656	.3535		
Total	19	298.858			

TABLE 10.9 *Computation of Sums of Squares from Effect Estimates*

Effect	Effect Estimate	Sum of Square
M (moisture)	−6.50	$5(-6.50)^2 = 211.25$
R (resin)	3.20	$5(3.20)^2 = 51.20$
M x R	−2.48	$5(-2.48)^2 = 30.752$

10.4 ADDITIONAL EXPERIMENTAL STRATEGIES

Section 10.3 describes one of the more simple and straightforward types of experiments. Yet even such a simple example provided an illustration of some of the main features of experimental design. The example also provided an illustration of the way in which previously developed process knowledge forms the basis of experimental design work. In particular, the 2^2 experiment examined in the previous section allowed a consideration of

- the necessity of understanding factor interactions,
- the management of other, influential factors when conducting experiments,
- the role of randomization in conducting experiments,
- the use of graphing procedures for describing experimental results,
- the statistical comparison of factor effects with mean squared error, and
- the need to confirm, in practice, experimental results.

However, the experimental strategies of complete randomization and of conducting runs for all possible combinations of factors cannot be generalized to all experimental conditions encountered in an industrial setting. In practice, experimental strategies will need to be expanded in several ways. Although a complete discussion of different types of experiments useful in industry is beyond the intended scope of this chapter, a brief description of other topics in the area of designed experiments is presented in order to provide direction for further study in this area.

10.4.1   **Studying Many Factors with a Small Number of Experimental Runs**

It is often the case, particularly in the early stages of the development of process knowledge, that investigation of a long list of possible factors affecting a critical process characteristic is required. For example, in the investigation of the gloss obtained on paper run through a supercalender, we wanted to study the effect on gloss of the following seven factors:

1.   moisture content,
2.   speed,
3.   temperature 1,
4.   temperature 2,
5.   roller pressure,
6.   steam pressure, and
7.   machine.

If two levels of each factor are to be investigated and the same approach is used for deciding on the number of runs as was used in the previous experiment with two factors, there would be a total of $2^7 = 128$ different experimental runs to be performed. Although the number of values for each included factor is only two, just one replication of all combinations of the seven factors requires a prohibitively large number of runs. Fortunately, it is not necessary to complete all of these runs to discover useful process information. An often-used approach is to perform some fraction of all possible runs (generally without replication) in order to identify which of the factors seem to have the biggest impact on the response variable (gloss in the previous example). Once the variables that have the largest impact have been identified, further experimentation on this smaller set of factors may be performed, if required. The type of designs employed in this kind of situation are often referred to as *fractional factorial designs*.

A primary reason that fractional factorial designs can serve to answer required process questions is that doing all possible runs (a *full* factorial experiment) for a large number of variables provides information for some effects that in all likelihood are not useful. Specifically, the capability to describe effects that are higher order interactions is often not worth the cost or time. When studying a small number of variables, say, two or three, a sufficient number of runs should be conducted to examine the two-factor interactions. In many processes, two-factor interactions have a strong effect and process improvement requires their recognition and use in process setup and control strategies. When experimenting with three variables, say, A, B, and C, the number of interactions grows two ways. There are now more two-factor interactions (AB, AC, and BC) and one three-factor interaction, ABC. Any of these interactions could have an effect on the process and needs consideration in completing process knowledge. When an experiment involves changing even more variables, there exists the possibility of observing more interactions as well as more higher order interactions. It is the higher order interactions, three and higher, that are thought of as providing superfluous information, at least for some processes and for some experimental objectives. With the agreement that it is not necessary to study all possible interactions, fewer runs become statistically feasible.

Fractional factorial experiments are experiments that deliberately complete only a part, a fraction, of all possible runs. When such an experimental strategy is suggested, the experimenter must consider two critical issues:

1. *What happens to the integrity and quantity of data, and thus information, provided by fractional factorials?* Because only a fraction of all possible runs is completed, the integrity of data results is compromised. Something is gained (time and cost) but something is lost (complete information on all effects). When fractional factorial experiments are conducted, some effects are "confounded" with other effects, meaning that with the data generated from the experiment, it will not be possible to separate the effects on the response of different individual factors from the effects created by interactions of other factors.

2. *How does data analysis change with the use of fractional factorials?* Data analysis for the fractional factorial experiment proceeds in much the same fashion as for the full factorial with the understanding that some effects are confounded with other effects. The plots and effect estimates described in connection with the 2^2 experiment can be used to describe those effects of interest.

10.4.2   Experimenting Without Complete Randomization

The word "design" in the statistical design of experiments has a particular definition and use. Specifically, it refers to the planned number of runs, the run sequence, the randomization pattern, other conditions on testing the recommended factors, and the particular combinations of factors being tested. The pattern in the data collection must be recognized by the experimenter since correct data analysis will need to take this pattern into account. Complete randomization, as previously discussed, means that all combinations of factors to be run are written and then randomized by some random mechanism. That approach is not possible in all circumstances. For example, it is often true that one or more of the experimental factors is very expensive or time consuming to change. In such circumstances, practicality dictates that complete randomization not be done. Instead, a statistical design must recognize the restriction on the factor or factors. Recognition of the restriction means that the design must address how to complete the runs inside the restriction and how to analyze appropriately the data collected from the experiment.

Experiments in which one or more factors are changed less frequently than for a completely randomized design are called *split-plot experiments*. As a practical matter, most industrial experiments are of this type. It is important to recognize that such circumstances occur, to know that valid experiments can be conducted with the imposed restriction, to know that there are recommended data analysis procedures available, and that these procedures can be learned and practiced by individuals having quantitative skills.

10.4.3   Experimenting in the Presence of Uncontrolled Variation

Many statisticians recommend that industrial experimentation only be done when the process under investigation is in control. The reasoning behind this judgment is that if the process knowledge of how to run the process consistently does not exist, it is not likely that the knowledge of how to apply the results of an experiment will be in place either. However, processes are sometimes out of control because of the effects of known, but uncontrollable (and possibly unmeasureable), variables. An example would be a process that uses wood products whose properties vary. In such situations, it may be necessary to optimize an output characteristic in the presence of these changing conditions. In other situations, a process may run consistently for a period of time, say, a shift, a week, or even longer. But, at times, conditions and circumstances change because of effects created by known or unknown variables. Maybe the result is that the process average shifts to a new level. In many of these circumstances, it is possible to conduct productive experiments and determine the effect of some variables, although in the long term the process is not stable.

Several experimental approaches might be considered in these types of situations. Three such approaches are

1. block designs,
2. analysis of covariance, and
3. robust designs.

An example from the manufacture of circuit boards is provided next to illustrate processing situations for which the preceding designs are useful.

An experimental objective is to learn about the effects that changes in flux type, solder temperature, and conveyor speed have on the average number of defects on circuit boards. By knowing the individual and collective effects of these three factors on defects, it may be possible to improve the process by selecting and setting values for these factors that will result in fewer defects than are now observed. This experimental objective sounds much like previous ones discussed. However, in the present instance, the environment and processing circumstances complicate the objective in very practical ways.

We need to determine the relative magnitude of these effects in an active and ongoing environment where there are also effects on the number of defects due to the following:

1. *Controlled variables that are not be included in the experiment.*
2. *Line or plant variations created by uncontrolled variables.* For example, purity of materials, setup conditions, people (style and experience), schedule (mix and sequence), and other, similar variables, affect the average number of defects per board in different ways and with different severity from time to time. The state or condition of these variables is not always well known. It is not possible to predict when and how they become active and influential. There is little confirmed knowledge regarding the specific effects which these factors, individually or collectively, have on board defects, either separate from or interacting with the managed process variables, although results are fairly stable once a run is "broken in."
3. *Unknown or unconsidered sources.* It is known that "conditions" seem to last for fairly lengthy periods of time, varying from a day or two up to a week or so before conditions change. All of these, in very practical ways make up the environment in which the line runs.

The last two categories create concern when conducting experiments having valid results. The experiment will have to be run in this environment because results will be expected to be applied to this line in these circumstances. It is known that the average number of defects changes in unexplained ways from run to run, but it is thought that specific values for flux type, solder temperature, and conveyor speed can be found that tend to result in the fewest number of defects, regardless of which stage, step, or circumstance of the "conditions" are active. This knowledge would be useful in identifying process refinement possibilities. And by conducting the experiment within the reality of the environment within which the line runs, experimental validity is improved.

A run plan will have to be defined and tested for responsiveness not only to the above experimental objective but also for process practicality. The analysis of the data from the experiment must take into account the specific nature of the run assignment. It may be that settings of flux type, solder temperature, and conveyor speed can be used to make the process somewhat robust with respect to the collective effect of these nameless but real variations. It would be useful to find an operating value for each of the three variables that would tend to result in a small number of defects no matter what conditions prevail due to other conditions. Statistical design would propose the following run plan: Within a fixed period of time, during which it is thought that the environment will be relatively homogeneous, conduct a complete set of randomized runs in the three variables and record the results as usual. If each of the three factors is to be tested at two levels, there will be eight runs in one complete replication. These experimental runs are then repeated (with a different randomization) at another time during which it is thought that environmental variables are likely to be different from what they were in the first period, but will be relatively homogenous in their behavior throughout this second selected time period. Absolute homogeneity, of course, is not possible nor is it needed. Relatively homogeneous means that during a shorter period of time the environment remains quieter than if a considerably longer period of time were considered.

Recommended data approaches are available for these designs. This experimental approach is called *"blocking,"* an accurate and descriptive name. The image of a "block" captures the idea of relative homogeneity over a set of conditions or circumstances. All planned runs, eight in this example, are randomized within the block. There would be as many blocks as the user group thought necessary to detect differences across the conditions and circumstances represented by different blocks.

Time has been used as the blocking variable in the previous example. But the blocking idea is much more general. The block for example could be material lots, equipment configurations, or other conditions defying simple description as to cause and effect. Blocking can also be used in other ways by experimenters. Results and learning from an experiment are strengthened if the results can be replicated across several different environments that differ in various ways, perhaps known in some respects and unknown in others.

Another tactic being used more and more frequently is for experimenters to construct artificial environments that represent extreme sets of conditions for variables that cannot be managed. Product or process factors are then tested in these respective environments. The objective of the data analysis is then to select values of these factors that make the experimental results impervious or "robust" with respect to the effects of these uncontrollable environmental variables. This type of an experimental strategy is called *robust design*.

Another processing circumstance that is often encountered by experimental users is one in which we need to know the effects of one, two, or more factors on one or more response variables but a process variable, known to be influential in affecting the response, is not controllable and will not be controlled during actual operation. This variable is known to change values in the short term and in fact will be different for every experimental combination tested. What distinguishes this situation from the earlier ones discussed is that the variable in question can be measured, even though it cannot be controlled. For example, it may be known that a specific material property, which can be measured, affects the results of the response variable and it is known that this material property changes significantly over short periods of time. We need to learn the effects on the response of identified, experimental factors although the material property will also vary and thus create variations in the response variable. Experimental design procedures recognize this situation, state a suggested design methodology to take the extraneous variable into account during the course of the experiment, and then provide means to statistically adjust out the effects of the variable and proceed with the analysis of the experimental variables in determining their effect on the response variable or variables. The statistical term for this type of experimental analysis is *analysis of covariance*.

10.5   SUMMARY AND DESIGN REFERENCES

The intent of this chapter was to illustrate the integration of the use of designed experiments in an industrial setting with the techniques for process study discussed earlier in the book. Both techniques provide useful methods for studying and improving the management of industrial processes. Experiments in industry are likely to be ineffective when the design of those experiments is not informed by the kind of process knowledge generated from ongoing process study. Each step in designing an experiment rests on a foundation of process knowledge. In particular, the following steps in designing an experiment have been discussed:

1. Define experimental objectives.
2. Select response variable(s).
3. Specify how and when the response(s) will be measured.
4. Determine experimental factors.
5. Select factor levels.
6. Determine the number of replications required.
7. Decide on the run order of the experiment.
8. Decide on how to manage other influential factors.

The required process knowledge for completing these steps has also been described. Experimental design serves as a complement to other process improvement work. Designed experiments can be used to

1. confirm the importance of certain process variables,
2. identify variables that must be managed to different tolerances, and
3. support the creation of knowledge about the multivariate generation of variation.

The overview in this chapter on designed experiments is clearly insufficient for confidently knowing how to design and analyze an industrial experiment. The books in the following list provide good introductions to the study of designed experiments. The ordering of the list of books corresponds to a suggested order in which these books might be studied.

1. Box, George E. P., Hunter, William G., and Hunter, J. Stuart. *Statistics for Experimenters*. New York, John Wiley & Sons, 1978.
2. Moen, Ronald D., Nolan, Thomas W., and Provost, Lloyd, P. *Improving Quality through Planned Experimentation*. New York, McGraw-Hill, 1991.
3. Anderson, Virgil L., and McLean, Robert A. *Design of Experiments: A Realistic Approach*. New York, Marcel Dekker, 1974.
4. Hick, Charles R. *Fundamental Concepts in the Design of Experiments*, 4th ed. New York, Saunders College Publishing, 1993.
5. Montgomery, Douglas C. *Design and Analysis of Experiments*, 3rd ed. New York, John Wiley & Sons, 1991.

10.6   PRACTICE PROBLEMS

10.6.1 In a batch process, one of the critical characteristics is yield. The operation under study involves a biochemical process and past yields have been erratic. During recent processing of one such batch, the size tank typically used in step 2 of the process was not available so a larger tank was used. The yield for that batch was markedly higher than that which had been experienced in the past. Whether or not the change in the tank size was the reason for the increased yields was not clear. However, speculation about why the change in tanks might have increased yields identified two possible reasons for such an increase:

1. The larger tank might have meant that the agitation occurring during this phase may have increased because of the additional space in the tank.
2. The growth in the biological mechanism occurring at this stage may proceed in a more efficient fashion with the additional air space provided by using a larger tank.

In an attempt to isolate which, if either, of these causes might produce increased yields, a small experiment was run. Two different agitation speeds and two different tank sizes were investigated in this experiment. The yields at each of the four combinations of speed and tank size settings are provided in the table:

Speed	Tank Size	Yield
Low	Small	87.2
Low	Large	88.4
High	Small	88.6
High	Large	90.9

a. Does yield appear to be affected by agitation speed and/or tank size? Construct an interaction plot to describe the effects of the two factors on batch yield.
b. In view of the fact that yields on the process have been reported to be erratic, what kind of follow-up work would you suggest on this process?

10.6.2 The experimental data provided below were generated from an experiment conducted to better understand a reaction in drug manufacturing. The objectives of this experiment were to discover process settings that could be used to increase yield and manage reaction time.

Two response variables were identified for the experiment:

   $Y1$, yields as a percentage of theoretical yield
   $Y2$, Reaction time minus 200 minutes (as judged by FTIR).

Two process factors thought to have an impact on these two responses were the mole ratio of compound A to compound B and the addition time of chemical T. Initially, two levels of these two factors were to be investigated:

Factors	*Levels*	
1. Mole ratio A/B in initial solution (R)	.8(−1)	1.2(+1)
2. Addition time in minutes of a chemical (T)	20(−1)	90(+1)

Temperature and catalyst quality were held constant for all experimental runs. Each factor combination was replicated two times. The responses recorded at these runs are given:

Observation	R	T	Y1	Y2
1	−1	−1	61.8	38
2	−1	−1	63.2	43
3	−1	+1	64.4	70
4	−1	+1	65.8	77
5	+1	−1	77.9	41
6	+1	−1	73.1	33
7	+1	+1	74.2	75
8	+1	+1	77.4	69

a. Does variation appear to be consistent at the different levels of the two process factors?

b. What effect, if any, do mole ratio and addition time have on yield? On reaction time? Draw appropriate plots to support your conclusions.

c. Are the effects noted in part b statistically significant?

d. One of the controlled factors, temperature, often varies by as much as 30 degrees in process operation. What statements can be made about the validity of your conclusions in the presence of this variation in temperature?

e. What actions would you recommend based on the results of this experiment?

Appendix A:
Probability Models

Much of the statistical analysis of data relies on the use of theoretical models. These models are used for comparing observed behavior to behavior that might be expected if the experimental situation could be described by the chosen theoretical model. Keep in mind that any model used in this fashion is only that, a model, and cannot be expected to mimic reality. However, such models do serve as useful benchmarks for evaluating process behavior. Deciding whether a model is useful relies on understanding the assumptions on which the model is based and then determining whether the experimental situation is adequately described by these assumptions. The three sections of this appendix describe three different models that are used extensively in industrial statistics applications. Descriptions of the models and discussion about the appropriate use of the models are provided.

A.1   THE NORMAL PROBABILITY MODEL

The normal probability model is a widely used model for describing variables data, not just in industrial applications, but throughout most sciences. Its broad usefulness derives from the following two characterizations of this model:

1. It describes the distribution of measurements subject to numerous, small, independent sources of variation,
2. As discussed in Section A.1.4, it can be mathematically shown that the distribution of sample means, for large enough sample sizes, has approximately a normal distribution.

Because of its wide applicability, the normal model serves as a basis for the control chart constants used to construct R and X-bar charts as well as s and

X-bar charts (Chapter 5, Section 5.3). It is also often used as a basis for describing process capability (Chapter 5, Section 5.4). On the basis of the large sample properties described in point 2, above, three standard deviations are used for limits on most control charts. The following sections describe the properties of the normal probability model, how this model is used for characterizing distributions, and why the normal model can be used for describing large sample properties of sample averages.

A.1.1   Characterizing the Normal Probability Model

As stated earlier, the normal probability model is used to describe the behavior of variables data. Normally distributed data can assume infinitely many values on a line interval. The approach used to assign a probability of occurrence to these numbers is similar in concept to the frequency histogram discussed in Chapter 5. A smooth curve is drawn to represent the probability distribution. Areas under this curve between two values, say, *a* and *b*, represent the proportion of measurements that would be expected to fall between the numbers *a* and *b*. The following formula provides a mathematical description of a normal probability curve:

$$f(x) = \frac{1}{\sqrt{2\pi\hat{\sigma}^2}} \, \exp\left(-\frac{(x - \mu)^2}{2\hat{\sigma}^2}\right)$$

This curve has been plotted in Figure A.1. The total area under the curve is one, and areas under portions of the curve correspond to probabilities. For example, the area under the curve between $\mu - 1\sigma$ and $\mu + 1\sigma$ is 0.68. Thus, if a distribution of measurements is normally distributed, the probability of choosing a value at random from that distribution and having it fall between $\mu - 1\sigma$ and $\mu + 1\sigma$ is 0.68. Stated somewhat differently, for a set of normally distributed data, about 68% of the measurements will lie between $\mu - 1\sigma$ and $\mu + 1\sigma$. About 95% of measurements from a normal distribution will fall between $\mu - 2\sigma$ and $\mu + 2\sigma$ and almost all or 99.7% between $\mu - 3\sigma$ and $\mu + 3\sigma$.

From the equation and graph of the normal curve, note that three pieces of information are used to specify normally distributed values:

1. the shape of the distribution, as described by the curve, *f*(*x*),
2. μ, the mean of the distribution, and
3. σ, the standard deviation of the distribution.

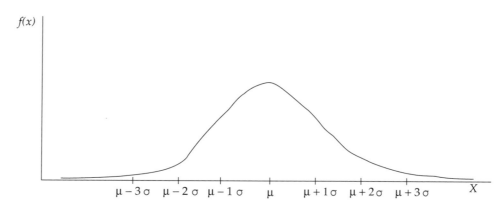

FIGURE A.1 *Normal Probability Curve*

A.1.2   Tabulated Areas of the Normal Probability Distribution

Determining proportions or percentages for normally distributed data requires finding areas under the normal curve. The equation for the normal distribution is dependent on the numbers μ and σ. If tabulated areas under normal curves were to be created for every possible value of these two numbers, an infinite number of tables would be required. A different approach is taken; one table is developed that describes areas under a normal curve lying within a specified number of standard deviations from the mean. The resulting table is called the *standard normal table*. This is Table D.1 in Appendix D.

Before describing how to use this table, a few pieces of information about the normal distribution will be useful.

1. The normal distribution is symmetrical about its mean. Consequently half of the area under the curve lies to the left of the mean and half to the right.
2. Because of the symmetry of the normal curve, the table of areas is simplified by only listing areas between the mean and a specified number, z, of standard deviations to the right of the mean. An area to the left of the mean can be determined by calculating the corresponding and equal area to the right of the mean. Figure A.2 illustrates these two equal areas.

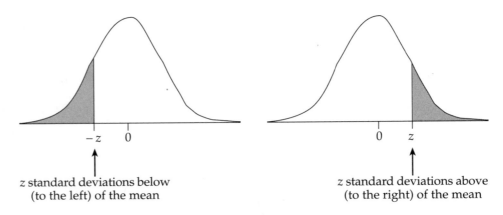

z standard deviations below
(to the left) of the mean

z standard deviations above
(to the right) of the mean

FIGURE A.2 *The Symmetry of the Normal Curve*

The information in Table D.1 is given in the form of the proportion of values from a normal distribution that lies more than z standard deviations from the average, where z is a number between –4.0 and +4.0. Note that z, correct to the nearest tenth, is recorded in the left-hand column (Figure A.2). The second decimal place for z, corresponding to hundredths, is given across the top row. To illustrate the use of the table, suppose that a process characteristic has been shown to be in control, is approximately normally distributed, has an estimated mean of 34.35, and an estimated standard deviation of 2.27. We want to determine the percent of material produced by the process whose value for this characteristic would lie above 40. The following steps explain how to determine this percentage:

Step 1

> Sketch a normal distribution and label it. In particular, locate the estimated process average at the center of the curve and the value of 40 at the appropriate point on the vertical axis. Shade the area under the curve to the right of 40. Note that this corresponds to the values that are larger than 40. For the current example, this curve is displayed in Figure A.3.

Step 2

> Calculate z, the number of standard deviations that 40 is from the mean

$$z = \frac{40 - \hat{\mu}}{\hat{\sigma}} = \frac{40 - 34.35}{2.27} = 2.49$$

It is useful to include this information on the normal curve drawn in step 1.

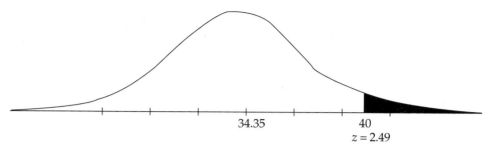

FIGURE A.3 *An Example of a Normal Curve for Computing Probability*

Step 3

Use the table of the standard normal distribution to find the proportion of material above 40. The z value for 40 is a positive number. The proportion of observations that would be more than +2.49 standard deviations from the mean is found by finding the proportion associated with this z value in the table. In particular, in the column labeled $|z|$, find the number 2.4. (Note that 2.4 represents the integer and tenths place of the number 2.49.) The number in this row that appears under the value x.x9 will give the proportion of observations larger than 2.49. (Note that the x.x9 column is used since 9 is the number in the hundredths place of 2.49.) The number found in the z table is .0064. So the percentage of measurements that could be expected to fall above 40 would be .0064 $\times$ 100% = .64%.

A.1.3 Examples Illustrating the Use of the Normal Distribution

In the two examples that follow, the processes being characterized have been shown to be operating in control, and histograms constructed from process data appear to have an approximate normal distribution.

Example 1

For a process with an estimated mean of 3.6 and an estimated standard deviation of .6, what percentage of measurements will be smaller than 2.8?

The idealized normal curve is contained in Figure A.4. It should be noted that the reported process mean, 3.6, is at the center of the curve. The curve approaches the horizontal axis at about 3 standard deviations away from this average, or in this example, at about 3.6 – 3(.6) = 1.8 and 3.6 + 3(.6) = 5.4.

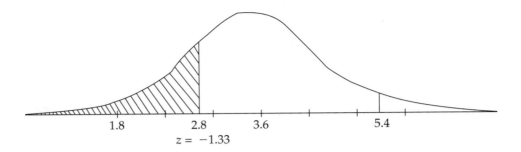

FIGURE A.4 *Normal Curve for Example 1*

The z value for the number 2.8 is found by:

$$z = \frac{2.8 - \hat{\mu}}{\hat{\sigma}} = \frac{2.8 - 3.6}{.6} = -1.33$$

The area under the normal curve to the left of –1.33 has been shaded in Figure A.4. Looking up the z value of 1.33 in the normal table results in a value of .918. So the percentage of measurements smaller than 2.8 is 9.18%.

Example 2
For a process with an estimated mean of 1180 and an estimated standard deviation of 83, what percentage of measurements will be within 1000 and 1400?

The z values for 1000 and 1400 are:

$$z_L = \frac{1000 - \hat{\mu}}{\hat{\sigma}} = \frac{1000 - 1180}{83} = -2.17$$

$$z_U = \frac{1400 - \hat{\mu}}{\hat{\sigma}} = \frac{1400 - 1180}{83} = 2.65$$

Looking up the z values for –2.17 and 2.65 in the normal table results in values of .0150 and .0040, respectively. Therefore, the two areas shaded in Figure A.5 represent a total area of .0150 + .0040 = .0190. The unshaded area in Figure A.5 is the proportion of measurements that falls between 1000 and 1400. Since the total area under the curve is 1, this area is 1.0 – .0190 = .9810. In other words, the percentage of measurements that lies between 1000 and 1400 is 98.1%.

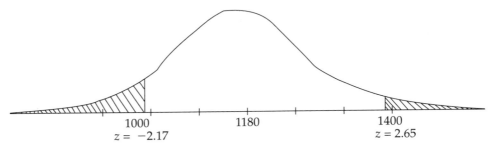

FIGURE A.5 *Normal Curve for Example 2*

A.1.4   The Central Limit Theorem

The central limit theorem states that under fairly general conditions, the means of random samples drawn from any distribution will have an approximately normal distribution. This is a very powerful statement and supports the practical application of normal probability theory to the analysis of process data. Although the proof of this theorem is beyond the scope of this text, the following example illustrates the result. For this example, the distribution from which a sample is to be drawn will be a very simple one; this distribution is generated by rolling a die. When rolling a die, one of six possible outcomes might occur: a 1, 2, 3, 4, 5, or a 6. Because each of these outcomes is equally likely to occur, the probability of any one of the outcomes is 1/6. A probability distribution for this experiment has been drawn in Figure A.6. The possible results of the experiment are marked on the horizontal axis and a scale for probabilities associated with these outcomes is drawn on the vertical axis. The probability associated with each outcome is plotted. The points which describe the probabilities associated with the six possible outcomes have been connected by lines.

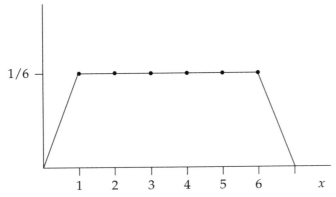

FIGURE A.6 *Probability Distribution for the Die Rolling Experiment*

Note that the probability of observing, say, a 2 when rolling a die is described by the fact that the distance of the point above the number 2 is $1/6$.

Now suppose that a sample is generated by rolling a die two times. The possible samples that could occur, along with the sample means for these samples, are listed in Table A.1. A probability distribution of the sample averages is drawn in Figure A.7. Thirty-six different samples might occur. A sample mean of 2.5 occurs in 4 of these 36 samples. In other words, the frequency or probability of a sample mean of 2.5 occurring in two rolls of a die is $4/36$. So the point above the number 2.5 in Figure A.7 has a height of $4/36 = 1/9$. The rest of the distribution was constructed using the same reasoning.

TABLE A.1 *Possible Samples from Rolling a Die Twice*

Sample	X_1	X_2	$\bar{X}$	Sample	X_1	X_2	$\bar{X}$
1	1	1	1.0	19	4	1	2.5
2	1	2	1.5	20	4	2	3.0
3	1	3	2.0	21	4	3	3.5
4	1	4	2.5	22	4	4	4.0
5	1	5	3.0	23	4	5	4.5
6	1	6	3.5	24	4	6	5.0
7	2	1	1.5	25	5	1	3.0
8	2	2	2.0	26	5	2	3.5
9	2	3	2.5	27	5	3	4.0
10	2	4	3.0	28	5	4	4.5
11	2	5	3.5	29	5	5	5.0
12	2	6	4.0	30	5	6	5.5
13	3	1	2.0	31	6	1	3.5
14	3	2	2.5	32	6	2	4.0
15	3	3	3.0	33	6	3	4.5
16	3	4	3.5	34	6	4	5.0
17	3	5	4.0	35	6	5	5.5
18	3	6	4.5	36	6	6	6.0

The curve drawn in Figure A.7 is referred to as the sampling distribution of the sample mean. It describes the frequency with which various possible values of means from samples of size two are likely to occur. This curve looks similar to that of a normal curve. Figure A.7 also contains a histogram of the sampling distribution of the mean for samples of size three. It should be noted that as the sample size gets larger, the histograms of the sampling distributions of the means begin to look more and more like the normal curve drawn in Figure A.1.

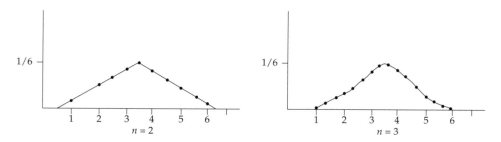

FIGURE A.7 *Sampling Distributions for the Sample Mean*

This simple example illustrates that when sampling from a distribution as pictured in Figure A.6, even for samples as small as size two, a normal distribution can serve as a useful model of the distribution of sample means. The remarkable fact about the central limit theorem is that this result does not depend on the shape of the distribution from which the sample is drawn. Figure A.8 contains two different distributions from which samples are taken. The sampling distributions of sample means for samples of size two, five, and thirty are also provided in Figure A.8. In both cases, for samples of size five and thirty, histograms of the sampling distributions resemble a normal distribution.

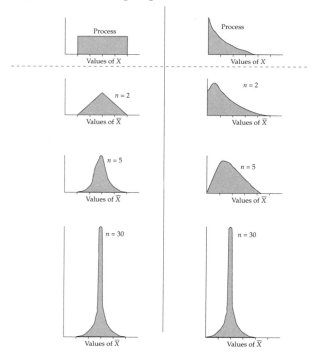

FIGURE A.8 *Illustrations of the Central Limit Theorem*

A.2   # THE BINOMIAL PROBABILITY MODEL : A THEORETICAL BASIS FOR *p* AND *np* CHARTS

A.2.1   ## The Coin Tossing Experiment

An experiment in which a coin is tossed repeatedly, say, *n* times, and the number of heads observed in the *n* tosses is counted provides an illustration of a binomial experiment. The elements of this experiment that make it a useful example of a binomial probability model are as follows:

1. The same operation, tossing a coin, is repeated *n* times. The *n* trials (tosses) are said to be independent since the outcome on one toss does not affect and is not affected by the outcomes on the other tosses.
2. Each operation can result in one of two outcomes, heads or tails. (The prefix bi- refers to the fact that there are two possible outcomes.) The probability of observing a head on any one trial remains the same from trial to trial. If the coin is a fair one, the probability of heads on any one trial would, of course, be $P = 1/2$. Typically, the outcome of interest ("heads" in this situation) is called a "success."
3. Of interest in the experiment is the number of heads occurring in the *n* trials.

A binomial experiment has all three elements as its theoretical basis. In short, these elements are

1. *n* independent trials
2. each trial may result in one of two outcomes, "success" or "failure." The trials are identical in the sense that the probability of success, *P*, stays the same from trial to trial, and
3. the number of successes in the *n* trials is counted.

A binomial experiment might be considered as a possibility for describing the results of a basketball player shooting 20 free throw shots over the course of a ball game. The three elements of the binomial model will be examined in order to determine whether these elements apply to the physical situation of shooting free throws.

1. The number of trials is $n = 20$. However, before using the binomial model it will be necessary to decide whether the assumption of independent trials is reasonable in this situation. If it is thought that the basketball player might be less likely to make subsequent shots if she misses some of the first shots, then the assumption of independence is not valid. Similarly, if several of the early shots result in "success" and so the player gains confidence and is thus more likely to make subsequent shots, the trials cannot be thought of as independent. In this case the assumption of independence would mean

that the likelihood of success on later trials is not affected by the results of the earlier trials.

2. The probability of success on the trials, *P*, remains the same from one trial to the next. In this case, this assumption would amount to assuming that the player does not benefit in later shots from the practice obtained in the early shots. Nor does the player get tired as she shoots the 20 shots. The chance of her making any one shot stays the same throughout the experiment.

3. The player, coach, teammates, and spectators are interested in the number of shots the player makes in the 20 trials.

A.2.2   The Binomial Model Applied to *p* and *np* Charts

The binomial model is the basis on which the control limits for *p* and *np* charts are calculated. Whether attributes data are to be plotted on a *p* or an *np* chart, the physical situation in both cases is that a subgroup of *n* items is selected from a process and the number of nonconforming items ("successes") is counted. If a chart includes, say, *k* = 20 subgroups each of size *n*, then 20 separate binomial experiments are being considered. For each one of these experiments, there exists the underlying assumption that the three elements of the binomial model provide an adequate description of the data collection situation. It is worthwhile to examine this assumption in the present case:

1. There are *n* independent trials. For the attributes data being considered, *n* would refer to the subgroup size. (Keep in mind that, at this point in the discussion, only the results expected in one subgroup are being discussed.) The assumption of independence must be considered. Suppose that in the physical situation under study, nonconforming items occur in a systematic pattern; maybe for some processing reason the process produces a string of five nonconforming items followed by several hundred conforming items. Thus, if it is known that the first five items in a subgroup of *n* = 100 successive items are nonconforming, it is highly unlikely that any of the remaining 95 items are nonconforming. Consequently, the assumption of independence in the 100 trials is not valid.

   The binomial model will best describe situations in which there is not a systematic pattern in the nonconforming items produced by a process; rather the nonconforming items occur in a random, nonsystematic fashion.

   In the preceding statement, it is not stated that a *p* or *np* chart would not be used since the assumptions of binomial model are violated. The systematic pattern described may not be known prior to data collection. Thus, a *p* chart might be constructed and then a peculiar pattern on the chart would provide evidence of the systematic behavior described above.

2. The probability, *P*, of an item being nonconforming (a "success" in the language of probability) stays the same from trial to trial. Suppose a subgroup contained items from two or more machines and the machines produced quite different levels of nonconforming items. Then, the chance of an item being nonconforming would be different depending on the machine from which the item came. Stated differently, the probability of an item being nonconforming would not be the same from one item to the next.

    In planning data collection for a study, we recommend that the preceding situation be avoided. Subgroups should be collected so that the material within a subgroup is as homogeneous as possible. If this strategy is not followed, the use of the binomial model for constructing a *p* or *np* chart is questionable. However, there is a more compelling, process-based, reason for not including, say, separate machines within the same subgroup; if different subgroups contain items from different machines, then differences between machines are not as easily detected than if the items from different machines are in separate subgroups.

3. For constructing either a *p* or an *np* chart, the number of nonconforming items in the subgroup of size *n* are counted.

A.2.3   Assigning Probabilities to Outcomes of a Binomial Experiment

The limits on a *p* or *np* chart (as with any chart) are determined based on what is likely to happen under a certain set of assumptions. For the binomial experiment, the probability of observing *r* successes in *n* trials can be calculated from the following formula:

$$P(r) = \frac{n!}{r!(n-r)!} P^r (1-P)^{n-r} \qquad \text{for } r = 0, 1, \ldots, 20 \quad (A.1)$$

where $n! = n(n-1)(n-2) \ldots 3.2.1$ (Note that 0! is defined to be 1.)

For example, suppose that the basketball player will shoot *n* = 20 free throws, and that the probability of hitting any one shot is *P* = 0.8 for each shot. Then, the probability that she will hit 15 of the 20 throws is:

$$P(15) = \frac{20!}{15!15!} (0.8)^{15} (0.2)^5$$

$$= 15,504(0.8)^{15}(0.2)^5$$

$$= .1746$$

TABLE A.2 *Binomial Probability Distribution for* n = 20 *and* P = .8

r	p	P(r)	r	p	P(r)
0	0	.0000	11	.55	.0074
1	.05	.0000	12	.60	.0221
2	.10	.0000	13	.65	.0546
3	.15	.0000	14	.70	.1091
4	.20	.0000	15	.75	.1746
5	.25	.0000	16	.80	.2182
6	.30	.0000	17	.85	.2053
7	.35	.0000	18	.90	.1369
8	.40	.0001	19	.95	.0577
9	.45	.0005	20	1.00	.0115
10	.50	.0020			

The probability of observing *r* successes could be calculated for every possible value of *r* between 0 and 20. These calculations have been made and the results reported in Table A.2. This summary of all possible values of the counts of successful free throws and the probability of observing each of the possible values is referred to as a *probability distribution*. The different values of *r* refer to the number of successes that may occur in *n* trials. At times it might be more useful to talk of the proportion of successes in *n* trials. To be consistent with the notation of Chapter 3, the lowercase letter *p* is used to denote this proportion. The possible values for *p* are also listed in Table A.2. Note that if the probability of observing, say, 15 successes in 20 trials is .1746, then the probability of observing $p = 15/20 = .75$ is also .1746.

Another way of displaying the information in Table A.2 would be to draw a histogram of the probability distribution. Such a picture appears in Figure A.9.

A.2.4 Characterizing a Binomial Distribution

The probability distribution in Figure A.9 provides a reference for considering three methods of characterizing a probability distribution. These methods are

1. describing a typical value (or *expected value*) of the distribution,
2. describing the spread (or *variance*) of the distribution, and
3. describing the shape of the distribution.

The mathematical definition of the *expected value* (EV) for the binomial experiment is:

$$EV = \sum_{r=0}^{n} r \times P(r) \quad (A.2)$$

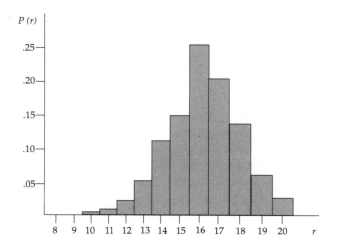

FIGURE A.9 *Binomial Distribution for* n = 20 *and* P = .8

Note that the formula for the expected value simply weights each possible outcome of the binomial experiment by the probability of observing that outcome. Understanding the information about the probability distribution contained in the expected value is aided by noting that the expected value is like a long-run average value. In other words, if the experiment of shooting 20 free throws were repeated a great many times, and at the end of each 20 throws a count of the number of successful shots was made, and then these numbers of successful free throws were averaged, this average would be close to the expected value of the experiment.

The expected value of a probability distribution is also called the mean of the distribution and is denoted by a Greek lowercase mu, μ.

For the binomial experiment it is not necessary to calculate the probability of each possible outcome and then use the preceding formula to find the expected value. It can be shown mathematically that:

$$\text{EV} = \sum_{r=0}^{n} rP(r) = \sum_{r=0}^{n} r \frac{n!}{r!(n-r)!} P^r (1-P)^{n-r} = nP$$

For the experiment of shooting 20 free throws, the expected value of the number of successful shots would be:

$$\text{EV} = nP = 20(0.8) = 16$$

Note that this expected value falls at the center, or balance point, of the probability distribution in Figure A.9.

The *variance* of a probability distribution is a number that captures the spread or dispersion in the likely outcomes of an experiment. The variance (VAR) is mathematically described by the following formula:

$$\text{VAR} = \sum_{r=0}^{n}(r - \text{EV})^2 P(r) \quad \text{(A.3)}$$

The variance of a probability distribution is also denoted by σ^2. The square root of the variance, denoted by σ, is called the *standard deviation*.

An examination of Equation (A.3) provides some insight into how the variance of a distribution describes the spread in a probability distribution. From this equation we can see that the squared distance of each possible outcome of an experiment from the expected outcome is weighted by the probability that that outcome is observed. Thus, if values close to the expected value are likely to occur, the contribution of these values to the variance will be small. On the other hand, if values far from the expected value have a high probability of occurring, the variance will be large. A further description of the information contained in the variance will be provided after some ideas about the shape of the distribution are in place. For now, suffice it to say that large variances indicate a higher probability of observing values far from the expected value than do smaller variances.

The variance of a binomial distribution can be shown mathematically to be:

$$\text{VAR} = \sum_{r=0}^{n}(r - \text{EV})^2 P(r) = nP(1 - P)$$

For the experiment of shooting 20 free throws, the variance of the number of successful throws would be:

$$\hat{\sigma}^2 = \text{VAR} = 20(.8)(1 - .8) = 3.2$$

and the standard deviation, σ, would be:

$$\hat{\sigma} = \sqrt{3.2} = 1.789$$

The third method of characterizing a probability distribution is by the shape of the distribution. The histogram in Figure A.9 provides one means of describing the shape of the distribution for the binomial experiment. However, as discussed in the following section, an important property of the shape of a binomial distribution is that, for sufficiently large n, the shape of the distribution will be close to that of a normal.

A.2.5   ## Approximate Normality of the Binomial Distribution

For large sample sizes, the shape of the binomial distribution will resemble that of a normal distribution. This statement concerning the shape of the binomial distribution can be proven mathematically. Although it is beyond the scope of this text to provide a proof of this result, the result is illustrated by graphing the probability distribution for the number of free throws successfully made in $n = 25$ tries (Figure A.10). Whereas the probability distribution when n was 20 showed a slight left-skew (one which tails off on the left side of the graph), the distribution in Figure A.10 is more symmetric about its mean ($\mu = nP = 20$.) In other words, with increasing sample size, n, the distribution looks more and more like that of a normal distribution. When both nP and $n(1 - P)$ are larger than five, a binomial distribution can be approximated by a normal distribution.

Based on the approximate normality of the binomial distribution, it is possible to provide further descriptions of the binomial distribution in terms of the mean and standard deviation of the distribution. For example, almost all of the measurements (99.7%) from a normal distribution fall within three standard deviations of the mean. For a binomial distribution with a sufficiently large n [i.e., both nP and $n(1 - P)$ larger than 5] this would mean that 99.7% of the time one would expect to observe a value of p in a sample of size n falling between the values:

$$P - 3\sqrt{\frac{P(1 - P)}{n}} \quad \text{and} \quad P + 3\sqrt{\frac{P(1 - P)}{n}}$$

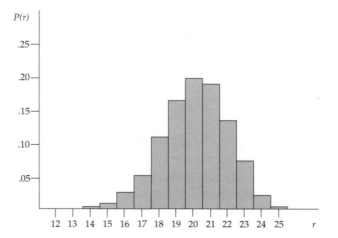

FIGURE A.10   *Binomial Distribution for* n = 25 *and* P = .8

It is this property of the binomial distribution on which the 3σ limits for the *p* and *np* chart are based.

A.3 THE POISSON PROBABILITY MODEL: A THEORETICAL BASIS FOR *c* CHARTS

The Poisson probability model often provides a good method for describing the distribution of the number of rare events that occur infrequently in time or space. For example, a Poisson model may be used to describe probabilities associated with the following counts of occurrences:

* the number of defects in a unit area of glass, textile, or paper,
* the number of insulation breaks in a certain type of wire,
* the number of industrial accidents in a given unit of time, or
* the number of errors a typist makes in typing a page.

The Poisson probability model serves as the basis for constructing *c* and *u* charts. The following section contains a characterization of the Poisson model; the next section describes its use in constructing *c* charts.

A.3.1 Characterizing the Poisson Probability Model

In each of the preceding examples, there is a given unit of inspection, call it A, which is being considered. This unit of inspection may be a specified amount of material (say, a roll of insulated wire) or a specified amount of time (maybe a month.) Counts of occurrences in a specified unit of inspection can be described by a Poisson probability model if the following four characterizations of the mechanisms creating the occurrences are true:

1. The number of occurrences in one unit of inspection, say, A1, and another unit, A2, are independent.
2. If the unit of inspection, A, is divided into many small increments, the probability of observing a single event is approximately proportional to the size of the increment.
3. The chance of observing more than one event in a small increment is negligible.
4. The probability of observing some number of counts in an inspection unit is constant, in other words does not change across time or condition.

For example, if the counts of insulation breaks in a roll of wire followed a Poisson probability model, a practical interpretation of the preceding four characterizations of the process creating wire rolls would be:

1. If there are a large number of breaks on one roll of wire, that does not mean that a large or small number of breaks on the next roll is more or less likely to occur.
2. If a small piece of wire of length t were snipped of the end of a roll, the probability that it would have a defect would be about λt, where λ is some constant. Noted that because any small piece of wire might have a break, there is an opportunity for a great many breaks, although the chance of observing a large number of breaks may be quite small.
3. The chance that this small piece of wire would have more than one break is very small.
4. The number of breaks that one might expect to observe on rolls of wire would not change throughout the time of production.

The probability of observing a count of occurrences, C, to be equal to c when the counts behave as described by the four characterizations just discussed can be calculated by:

$$P(C = c) = \frac{e^{-\lambda}\lambda^c}{c!}, \qquad \text{where } c \text{ might be } 0, 1, 2, 3,\ldots \quad (A.4)$$

Note that the constant λ is the same as the constant defined in characterization 2 above.

Just as with the binomial experiment, it is useful to characterize the Poisson model with an expected value and a variance. The expected number, or expected value, of occurrences in a given unit of inspection, say, A, is $EV = \lambda$, and the variance of the number of counts is $VAR = \lambda$, the same quantity as the expected value.

A.3.2   The Use of the Poisson Model for Charting Count Data

The Poisson probability model is used to describe the expected behavior of counts of occurrences from a process that is operating in a stable fashion. Stable behavior would mean that occurrences happen "at random," in other words, that occurrences are not more likely at some times or locations than at others. For example, suppose that the process for making insulator wire is being evaluated by counting the number of breaks in the wire. An inspection unit is taken to be a roll of wire. Stable process behavior would be taken to mean that there are common cause sources that affect the formation of breaks in the wire, and that these breaks might occur at any time and on any roll. The presence of special causes, those causes of breaks which are present in the manufacture of some rolls and not others, would mean that some rolls would be subject to more or fewer breaks than is typically seen.

Suppose that k rolls of wire are taken from the wire process at k distinct points in time. These k rolls would form k subgroups for evaluating process behavior. The centerline and control limits for a c chart are calculated under the assumption that the process is operating in a stable or consistent fashion. If the process were stable, then the average number of breaks on the 20 rolls would provide an estimate of the expected number (λ) of breaks. This estimate, $\bar{c}$, is the centerline for the c chart. Control limits for a chart are based on the assumption that for a controlled process most observations will fall within three standard deviations (3σ) from the centerline. Since the variance of a Poisson probability model is the same as the expected value, the standard deviation is estimated to be the square root of $\bar{c}$. The resulting upper and lower control limits are as given in Chapter 4:

$$\text{UCL}_c = \bar{c} + \sqrt{\bar{c}}$$

$$\text{LCL}_c = \bar{c} - \sqrt{\bar{c}}$$

The rationale for adding and subtracting three standard deviations from the centerline is that for normally distributed data, almost all of the observations will fall within 3σ of the mean. For large enough values of λ, the shape of a Poisson distribution is very similar to that of a normal distribution. Figure A.11 contains two Poisson probability distributions, one for $\lambda = 5$ and one for $\lambda = 2$. The probability distribution for $\lambda = 2$ is skewed to the right. The distribution for $\lambda = 5$ is also skewed, but not as much as that for $\lambda = 2$. As λ becomes larger and larger, the distribution becomes more and more symmetrical and in fact more and more normal.

The two distributions in Figure A.11 also provide a way of observing the implications of using 3σ limits when the average count ($\bar{c}$) is small. For both distributions, the mean minus 3σ falls below zero, so the probability is zero of observing a count below this value. The value of the mean plus 3σ has been noted on each of the two distributions. As can be seen from these graphs, the chance of observing a count above the mean plus 3σ is very small in both cases, even though the two distributions are not symmetrical. Calculating these probabilities from the probability formula Equation (A.4), results in a probability of about .005 for both distributions. For normal distributions, the probability of observing a value that is larger than the mean plus 3σ is .0015. The fact that the difference in these two probabilities, .005 and .0015, is so small means that even if the expected number of counts is fairly small, 3σ limits still provide a reasonable method of placing control limits on a c chart.

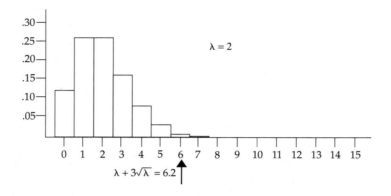

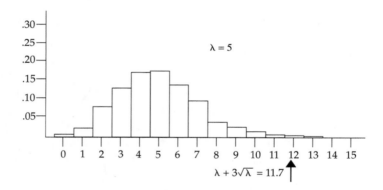

FIGURE A.11 *Poisson Probability Distributions*

Appendix B:
Runs Tests

In addition to points outside of the control limits, a systematic pattern in process data also indicates that special causes of variation are acting on a process. These systematic, nonrandom patterns in a set of time-ordered data are often characterizied by the number or length of "runs" that exist in the plotted data points. A *run* is defined to be a sequence of one or more consecutive data points on the same side of the centerline of a control chart. Runs tests are used to determine whether or not an identified pattern in a data set is likely to be due to special causes of variation. Many different types of runs tests exist to analyze a set of data for various nonrandom patterns. Seven commonly used runs tests are presented in this appendix. However, the simultaneous application of all seven (or even more than two) runs tests on the same set of plotted points makes the chance of detecting a "special cause" when one does not exist too likely to occur; as a consequence, too much time would be devoted to searching for special causes. Given that there is a chance for any single runs test to erroneously indicate a nonrandom pattern, application of multiple runs tests merely increases the chance of drawing a wrong conclusion concerning the stability of a process. To avoid this possibility, this book has recommended the consistent use of one runs test, the Rule of Seven, to all control charts except the moving range and moving average charts. However, just using one runs test to judge process stability is not the same thing as saying that any other patterns in the plotted points should be ignored. Other kinds of systematic patterns, such as a "see-saw" or "up-and-down" pattern in the data, may lead to useful process speculations about the underlying causes generating that behavior. A different subgrouping of the data or further data collection may be undertaken to confirm the existence or not of the speculated source of variation. Knowledge of the process, factors impacting the operation of the process, and a carefully planned data collection strategy should provide the necessary guidance in studying the patterns in a set of data.

Having stated that the use of multiple runs tests is not recommended, it should be noted that the use of multiple runs tests, or rules, is common practice in some places and is often provided by software packages. Since many different runs tests are in common usage, for completeness, these runs tests are described in this appendix.

If the data come from a stable, in-control process, then the plotted points will vary at random around the centerline and within the upper and lower control limits. In fact, since many sampling distributions approach the normal, bell-shaped curve (see Appendix A), a stable, predictable process should generate data points that are within ±3 standard deviations from the average with most of the points randomly falling near to the centerline and a few points scattered closer to the control limits. Thus, many runs tests are based on a division of the space between the control limits into zones that represent distances of ±1 standard deviation, ±2 standard deviations, and ±3 standard deviations, respectively, from the centerline. Figure B.1 illustrates this breakdown of the area between the control limits that is used in several of the following runs tests.

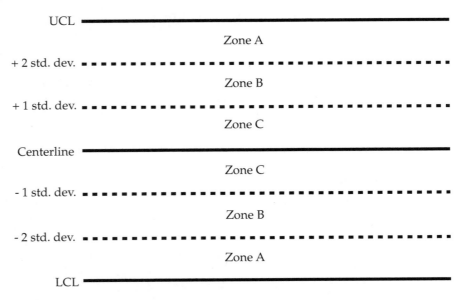

FIGURE B.1 *Zones for Applying Runs Tests*

B.1 SEVEN SUCCESSIVE POINTS IN OR BEYOND ZONE C ON SAME SIDE OF CENTERLINE (OR THE RULE OF SEVEN)

This runs test was introduced in Chapter 3 and has been the one runs test used throughout this book. Whenever seven or more data points fall successively on the same side of the centerline, there is evidence that a special cause is acting on the process. A trend in a process, a gradual change in a process average, or a planned change in order to shift a process average are all possible reasons that one or more runs in the data may contain seven or more points. To implement this runs test:

1. Identify all of the runs in the plotted values.
2. Count the number of points in the longest run.
3. If the number of points in the longest run is greater than or equal to seven, there is evidence of nonrandom influences acting on the process.

Figure B.2 illustrates process values plotted over time in which, based on this test, provide evidence of special causes of variation impacting the process. The last of the four runs in the plotted points is a run of length nine.

This runs test has seven as the critical value for the number of data points in the longest run; i.e., a signal for a nonrandom pattern occurs when the longest run has seven or more data points. Other books or software often suggest that eight be used as the critical value. The test discussed in B.2 also indicates the occurrence of a special cause(s) based on the length of the longest run.

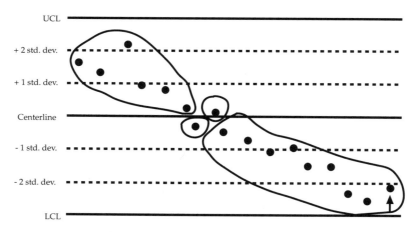

FIGURE B.2   *Eight Successive Points In or Beyond Zone C on Same Side of Centerline*

B.2   TEST FOR LENGTH OF LONGEST RUN

This test for nonrandom, systematic patterns in a set of data is very similar to the test discussed in Section B.1. To implement the test for the length of the longest run:

1. Identify all of the runs in the plotted values.
2. Find the longest run above or below the centerline.
3. Count the number of points in the longest run.
4. Count the total number of points above the centerline and the number of points below the centerline. Let *s* be the smaller of the two counts and *r* be the larger of the two counts.
5. Using Table B.1 and the values of *s* and *r*, find the critical value for the length of a longest run.

6. If the length of the longest run in the plotted points is greater than or equal to the critical value, there is evidence of nonrandom influences acting on the process. Otherwise, there is no evidence of nonrandom influences.

TABLE B.1   *Critical Values for Length of Longest Run*

r / s	5	6	7	8	9	10	11	12	13	14	15	16	17	18	19	20
5	5															
6	6	6														
7	6	6	6													
8	7	7	7	7												
9	8	7	7	7	7											
10	8	8	7	7	7	7										
11	9	8	8	8	7	7	7									
12	10	9	9	8	8	8	8	8								
13	10	10	9	9	8	8	8	8	8							
14	11	10	10	9	9	8	8	8	8	8						
15	11	11	10	10	9	9	9	8	8	8	8					
16	12	11	11	10	10	9	9	9	8	8	8	8				
17	13	12	11	11	10	10	9	9	9	9	8	8	8			
18	13	12	12	11	11	10	10	9	9	9	9	9	9	9		
19	14	13	12	12	11	11	10	10	9	9	9	9	9	9	9	
20	15	14	13	12	11	11	10	10	10	9	9	9	9	9	9	9

This table is adapted from Takashima, M. (1955), "Tables for Testing Randomness by Means of Length of Runs" in *Bulletin of Mathematical Statistics*, Volume 6, pp. 17–23.

The test for the length of the longest run is applied to the data plotted in Figure B.2. The four runs in the plotted values have been identified and circled. The longest run occurs below the centerline and has nine consecutive points. Seven points fall above the centerline and 10 points fall below the centerline; thus, $s = 7$ and $r = 10$. Based on Table B.1, the critical value for the length of the longest run for the stated values of s and r is seven. Since the length of the longest run is nine, there is evidence that a special cause is acting on the process and a nonrandom pattern exists in the data. Thus, the same conclusion is reached whether the Rule of Seven is used or the test for the length of the longest run is applied. However, this is not always the case, as illustrated by the values plotted in Figure B.3. In this chart, the length of the longest run is eight. Hence, one would conclude that a nonrandom influence is acting on the process based on the Rule of Seven. The total number of points that falls below the centerline is 12 and the number of points above the centerline is 6, so the critical value from Table B.1 for the length of the longest run is nine. Therefore, according to the test for the length of the longest run, there is no evidence of nonrandom patterns in the data. The test for the length of longest run is more conservative than the Rule of Seven. When there are only random influences present, the probability of finding one or more runs in the data with lengths greater than or equal to the critical value (from Table B.1) is less than .05.

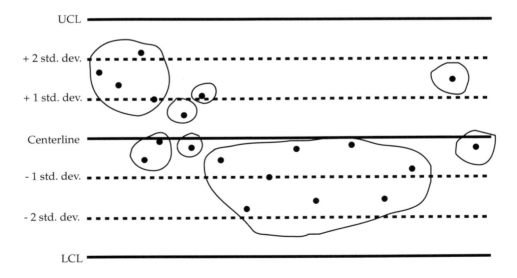

FIGURE B.3 *Illustration of Test for Length of Longest Run*
$r = 12$, $s = 7$, critical value = 9

B.3   TEST FOR TOO FEW RUNS

When a process is changing over time, a plot of the data collected from the process often shows that the plotted points do not fluctuate randomly above and below the centerline. The test for too few runs is based on probability calculations of the expected number of times that data from a stable process will fluctuate from below the centerline to above the centerline and vice versa. If there is an insufficient number of runs in the data, then there is evidence that a special cause of variation is acting on the process. To implement the test for too few runs:

1. Identify the runs in the data points plotted over time.
2. Count the total number of runs in the plotted data.
3. Count the total number of points above the centerline and the number of points below the centerline. Let s be the smaller of the two counts and r be the larger of the counts.
4. Using Table B.2 and the values of s and r, find the critical value for the number of runs.
5. If the actual number of runs in the plotted points is less than or equal to the critical value, there is evidence of nonrandom influences acting on the process. Otherwise, there is no evidence of nonrandom influences.

TABLE B.2 *Critical Values for Too Few Runs*

r / s	5	6	7	8	9	10	11	12	13	14	15	16	17	18	19	20
5	3															
6	3	3														
7	3	4	4													
8	3	4	4	5												
9	4	4	5	5	6											
10	4	5	5	6	6	6										
11	4	5	5	6	6	7	7									
12	4	5	6	6	7	7	8	8								
13	4	5	6	6	7	8	8	9	9							
14	5	5	6	7	7	8	8	8	9	10						
15	5	6	6	7	8	8	9	9	10	10	11					
16	5	6	6	7	8	8	9	10	10	11	11	11				
17	5	6	7	7	8	9	9	10	10	11	11	12	12			
18	5	6	7	8	8	9	10	10	11	11	12	12	13	13		
19	5	6	7	8	8	9	10	10	11	12	12	13	13	14	14	
20	5	6	7	8	9	9	10	11	11	12	12	13	13	14	14	15

This table is adapted from Swed, F. S., and Eisenhart, C. (1943), "Tables for Testing Randomness of Grouping in a Sequence of Alternatives," in *Annals of Mathematical Statistics*, Volume 14, pp. 66–87.

The chart in Figure B.3 shows eight runs in the plotted data points with 12 points below the centerline and 7 points above the centerline. Thus, r is equal to 12 and s is equal to 7. From Table B.2, the critical value for too few runs is six. The probability of finding six or fewer runs when $r = 12$ and $s = 7$ is no larger than .05 when there are only random influences impacting the process. Therefore, for the points plotted in Figure B.3, one would conclude that there is no evidence of nonrandom patterns according to the test for too few runs.

B.4 TWO OUT OF THREE SUCCESSIVE POINTS IN ZONE A OR BEYOND*

Even though many of the runs tests are based primarily on the length of a run as the criterion for evidence of an out-of-control process, shorter runs that occur close to the upper or lower control limits are often used to indicate the presence of special causes acting on a process. The test for two out of three successive points in Zone A or beyond uses the distance that points fall from the centerline as the primary criterion for evidence of lack of control. According to this test, if two out of three consecutive points fall more than two standard deviations away from the centerline and the two points are on the same side of the centerline, then there is evidence of nonrandom patterns in the data. Consider the data plotted in Figure B.4. Even though the longest run in the data consists only of three points, there is evidence of nonrandom behavior in the data based on the test for two out of three successive points in zone A or beyond. To implement this runs test:

1. Compute the value of the centerline for the plotted data points.
2. Calculate the estimated standard deviation based on the data. Use the estimated standard deviation to determine the values for one, two, and three standard deviations above and below the centerline (i.e., determine the zone lines.)
3. Identify all points that fall in zone A or beyond on either side of the centerline.
4. If two out of three consecutive points, on the same side of the centerline, are beyond two standard deviations of the centerline, then there is evidence of nonrandom patterns in the data. Note that the third point can be anywhere on the control chart.

Adapted with the permission of AT&T © 1956. All rights reserved. AT&T Statistical Quality Control Handbook by Western Electric Co., Inc. (1956) Delmar Printing Co., Charlotte, NC , p. 26.

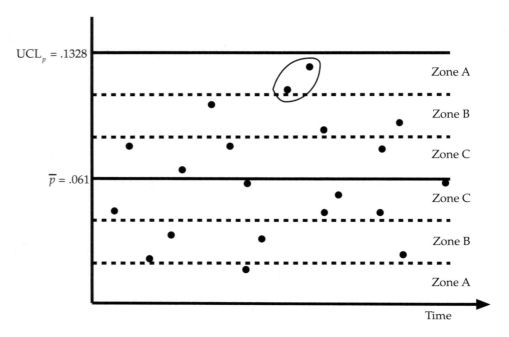

FIGURE B.4 *Two Out of Three Successive Points in Zone A or Beyond*

For the data plotted on the p chart in Figure B.4, the centerline value is .061 and the upper control limit is .1328. Hence, an estimate of the standard deviation for an in control process is:

$$\frac{1}{3} \ (\text{UCL} \ - \ \text{centerline}) \ = \ \frac{1}{3} \ (.1328 \ - \ .061) \ = \ .0239$$

Two standard deviations above the centerline is $.061 + 2(0.0239) = .1089$. There is no lower control limit, but the line for zone A below the centerline falls at $.061 - .0478 = .0132$. There are three points in Figure B.4 that fall in zone A or beyond. For the point in zone A below the centerline, the points that immediately precede and follow that data point do not fall in zone A below the centerline. Hence, this point does not provide a signal of nonrandom behavior. However, there are two consecutive points that fall in zone A above the centerline. Based on this test, since both points are on the same side of the centerline and they are more than two standard deviations above the centerline, there is evidence of special causes acting on the process.

B.5   FOUR OUT OF FIVE SUCCESSIVE POINTS IN ZONE B OR BEYOND*

In the previous runs test, short runs far away from the centerline are evidence of an out-of-control process. The test of four out of five successive points in zone B or beyond is based on the idea that longer runs somewhat closer to the centerline also signal the presence of special causes. This test states that if four out of five data points are on the same side of the centerline and are more than one standard deviation away from the centerline, then the data exhibit nonrandom behavior. As with the previous test, the location of the fifth data point is irrelevant (i.e., it can fall anywhere above or below the centerline). The test is performed on a set of points plotted in time order in the same manner as the runs test presented in Section B.4:

1. Compute the value of the centerline for the plotted data points.
2. Calculate the estimated standard deviation based on the data. Use the estimated standard deviation to determine the values for one, two, and three standard deviations above and below the centerline (i.e., determine the zone lines).
3. Identify all points that fall in zone B or beyond on either side of the centerline.
4. If four out of five points in succession, on the same side of the centerline, are beyond one standard deviation of the centerline, then there is evidence of nonrandom patterns in the data. Otherwise, there is no evidence of lack of control.

Figure B.5 illustrates the runs test for four out of five successive points in zone B or beyond. Note that this particular set of time-ordered data fails two different runs tests. There are four out of five successive points below the centerline that fall beyond one standard deviation from the centerline, and two out of three consecutive points fall in zone A below the centerline. Hence, based on either of the two runs tests, one would conclude that the data exhibit a nonrandom behavior pattern and would seek to identify and eliminate the special causes of this nonrandom behavior.

*Adapted with the permission of AT&T © 1956. All rights reserved. *AT&T Statistical Quality Control Handbook* by Western Electric Co., Inc. (1956) Delmar Printing Co., Charlotte, NC, p. 29.

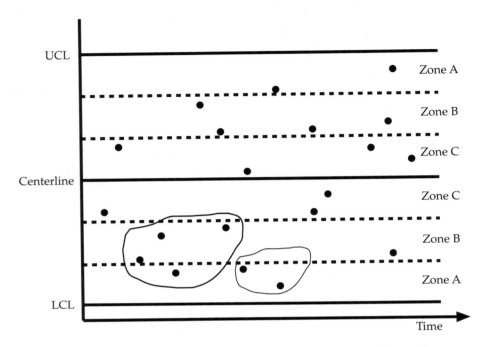

FIGURE B.5 *Example of Nonrandom Behavior in Data Plotted Over Time*

B.6   FIFTEEN OR MORE CONSECUTIVE POINTS IN ZONE C ON EITHER SIDE OF CENTERLINE*

The chart provided in Figure B.6 exhibits very little up-and-down variation in the plotted points. All of the points fall inside the upper and lower control limits and, based on the previous runs tests, there is no evidence of instability. All of the plotted points fall close to the centerline. Control charts, such as the one in Figure B.6, that have very few points away from the centerline also signal an unnatural pattern in the data. When data are collected from a stable process, we expect to see the plotted points varying randomly around the centerline with a few points close to the upper or lower control limits. When no points fall anywhere near the control limits, there is evidence of a nonrandom behavior in the data. This situation often occurs when the process measurements come from two very different distributions. When data are collected in such a way that samples or subgroups consistently contain data from different distributions,

*Adapted with the permission of AT&T © 1956. All rights reserved. *AT&T Statistical Quality Control Handbook* by Western Electric Co., Inc. (1956) Delmar Printing Co., Charlotte, NC, p. 29.

then the data are said to be *stratified*. For example, if two parallel production lines feed assembled engines into a common inspection point where the number of defective engines in each lot is counted, then the collected data are a composite of information on defective engines from the two different production lines. If one line produces defective engines at a much higher level than the other line and each subgroup of engines contains engines from both lines, then there is an "averaging out" of the differences between the proportion defective in each of the lines. Figure B.7 illustrates this phenomenon.

Whenever the amount of variation in the plotted points is small compared to the width of the upper and lower control limits, there is evidence of the subgroups consistently containing data from multiple processes. A runs test that identifies this stratification in the plotted data is the test for 15 or more consecutive points in zone C. The test states that if 15 or more consecutive points fall within one standard deviation above or below the centerline, then there is evidence to conclude that there is nonrandom variation in the plotted points (i.e., there is evidence of sampling from two different distributions). Figure B.8 illustrates this runs test.

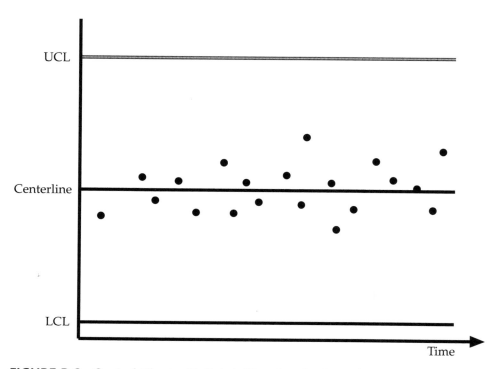

FIGURE B.6 *Control Chart with Points Hugging the Centerline*

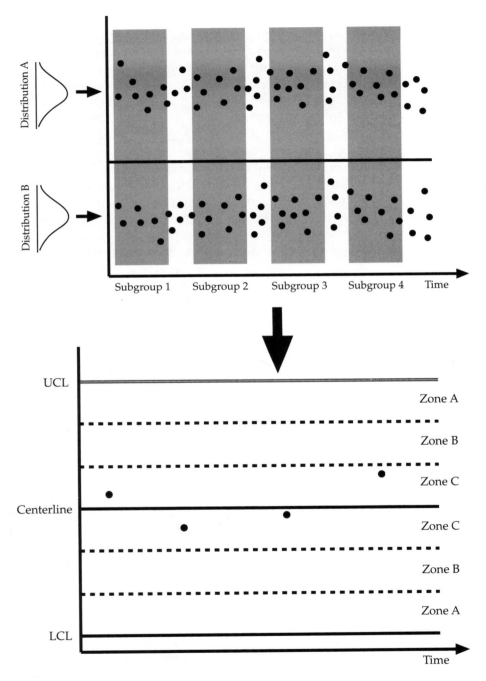

FIGURE B.7 *Illustration of Stratified Data*

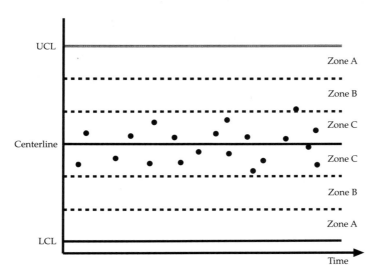

FIGURE B.8 *More Than 15 Consecutive Points in Zone C*

B.7 EIGHT CONSECUTIVE POINTS BEYOND ZONE C ON EITHER SIDE OF CENTERLINE*

In the previous section, we saw that when subgroups are taken so that each subgroup consistently contains measurements from different distributions, the control chart will appear to have the plotted points clustered tightly around the centerline. When data are collected from two different processes with different distributions, but the sampling is performed in such a way that each subgroup contains data from only one of the processes, it may be that very few points fall close to the centerline. For example, 25 shampoo bottles are filled simultaneously on one of two filling machines. After the filling stage, the filled bottles are inspected to ensure that the minimum amount of shampoo is in each bottle. Hence, each subgroup consists of 25 bottles all filled by the same machine. If the two filling machines operate at different levels and the measurements from the two machines are charted on the same control chart, there will be evidence of a mixture of distributions. As shown in Figure B.9, the first subgroup contains information only about the bottles from machine A, whereas the second subgroup contains data from machine B. Since the two machines fill at two different levels, the points fluctuate above and below the centerline, with no points near or at the centerline.

Whenever there is a mixture of multiple distributions and each subgroup captures information from only one of the distributions, there will be a pattern in the data where the points tend to fall beyond one standard deviation above and below the centerline. If eight or more consecutive points fall in either zone B or zone A on either side of the centerline (i.e., none of the points fall in zone C), then there is evidence of such a mixture in the data. This runs test is illustrated in Figure B.10.

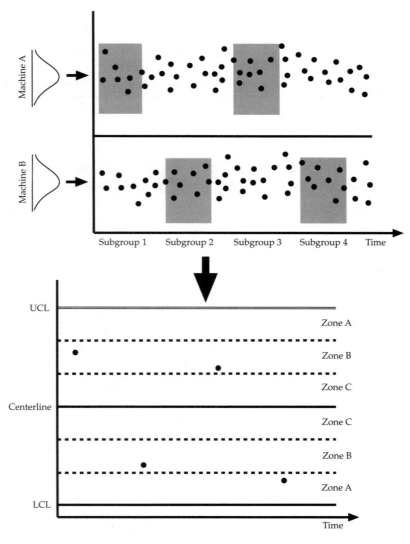

FIGURE B.9 *Mixture of Two Distributions*

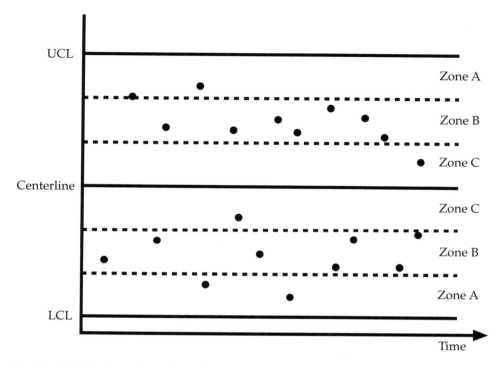

FIGURE B.10 *Runs Test for Eight or More Consecutive Points Beyond Zone C*

Appendix C:
Symbols for More Detailed
Process Flowcharts[*]

Chapter 2 provides examples of process flowcharts using the most useful and common flowcharting symbols—the diamond, the rectangle, and the oval circle. Many additional flowcharting symbols exist for depicting detailed information in a process flow diagram. The primary purpose for flowcharting a process is to understand the current process operation and its critical steps in order to identify key points in the process for study and measurement. Thus, the constructed flowchart should serve as a working document that portrays the important features of a process and that changes as additional process knowledge is gained or process changes are made. The following techniques for more detailed flowcharts might be useful in supporting that purpose.

---

[*] Dr. Kenneth Kirby, Department of Industrial Engineering, University of Tennesee, Knoxville, provided the ideas presented in this appendix.

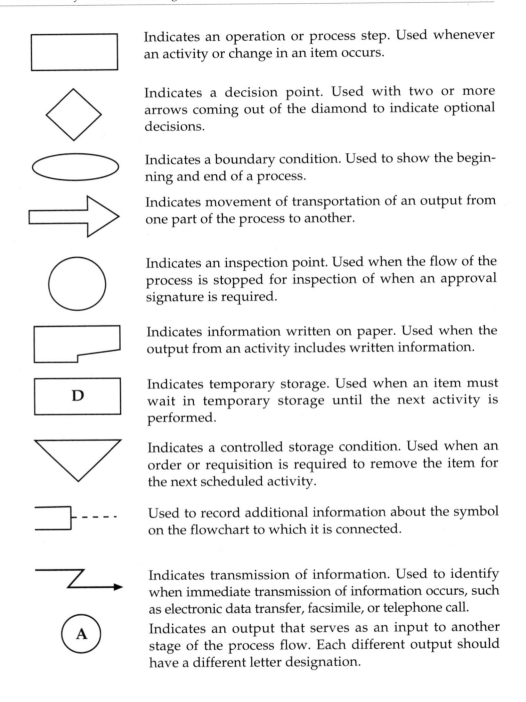

Indicates an operation or process step. Used whenever an activity or change in an item occurs.

Indicates a decision point. Used with two or more arrows coming out of the diamond to indicate optional decisions.

Indicates a boundary condition. Used to show the beginning and end of a process.

Indicates movement of transportation of an output from one part of the process to another.

Indicates an inspection point. Used when the flow of the process is stopped for inspection of when an approval signature is required.

Indicates information written on paper. Used when the output from an activity includes written information.

Indicates temporary storage. Used when an item must wait in temporary storage until the next activity is performed.

Indicates a controlled storage condition. Used when an order or requisition is required to remove the item for the next scheduled activity.

Used to record additional information about the symbol on the flowchart to which it is connected.

Indicates transmission of information. Used to identify when immediate transmission of information occurs, such as electronic data transfer, facsimile, or telephone call.

Indicates an output that serves as an input to another stage of the process flow. Each different output should have a different letter designation.

FIGURE C.1 *Flowcharting Symbols*

Appendix D: Statistical Tables

TABLE D.1 *Standard Normal Distribution*

Pz = the proportion of process output beyond a single specification limit that is z standard deviation units away from the process average (for a process that is in statistical control and is normally distributed).

$|z|$	x.x0	x.x1	x.x2	x.x3	x.x4	x.x5	x.x6	x.x7	x.x8	x.x9
4.0	.00003									
3.5	.00023									
3.0	.00135									
2.9	.0019	.0018	.0018	.0017	.0016	.0016	.0015	.0015	.0014	.0014
2.8	.0026	.0025	.0024	.0023	.0023	.0022	.0021	.0021	.0020	.0019
2.7	.0035	.0034	.0033	.0032	.0031	.0030	.0029	.0028	.0027	.0026
2.6	.0047	.0045	.0044	.0043	.0041	.0040	.0039	.0038	.0037	.0036
2.5	.0062	.0060	.0059	.0057	.0055	.0054	.0052	.0051	.0049	.0048
2.4	.0082	.0080	.0078	.0075	.0073	.0071	.0069	.0068	.0066	.0064
2.3	.0107	.0104	.0102	.0099	.0096	.0094	.0091	.0089	.0087	.0084
2.2	.0139	.0136	.0132	.0129	.0125	.0122	.0119	.0116	.0113	.0110
2.1	.0179	.0174	.0170	.0166	.0162	.0158	.0154	.0150	.0146	.0143
2.0	.0228	.0222	.0217	.0212	.0207	.0202	.0197	.0192	.0188	.0183
1.9	.0287	.0281	.0274	.0268	.0262	.0256	.0250	.0244	.0239	.0233
1.8	.0359	.0351	.0344	.0336	.0329	.0322	.0314	.0307	.0301	.0294
1.7	.0446	.0436	.0427	.0418	.0409	.0401	.0392	.0384	.0375	.0367
1.6	.0548	.0537	.0526	.0516	.0505	.0495	.0485	.0475	.0465	.0455
1.5	.0668	.0655	.0643	.0630	.0618	.0606	.0594	.0582	.0571	.0559
1.4	.0808	.0793	.0778	.0764	.0749	.0735	.0721	.0708	.0694	.0681
1.3	.0968	.0951	.0934	.0918	.0901	.0885	.0869	.0853	.0838	.0823
1.2	.1151	.1131	.1112	.1093	.1075	.1056	.1038	.1020	.1003	.0985
1.1	.1357	.1335	.1314	.1292	.1271	.1251	.1230	.1210	.1190	.1170
1.0	.1587	.1562	.1539	.1515	.1492	.1469	.1446	.1423	.1401	.1379
0.9	.1841	.1814	.1788	.1762	.1736	.1711	.1685	.1660	.1635	.1611
0.8	.2119	.2090	.2061	.2033	.2005	.1977	.1949	.1922	.1894	.1867
0.7	.2420	.2389	.2358	.2327	.2297	.2266	.2236	.2206	.2177	.2148
0.6	.2743	.2709	.2676	.2643	.2611	.2578	.2546	.2514	.2483	.2451
0.5	.3085	.3050	.3015	.2981	.2946	.2912	.2877	.2843	.2810	.2776
0.4	.3446	.3409	.3372	.3336	.3300	.3264	.3228	.3192	.3156	.3121
0.3	.3821	.3783	.3745	.3707	.3669	.3632	.3594	.3557	.3520	.3483
0.2	.4207	.4168	.4129	.4090	.4052	.4013	.3974	.3936	.3897	.3859
0.1	.4602	.4562	.4522	.4483	.4443	.4404	.4364	.4325	.4286	.4247
0.0	.5000	.4960	.4920	.4880	.4840	.4801	.4761	.4721	.4681	.4641

TABLE D.1   *Standard Normal Distribution (continued)*

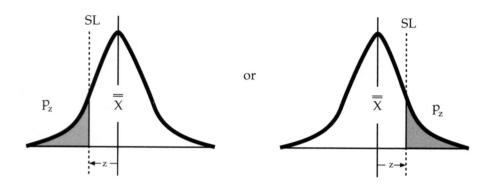

TABLE D.2   *Factors for Use with $\overline{X}$ and Range Charts*

Number of Observations in Subgroup	Factor for $\overline{X}$ Chart	Factors for Range Charts		Factor for Estimating σ
		LCL	UCL	
n	A_2	D_3	D_4	d_2
2	1.880		3.267	1.128
3	1.023		2.574	1.693
4	0.729		2.282	2.059
5	0.577		2.114	2.326
6	0.483		2.004	2.534
7	0.419	0.076	1.924	2.704
8	0.373	0.136	1.864	2.847
9	0.337	0.184	1.816	2.970
10	0.308	0.223	1.777	3.078
11	0.285	0.256	1.744	3.173
12	0.266	0.284	1.716	3.258
13	0.249	0.308	1.692	3.336
14	0.235	0.329	1.671	3.407
15	0.223	0.348	1.652	3.472
16	0.212	0.364	1.636	3.532
17	0.203	0.379	1.621	3.588
18	0.194	0.392	1.608	3.640
19	0.187	0.404	1.596	3.689
20	0.180	0.414	1.586	3.735
21	0.173	0.425	1.575	3.778
22	0.167	0.434	1.566	3.819
23	0.162	0.443	1.557	3.858
24	0.157	0.452	1.548	3.895
25	0.153	0.459	1.541	3.931

TABLE D.3 *Factors for Use with* X *and* s *charts*

Number of Observations in Subgroup n	Factor for $\bar{X}$ Chart A_3	Factors for s Charts		Factor for Estimating σ c_4
		LCL B_3	UCL B_4	
2	2.659		3.267	.7979
3	1.954		2.568	.8862
4	1.628		2.266	.9213
5	1.427		2.089	.9400
6	1.287	0.030	1.970	.9515
7	1.182	0.118	1.882	.9594
8	1.099	0.185	1.815	.9650
9	1.032	0.239	1.761	.9693
10	0.975	0.284	1.716	.9727
11	0.927	0.321	1.679	.9754
12	0.886	0.354	1.646	.9776
13	0.850	0.382	1.618	.9794
14	0.817	0.406	1.594	.9810
15	0.789	0.428	1.572	.9823
16	0.763	0.448	1.552	.9835
17	0.739	0.466	1.534	.9845
18	0.718	0.482	1.518	.9854
19	0.698	0.497	1.503	.9862
20	0.680	0.510	1.490	.9869
21	0.663	0.523	1.477	.9876
22	0.647	0.534	1.466	.9882
23	0.633	0.545	1.455	.9887
24	0.619	0.555	1.445	.9892
25	0.606	0.565	1.435	.9896

TABLE D.4 *Denominator Degrees of Freedom*
$F\alpha$ for $\alpha = .05$

V_1 / V_2	1	2	3	4	5	6	7	8	9	10	12	15	20	24	30	40	60
1	161.4	199.5	215.7	224.6	230.2	234.0	236.8	238.9	240.5	241.9	243.9	245.9	248.0	249.1	250.1	251.1	252.2
2	18.51	19.00	19.16	19.25	19.30	19.33	19.35	19.37	19.38	19.40	19.41	19.43	19.45	19.45	19.46	19.47	19.48
3	10.13	9.55	9.28	9.12	9.01	8.94	8.89	8.85	8.81	8.79	8.74	8.70	8.66	8.64	8.62	8.59	8.57
4	7.71	6.94	6.59	6.39	6.26	6.16	6.09	6.04	6.00	5.96	5.91	5.86	5.80	5.77	5.75	5.72	5.69
5	6.61	5.79	5.41	5.19	5.05	4.95	4.88	4.82	4.77	4.74	4.68	4.62	4.56	4.53	4.50	4.46	4.43
6	5.99	5.14	4.76	4.53	4.39	4.28	4.21	4.15	4.10	4.06	4.00	3.94	3.87	3.84	3.81	3.77	3.74
7	5.59	4.74	4.35	4.12	3.97	3.87	3.79	3.73	3.68	3.64	3.57	3.51	3.44	3.41	3.38	3.34	3.30
8	5.32	4.46	4.07	3.84	3.69	3.58	3.50	3.44	3.39	3.35	3.28	3.22	3.15	3.12	3.08	3.04	3.01
9	5.12	4.26	3.86	3.63	3.48	3.37	3.29	3.23	3.18	3.14	3.07	3.01	2.94	2.90	2.86	2.83	2.79
10	4.96	4.10	3.71	3.48	3.33	3.22	3.14	3.07	3.02	2.98	2.91	2.85	2.77	2.74	2.70	2.66	2.62
11	4.84	3.98	3.59	3.36	3.20	3.09	3.01	2.95	2.90	2.85	2.79	2.72	2.65	2.61	2.57	2.53	2.49
12	4.75	3.89	3.49	3.26	3.11	3.00	2.91	2.85	2.80	2.75	2.69	2.62	2.54	2.51	2.47	2.43	2.38
13	4.67	3.81	3.41	3.18	3.03	2.92	2.83	2.77	2.71	2.67	2.60	2.53	2.46	2.42	2.38	2.34	2.30
14	4.60	3.74	3.34	3.11	2.96	2.85	2.76	2.70	2.65	2.60	2.53	2.46	2.39	2.35	2.31	2.27	2.22
15	4.54	3.68	3.29	3.06	2.90	2.79	2.71	2.64	2.59	2.54	2.48	2.40	2.33	2.29	2.25	2.20	2.16
16	4.49	3.63	3.24	3.01	2.85	2.74	2.66	2.59	2.54	2.49	2.42	2.35	2.28	2.24	2.19	2.15	2.11
17	4.45	3.59	3.20	2.96	2.81	2.70	2.61	2.55	2.49	2.45	2.38	2.31	2.23	2.19	2.15	2.10	2.06
18	4.41	3.55	3.16	2.93	2.77	2.66	2.58	2.51	2.46	2.41	2.34	2.27	2.19	2.15	2.11	2.06	2.02
19	4.38	3.52	3.13	2.90	2.74	2.63	2.54	2.48	2.42	2.38	2.31	2.23	2.16	2.11	2.07	2.03	1.98
20	4.35	3.49	3.10	2.87	2.71	2.60	2.51	2.45	2.39	2.35	2.28	2.20	2.12	2.08	2.04	1.99	1.95
21	4.32	3.47	3.07	2.84	2.68	2.57	2.49	2.42	2.37	2.32	2.25	2.18	2.10	2.05	2.01	1.96	1.92
22	4.30	3.44	3.05	2.82	2.66	2.55	2.46	2.40	2.34	2.30	2.23	2.15	2.07	2.03	1.98	1.94	1.89
23	4.28	3.42	3.03	2.80	2.64	2.53	2.44	2.37	2.32	2.27	2.20	2.13	2.05	2.01	1.96	1.91	1.86
24	4.26	3.40	3.01	2.78	2.62	2.51	2.42	2.36	2.30	2.25	2.18	2.11	2.03	1.98	1.94	1.89	1.84
25	4.24	3.39	2.99	2.76	2.60	2.49	2.40	2.34	2.28	2.24	2.16	2.09	2.01	1.96	1.92	1.87	1.82
26	4.23	3.37	2.98	2.74	2.59	2.47	2.39	2.32	2.27	2.22	2.15	2.07	1.99	1.95	1.90	1.85	1.80
27	4.21	3.35	2.96	2.73	2.57	2.46	2.37	2.31	2.25	2.20	2.13	2.06	1.97	1.93	1.88	1.84	1.79
28	4.20	3.34	2.95	2.71	2.56	2.45	2.36	2.29	2.24	2.19	2.12	2.04	1.96	1.91	1.87	1.82	1.77
29	4.18	3.33	2.93	2.70	2.55	2.43	2.35	2.28	2.22	2.18	2.10	2.03	1.94	1.90	1.85	1.81	1.75
30	4.17	3.32	2.92	2.69	2.53	2.42	2.33	2.27	2.21	2.16	2.09	2.01	1.93	1.89	1.84	1.79	1.74
40	4.08	3.23	2.84	2.61	2.45	2.34	2.25	2.18	2.12	2.08	2.00	1.92	1.84	1.79	1.74	1.69	1.64
60	4.00	3.15	2.76	2.53	2.37	2.25	2.17	2.10	2.04	1.99	1.92	1.84	1.75	1.70	1.65	1.59	1.53
120	3.92	3.07	2.68	2.45	2.29	2.17	2.09	2.02	1.96	1.91	1.83	1.75	1.66	1.61	1.55	1.50	1.43
∞	3.84	3.00	2.60	2.37	2.21	2.10	2.01	1.94	1.88	1.83	1.75	1.67	1.57	1.52	1.46	1.39	1.32

Numerator Degrees of Freedom

TABLE D.5 Denominator Degrees of Freedom

$F\alpha$ for $\alpha = .01$

Numerator Degrees of Freedom

v_2 \ v_1	1	2	3	4	5	6	7	8	9	10	12	15	20	24	30	40	60
1	4,052	4,999	5,403	5,625	5,764	5,859	5,928	5,982	6,022	6,056	6,106	6,157	6,209	6,235	6,261	6,287	6,313
2	98.50	99.00	99.17	99.25	99.30	99.33	99.36	99.37	99.39	99.40	99.42	99.43	99.45	99.46	99.47	99.47	99.48
3	34.12	30.82	29.46	28.71	28.24	27.91	27.67	27.49	27.35	27.23	27.05	26.87	26.69	26.60	26.50	26.41	26.32
4	21.20	18.00	16.69	15.98	15.52	15.21	14.98	14.80	14.66	14.55	14.37	14.20	14.02	13.93	13.84	13.75	13.65
5	16.26	13.27	12.06	11.39	10.97	10.67	10.46	10.29	10.16	10.05	9.89	9.72	9.55	9.47	9.38	9.29	9.20
6	13.75	10.92	9.78	9.15	8.75	8.47	8.26	8.10	7.98	7.87	7.72	7.56	7.40	7.31	7.23	7.14	7.06
7	12.25	9.55	8.45	7.85	7.46	7.19	6.99	6.84	6.72	6.62	6.47	6.31	6.16	6.07	5.99	5.91	5.82
8	11.26	8.65	7.59	7.01	6.63	6.37	6.18	6.03	5.91	5.81	5.67	5.52	5.36	5.28	5.20	5.12	5.03
9	10.56	8.02	6.99	6.42	6.06	5.80	5.61	5.47	5.35	5.26	5.11	4.96	4.81	4.73	4.65	4.57	4.48
10	10.04	7.56	6.55	5.99	5.64	5.39	5.20	5.06	4.94	4.85	4.71	4.56	4.41	4.33	4.25	4.17	4.08
11	9.65	7.21	6.22	5.67	5.32	5.07	4.89	4.74	4.63	4.54	4.40	4.25	4.10	4.02	3.94	3.86	3.78
12	9.33	6.93	5.95	5.41	5.06	4.82	4.64	4.50	4.39	4.30	4.16	4.01	3.86	3.78	3.70	3.62	3.54
13	9.07	6.70	5.74	5.21	4.86	4.62	4.44	4.30	4.19	4.10	3.96	3.82	3.66	3.59	3.51	3.43	3.34
14	8.86	6.51	5.56	5.04	4.69	4.46	4.28	4.14	4.03	3.94	3.80	3.66	3.51	3.43	3.35	3.27	3.18
15	8.68	6.36	5.42	4.89	4.56	4.32	4.14	4.00	3.89	3.80	3.67	3.52	3.37	3.29	3.21	3.13	3.05
16	8.53	6.23	5.29	4.77	4.44	4.20	4.03	3.89	3.78	3.69	3.55	3.41	3.26	3.18	3.10	3.02	2.93
17	8.40	6.11	5.18	4.67	4.34	4.10	3.93	3.79	3.68	3.59	3.46	3.31	3.16	3.08	3.00	2.92	2.83
18	8.29	6.01	5.09	4.58	4.25	4.01	3.84	3.71	3.60	3.51	3.37	3.23	3.08	3.00	2.92	2.84	2.75
19	8.18	5.93	5.01	4.50	4.17	3.94	3.77	3.63	3.52	3.43	3.30	3.15	3.00	2.92	2.84	2.76	2.67
20	8.10	5.85	4.94	4.43	4.10	3.87	3.70	3.56	3.46	3.37	3.23	3.09	2.94	2.86	2.78	2.69	2.61
21	8.02	5.78	4.87	4.37	4.04	3.81	3.64	3.51	3.40	3.31	3.17	3.03	2.88	2.80	2.72	2.64	2.55
22	7.95	5.72	4.82	4.31	3.99	3.76	3.59	3.45	3.35	3.26	3.12	2.98	2.83	2.75	2.67	2.58	2.50
23	7.88	5.66	4.76	4.26	3.94	3.71	3.54	3.41	3.30	3.21	3.07	2.93	2.78	2.70	2.62	2.54	2.45
24	7.82	5.61	4.72	4.22	3.90	3.67	3.50	3.36	3.26	3.17	3.03	2.89	2.74	2.66	2.58	2.49	2.40
25	7.77	5.57	4.68	4.18	3.85	3.63	3.46	3.32	3.22	3.13	2.99	2.85	2.70	2.62	2.54	2.45	2.36
26	7.72	5.53	4.64	4.14	3.82	3.59	3.42	3.29	3.18	3.09	2.96	2.81	2.66	2.58	2.50	2.42	2.33
27	7.68	5.49	4.60	4.11	3.78	3.56	3.39	3.26	3.15	3.06	2.93	2.78	2.63	2.55	2.47	2.38	2.29
28	7.64	5.45	4.57	4.07	3.75	3.53	3.36	3.23	3.12	3.03	2.90	2.75	2.60	2.52	2.44	2.35	2.26
29	7.60	5.42	4.54	4.04	3.73	3.50	3.33	3.20	3.09	3.00	2.87	2.73	2.57	2.49	2.41	2.33	2.23
30	7.56	5.39	4.51	4.02	3.70	3.47	3.30	3.17	3.07	2.98	2.84	2.70	2.55	2.47	2.39	2.30	2.21
40	7.31	5.18	4.31	3.83	3.51	3.29	3.12	2.99	2.89	2.80	2.66	2.52	2.37	2.29	2.20	2.11	2.02
60	7.08	4.98	4.13	3.65	3.34	3.12	2.95	2.82	2.72	2.63	2.50	2.35	2.20	2.12	2.03	1.94	1.84
120	6.85	4.79	3.95	3.48	3.17	2.96	2.79	2.66	2.56	2.47	2.34	2.19	2.03	1.95	1.86	1.76	1.66
∞	6.63	4.61	3.78	3.32	3.02	2.80	2.64	2.51	2.41	2.32	2.18	2.04	1.88	1.79	1.70	1.59	1.47

Appendix E:
Answers to Practice Problems

3.8.1

a. Centerline and control limits for the p chart are
$$p = .003045$$
$$UCL_p = .00827$$
$$LCL_p - none$$

b. Comments about how the data were obtained:

- No information on how the data were collected.
- Are the data time-ordered? Without this knowledge a runs test is inappropriate.
- No information on the way the data were subgrouped.
- Easily obtained data do not always provide sufficient information about the process.

c. Next course of action: PLAN

- Draw process flowchart in order to better define the process.
- Analyze the potential causes of defective cans by means of a cause and effect diagram.
- Develop a data collection strategy to provide desired information.

3.8.2

a. Centerline and control limits for the np chart are
$$\overline{np} = 15.56$$
$$UCL_{np} = 26.92$$
$$LCL_{np} = 4.20$$

b. The process appears to be in control since all points fall within the control limits and the chart passes the Rule of Seven runs test.

c. Estimated fraction nonconforming: $\bar{p} = .0778$.
d. This might be an important characteristic to consumers. However, the current method for counting nonconformities would not count these kinds of problems.
e. Based on the information provided, little more is known than that within-subgroup causes would be those causes active within a day, and between-subgroup causes would be causes that acted to make days different from each other. Additional information on when and how the samples of 200 bottles were selected could be sought to identify more specifically the causes captured within subgroups. Alternatively, causes thought to affect the process should be listed and sampling and subgrouping performed to understand the presence (or not) and the effect of these causes.

4.7.1

a. $\bar{u} = \dfrac{324}{101.5} = 3.192$

Week	n	u	UCL	LCL
1	8.4	2.86	5.041	1.343
2	6.0	2.33	5.380	1.004
3	7.2	3.75	5.190	1.194
4	10.8	3.61	4.823	1.561
5	9.6	4.69	4.922	1.462
6	6.0	2.83	5.380	1.004
7	7.5	2.67	5.149	1.235
8	9.9	2.63	4.896	1.488
9	8.7	2.07	5.009	1.375
10	11.2	4.11	4.794	1.590
11	7.2	4.03	5.190	1.194
12	9.0	2.11	4.979	1.405

4.7.2

a. $\bar{c} = 8.8$
   $UCL_c = 17.70$
   LCL_c – none

5.7.1

a. Range chart:
   $\bar{R} = 0.3933$
   $UCL_R = 0.8314$
   LCL_R – none

X-bar chart:
$$\bar{\bar{X}} = 2.7708$$
$$\text{UCL}_X = 2.9977$$
$$\text{LCL}_{\bar{X}} = 2.5439$$

5.7.2

a. $\text{UCL}_s = 0.786$
$\text{LCL}_s = 0.130$

b. $\text{UCL}_{\bar{X}} = 9.617$
$\text{LCL}_{\bar{X}} = 8.723$

c. $\hat{\sigma} = 0.47$
$\text{NT} = 6(0.47) = 2.82$
$\text{ET} = 3$

Process is capable since NT < ET.

d. 7.64%.

5.7.3

a. $\bar{R} = 6.286$
$\text{UCL}_R = 12.597$
$\text{LCL}_R - \text{none}$
$\bar{\bar{X}} = 19.4812$
$\text{UCL}_{\bar{X}} = 22.517$
$\text{LCL}_{\bar{X}} = 16.445$

b. $\hat{\sigma} = 2.48$

c. $\bar{\bar{X}} = 19.48$

d. $\text{NT} = 6(2.48) = 14.88$; $\text{ET} = 6$, not capable; 30.6% will fall outside of specifications.

7.7.1

b. $\overline{\text{MR}} = 20.83$
$\text{UCL}_{MR} = 68.05$

Not appropriate to complete individuals chart.

c. Not appropriate to estimate the process standard deviation.

8.5.1

b. $\bar{R} = 0.00411$
    $\text{UCL}_R = 0.01058$
    $\text{LCL}_R - \text{none}$

    $\bar{\bar{X}} = 0.05409$
    $\text{UCL}_{\bar{X}} = 0.05828$
    $\text{LCL}_{\bar{X}} = 0.04989$

    $\hat{\sigma}_w = 0.00242$

d. $\overline{\text{MR}} = 0.00659$
    $\text{UCL}_{MR} = 0.02153$
    $\overline{\text{MR}} / d_2 = 0.00584$
    $\text{UCL}_{\bar{X}} = 0.07161$
    $\text{LCL}_{\bar{X}} = 0.03657$

$$\hat{\sigma}_B = \sqrt{\left(\frac{\overline{\text{MR}}}{d_2}\right)^2 - \frac{\hat{\sigma}_w^2}{3}} = .00567$$

$$\hat{\sigma}_T = \sqrt{(.00567)^2 + (.00242)^2} = .00616$$

8.5.2

e. s_p^2 provides an estimate of measurement variation; $s_p^2 = 4226.43958$.

    $s_{\bar{X}}^2 = 3710.405$
    $\sigma_B^2 = 2865.117$
    $\sigma_T^2 = 7091.557$

59.6% of observed variation is due to measurement variation.

Bibliography

Over the years, the authors have found the following works to provide useful information for guiding process study. We have ourselves referred to each of these works on numerous occasions. Having learned from these books, we have internalized the thinking of the authors. Consequently, the ideas presented in this book reflect the teachings contained in the following list. The reader will also find the list useful in his or her own studies.

Burr, Irving W. *Statistical Quality Control*. Marcel, Dekker, New York, 1976.

Feigenbaum, Armand, V. *Total Quality Control*, 3rd ed. McGraw-Hill, New York, 1983.

Grant, Eugene L., and Leavenworth, Richard S. *Statistical Quality Control*, 6th ed. New York, McGraw-Hill, 1988.

Imai, Masaaki, *Kaizen*. Random House Business Division, New York, 1986.

Ishikawa, Kaoru, *Guide to Quality Control*, 2nd ed. Asian Productivity Organization, Tokyo, 1986.

Ishikawa, Kaoru, *What Is Total Quality Control?* Prentice-Hall, Englewood Cliffs, NJ, 1985.

Juran, J. M. *Managerial Breakthrough*, McGraw-Hill, New York, 1964.

Ott, Ellis R. *Process Quality Control*. McGraw-Hill, New York, 1975.

Wheeler, Donald J., and Chambers, David S. *Understanding Statistical Process Control*, 2nd ed. SPC Press, Knoxville, TN, 1992.

Index